Conference Proceedings of the Society for Experimental Mechanics Series

Series Editor

Kristin B. Zimmerman
Society for Experimental Mechanics, Inc.,
Bethel, USA

The Conference Proceedings of the Society for Experimental Mechanics Series presents early findings and case studies from a wide range of fundamental and applied work across the broad range of fields that comprise Experimental Mechanics. Series volumes follow the principle tracks or focus topics featured in each of the Society's two annual conferences: IMAC, A Conference and Exposition on Structural Dynamics, and the Society's Annual Conference & Exposition and will address critical areas of interest to researchers and design engineers working in all areas of Structural Dynamics, Solid Mechanics and Materials Research.

Ludovic Renson • Robert Kuether • Paolo Tiso • Drithi Shetty
Editors

Nonlinear Structures & Systems, Vol. 1

Proceedings of the 43rd IMAC, A Conference and Exposition on Structural Dynamics 2025

River Publishers

Published, sold and distributed by:

River Publishers
Broagervej 10
9260 Gistrup
Denmark

www.riverpublishers.com

ISBN 97887-438-0146-7 (Hardback)

ISBN 97887-438-0158-0 (eBook)

https://doi.org/10.13052/97887-438-0146-7

Conference Proceedings of the Society for Experimental Mechanics

An imprint of River Publishers

Preface

Nonlinear Structures & Systems represents one of twelve volumes of technical papers presented at the *43*rd IMAC, A Conference and Exposition on Structural Dynamics, organized by the Society for Experimental Mechanics, and held February 10-13, 2025. The full proceedings also include volumes on Dynamics of Civil Structures; Model Validation and Uncertainty Quantification; Dynamic Substructuring & Transfer Path Analysis; Special Topics in Structural Dynamics & Experimental Techniques; Computer Vision & Laser Vibrometry; Dynamic Environments Testing; Sensors & Instrumentation and Aircraft/Aerospace Testing Techniques; Topics in Modal Analysis & Parameter Identification I&II; Data Science in Engineering; and Structural Health Monitoring & Machine Learning.

Each collection presents early findings from experimental and computational investigations on an important area within Structural Dynamics. Nonlinearity is one of these areas.

The vast majority of real engineering structures behave nonlinearly. In many cases, nonlinearities can even be exploited to achieve novel functionalities. Therefore, it is necessary to consider nonlinear effects in all of the engineering design steps, both in the experimental analysis tools (so that the nonlinear parameters can be correctly identified) and in the mathematical and numerical models of the structure (to run accurate simulations). In so doing, it is possible to create models representative of the reality that can be used for better predictions once validated.

Several papers address theoretical and numerical aspects of nonlinear dynamics (covering rigorous theoretical formulations and robust computational algorithms) as well as experimental techniques and analysis methods. There are also papers dedicated to nonlinearity in practice where real-life examples of nonlinear structures are discussed.

The organizers would like to thank the authors, presenters, session organizers, and session chairs for their participation in this track.

Editors:
Ludovic Renson–*Imperial College London, UK*; Robert Kuether–*Sandia National Laboratories, USA*; Paolo Tiso–*ETH Zürich, Switzerland, Drithi Shetty – University of South Florida, USA.*

Contents

Chapter 1
Linear and Nonlinear Response of a Continuous-Interface BARC: FEA Prediction and Test

Sean Magoffin, Brooklyn Andrus, Matthew S. Allen, and Raymond Joshua

Abstract This paper presents the results of finite element modeling and linear and nonlinear testing of a BARC structure with a continuous interface (as opposed to the split-interface BARC). Before the actual hardware was constructed, blueprints were used to create a FEM in which the bolts were modeled in detail, including contact at the interfaces due the expected preload based on the bolt torque specs. This model was used to predict the linearized natural frequencies, to infer which modes might exhibit nonlinearity, and then quasi-static modal analysis was used to predict the amplitude dependent natural frequency and damping for the modes that were expected to exhibit the strongest nonlinearity. Once the FEM predictions were mostly complete, a BARC was constructed based on the blueprints and tests were performed to measure its linear modes and characterize those modes that exhibit nonlinear behavior. The linear modes measured in test are compared to those computed from the FEM about the preloaded state. Additionally, for those modes where significant nonlinearity was observed, quasi-static modal analysis (QSMA) was used to predict their amplitude dependent natural frequencies and damping ratios.

Keywords BARC · Quasi-static modal analysis · Nonlinearities · FEA

Introduction

The Box Assembly with Removable Component (BARC) structure was created in 2017 to provide a common benchmark structure to evaluate dynamic environment testing methods [1]. State of the art environmental testing methods do not consider nonlinearity, although one could argue that they seek to capture it by testing the component of interest at high enough levels that any nonlinearities would be observed. In any event, while many studies have tested and modeled the BARC over the past decade, very few focused on its nonlinear behavior. On the other hand, because the removable component consists of three parts bolted together, it can behave nonlinearly and experimental campaigns have noted this since it was introduced. This work presents a first effort to characterize the nonlinear behavior of the continuous-interface BARC structure, both experimentally and using nonlinear finite element simulations.

This study is a complement to the various studies that have used experimental modal analysis to extract the linear modes of the BARC and compared those with modes from finite element models. For example, Alvis and Schoenherr [2] used the BARC to perform a detailed study on the best practices for modeling bolted interfaces. Their study, as with others,

Honeywell Federal Manufacturing & Technologies, LLC operates the Kansas City National Security Campus for the United States Department of Energy / National Nuclear Security Administration under Contract Number DE-NA0002839. UUR – NSC-614-6620

Sean Magoffin · Brooklyn Andrus
Undergraduate Students, Brigham Young University, Department of Mechanical Engineering
e-mail: sm996@byu.edu; brookand@byu.edu

Matthew S. Allen
Professor, Brigham Young University, Department of Mechanical Engineering
e-mail: matt.allen@byu.edu

Raymond Joshua
Sr. Mechanical Engineer, Advanced Development Domain (ADD), Kansas City National Security Campus
e-mail: rjoshua@kcnsc.doe.gov

Ludovic Renson et al. (eds.), *Nonlinear Structures & Systems, Vol. 1*, Conference Proceedings of the Society for Experimental Mechanics Series, https://doi.org/10.13052/9788743801467-1

[3][4][5][6] focused on the low amplitude, linearized behavior. Several studies have had good success in creating finite element models that correlated very well with measurements of the natural frequencies and mode shapes of the BARC, although most have noted that care had to be taken in modeling the bolted connections appropriately. The authors found very few studies that quantified the nonlinear behavior of the BARC. Takeshita's thesis [7] and some precursor papers did explore this somewhat, in particular the randomly excited response of the BARC to a base input. No other detailed studies were found before this paper had to be submitted. In this work the BARC is tested and simulated in free-free conditions. Note also that this study focuses on the continuous-interface BARC, while many prior studies have focused on the split-interface BARC.

In undertaking this study, the authors began with blueprints of the BARC structure, and the finite element models were created from those before any hardware had been machined. The authors did have the benefit of comparing their results to other papers that had also modeled this structure, but other than that the predictions were blind in that the linear and nonlinear testing was not performed until after the finite element models had been created and debugged. The FEM included a simplified, three-dimensional model of the bolts, which were preloaded based on the torque specified in the assembly drawings. The FEM was created in Abaqus, which solved the nonlinear contact problem to obtain the preloaded state of the bolts and contact between the components. Linear modal analysis was performed about this preloaded state. This produced a prediction of the linearized natural frequencies of the BARC that didn't depend on model updating. Similarly, the nonlinearity in the BARC was predicted using quasi-static modal analysis [8], which consists of applying static loads in the shape of a mode of interest in order to predict its amplitude dependent natural frequency and damping ratio. This was also performed before any nonlinear tests had been performed. Hence, the paper presents an idea of the predictions that might be obtained with these tools based only on the nominal design of the components. These predictions are then compared with the actual test measurements.

The following section details the finite element model that was created, the validation steps that were performed, and the linear and nonlinear results obtained. Then Section 3 discusses the manufacturing of the actual BARC and the linear and nonlinear tests that were performed. The FEM and test results are then compared and conclusions are presented.

Creating a Finite Element Model of the BARC

A 3D finite element model (FEM) of the BARC was created using Abaqus from the provided drawings. The joints were modeled by creating simplified 3D bolts that were meshed with solid elements, as was found to work well in [2] as well as in the authors' prior work [9, 10]. The structure was preloaded by applying a bolt load in Abaqus to the middle surface of each bolt (see Figure 1) and then fixing the bolt length in the subsequent step. The specified torque values were converted to preload values using the plot of preload force versus torque found by Wall et al. in [11]. Furthermore, in [10] the best agreement with measurements was found when using a friction coefficient of 0.2, so this was used in Abaqus even though this is a lower value than typically expected for steel-on-steel contact.

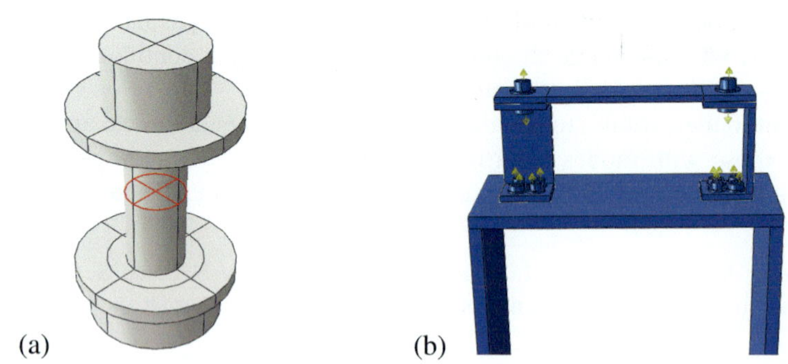

Fig. 1 (a) The surface that the bolt load is applied to is highlighted in red. (b) Yellow arrows represent the applied preload to those surfaces of each bolt.

The regions where contact was defined are shown in Figure 2. Notice that the interfaces between the bolts and the respective parts are tied in Abaqus instead of having contact defined because Jewell et al. [9] found these regions to contribute a negligible amount of damping and nonlinearity when similar joints were in a state of microslip.

The model was meshed with 3D, 8 node linear brick elements with reduced integration (C3D8R in Abaqus). In order to reduce the number of nodes in the model a separate part in Abaqus was created for the contact region on the box and is shown in Figure 4. These parts were then bonded together using a tie in Abaqus.

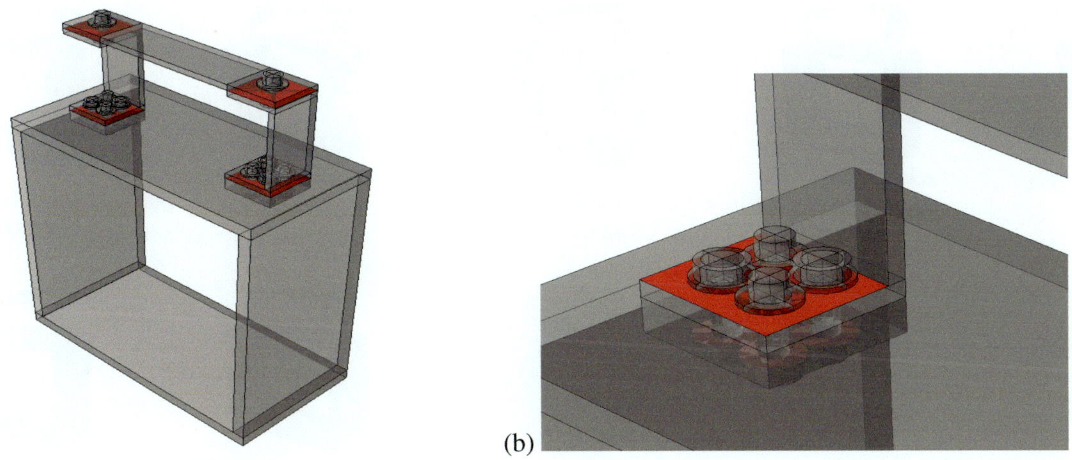

(a) (b)

Fig. 2 (a) surfaces where contact was defined and (b) example of where a tie was used between the bolt heads and their mating surface.

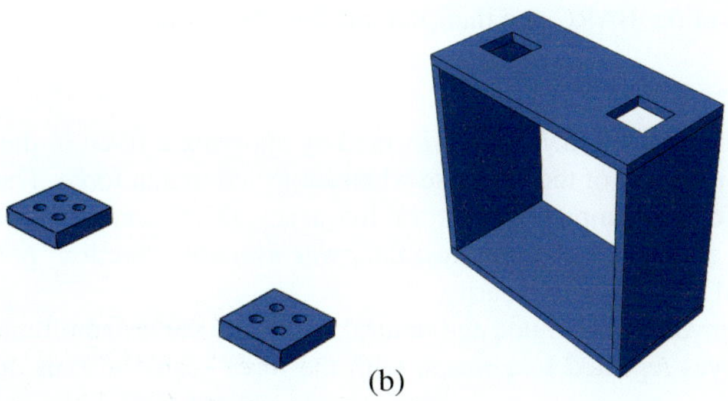

(a) (b)

Fig. 3 (a) Separate parts created for the contact region, which had a fine mesh so that the contact could be resolved. (b) Remaining part of the box assembly, which had a coarser mesh. The parts shown in (a) were connected to the box using tie constraints.

Splitting the box into three parts as shown allows one to use a coarser mesh on the box without losing the ability to mesh the contact region finely so that contact at the interface could be resolved. This technique brought the number of nodes from $\approx$ 218,000 in our initial model to $\approx$ 110,000, essentially cutting the size of the model in half. These models are denoted the 218k model and the 110k model respectively in the following. To verify this technique, modal analysis was performed on both models and the natural frequencies of the first ten elastic modes were compared. The maximum and average errors were 3.24% and 2.05% respectively between the two models. This confirmed that the two models were acceptably similar while also minimizing the number of nodes. The authors also created a third model, called the 190k model, which had a similar density of elements as the 110k model over the box, but a much finer mesh over the contact region. The 218k model was used to compute the linear natural frequencies and mode shapes in all results below, because that computation was fast with either model. In the following subsection the results of quasi-static modal analysis are shown for these models and the computational savings are discussed.

Abaqus then solved for the linear mode shapes and natural frequencies of the preloaded structure with free-free boundary conditions. Figure 4 shows the first ten resulting mode shapes and natural frequencies. The results were compared to [5, 12] and found to agree reasonably well with their published natural frequencies and mode shapes.

Nonlinear Predictions from the FEM

Once the authors were comfortable that the model was giving accurate results, the mode shapes were examined to determine which might induce the most slip in the joints and hence the most nonlinearity. Mode 2 was chosen as the first promising candidate because vibration in that mode shears the RC. Quasi-static modal analysis (QSMA) [8] was applied to the FEM

(a) Mode 1
391.77 Hz

(b) Mode 2
426.02 Hz

(c) Mode 3
523.77 Hz

(d) Mode 4
664.21 Hz

(e) Mode 5
977.38 Hz

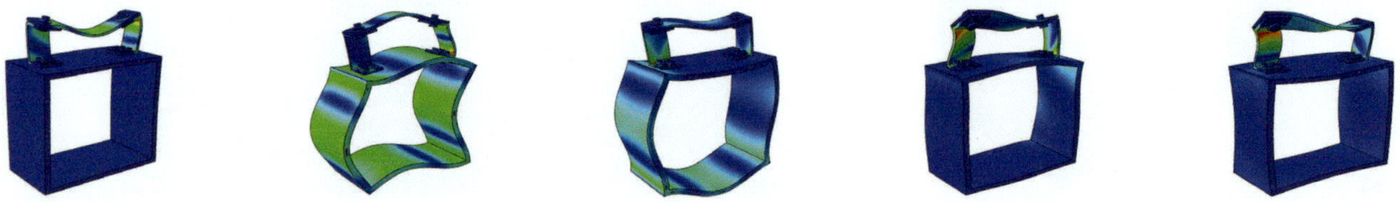

(f) Mode 6
1072.9 Hz

(g) Mode 7
1359.7 Hz

(h) Mode 8
1516 Hz

(i) Mode 9
1655 Hz

(j) Mode 10
1699 Hz

Fig. 4 First ten elastic modes of the BARC and their natural frequencies in Hz

to predict the nonlinearity in this mode. QSMA is performed by applying a force in the shape of the mode of interest, and solving for the nonlinear static response of the structure when subjected to that force. The static load displacement curve can then be used to compute the effective damping and natural frequency. Details on the QSMA process can be found in [8]. The predictions from QSMA were obtained before any test data was available. Section presents the test results, which reveal which modes actually exhibited the most nonlinearity.

Figure 5 shows the damping versus amplitude and natural frequency versus amplitude curves obtained for Mode 2 using the 218k model. The process was repeated four times, with the force scaled to four different amplitudes, denoted by the

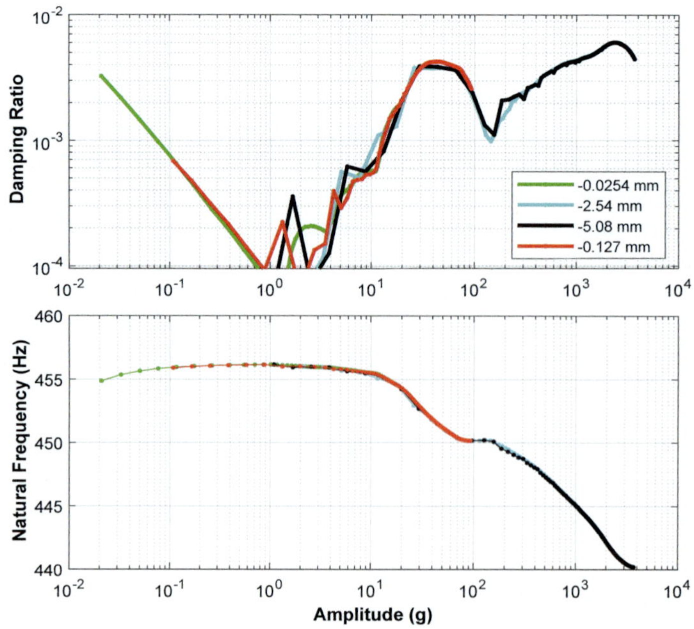

Fig. 5 Damping and Frequency vs. Amplitude for Mode 2. Each line represents a computation up to s different maximum force. These results are for the 218k model.

maximum displacement that each force would produce for the linear structure. These values ranged from 0.0254mm (0.001 in) to 5.08mm (0.2 in). The fact that the four curves agree with each other helps to verify that the behavior is physical and not an artefact of the numerical solution or function of the contact settings [9].

Figure 5 shows an increase in damping with amplitude as well as a decrease in frequency with amplitude. The damping increases with amplitude, up to a maximum value of about $\zeta = 0.005$, while the natural frequency decreases by about 15 Hz (3.3%). One would expect to be able to detect this level of damping and stiffness nonlinearity in a test. The QSMA predictions below about 1.0g are noisy, presumably because the mesh isn't fine enough to smoothly resolve the stick and slip transitions at this amplitude, so these should be ignored. At low amplitudes material damping would dominate so predictions are not needed for low amplitudes. At higher amplitudes the FEM shows microslip in the joints, which would likely be the source of the increase in damping observed here. Figure 7 shows the changing contact area and the slip in the joints as the analysis progresses. The mesh seems to be adequate to resolve a gradual change in the contact area and the sticking and slipping regions, providing additional confidence in the predictions.

Figure 6 compares the results between the 218k and 110k models. Although there is a shift in the frequency the trends between the two match well which helps validate this modeling technique. The 218k model, which had a single part for the box, took approximately 5.5 hours for the entire QSMA process while the 110k model, which had separate parts for the contact regions, finished in just under 3 hours. The fact that the results match suggests that one can reduce the computation time while keeping accuracy when examining a contact interface. This is especially helpful when one needs to run the model many times, for example to examine nonlinearity in several modes, or to cover a wide range of amplitude for a single mode. The frequency and damping for the 190k model were also nearly identical to these, so they were not shown here for brevity. The contact status for the 190k model was shown in Figure 7, where one can see that the overall areas of slip and stick were similar between the two models.

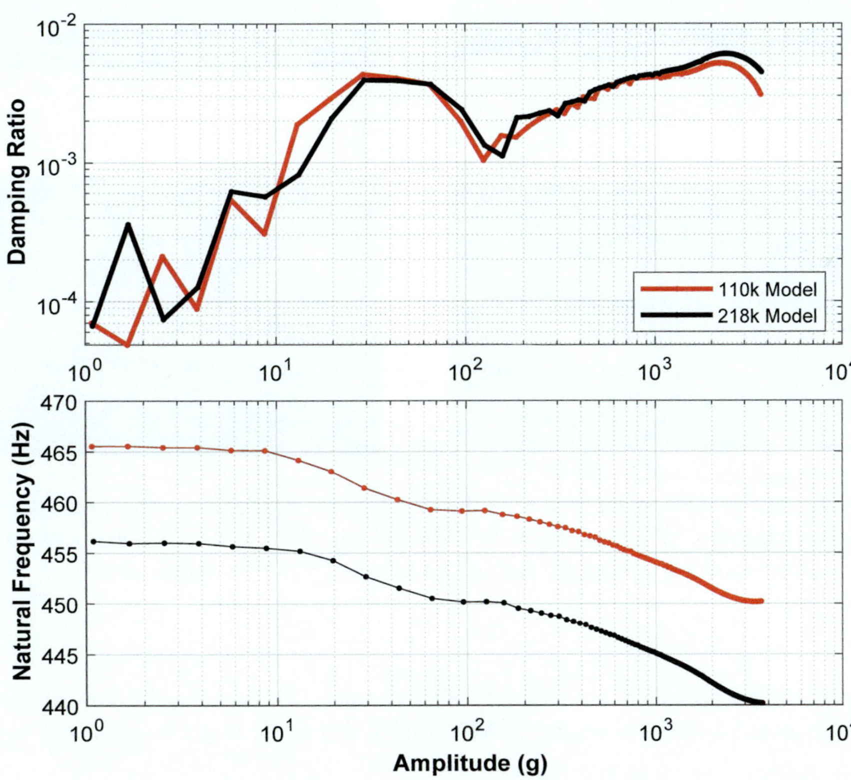

Fig. 6 Damping and Frequency vs. Amplitude for mode 2. The red line is from the 110k model and the black line is from the 218k model.

The authors also suspected that Mode 6 might exhibit nonlinearity, as the deformation of the removable component tends to want to open and close the bolted joints when the structure vibrates in that mode. QSMA was also attempted on this mode. Force displacement curves were obtained for a force applied in the shape of Mode 6 in both the positive and negative directions (i.e. a force that drives the removable component upward and downward respectively). These curves were compared and found to be significantly different. As a result, the standard QSMA approach could not be used on this

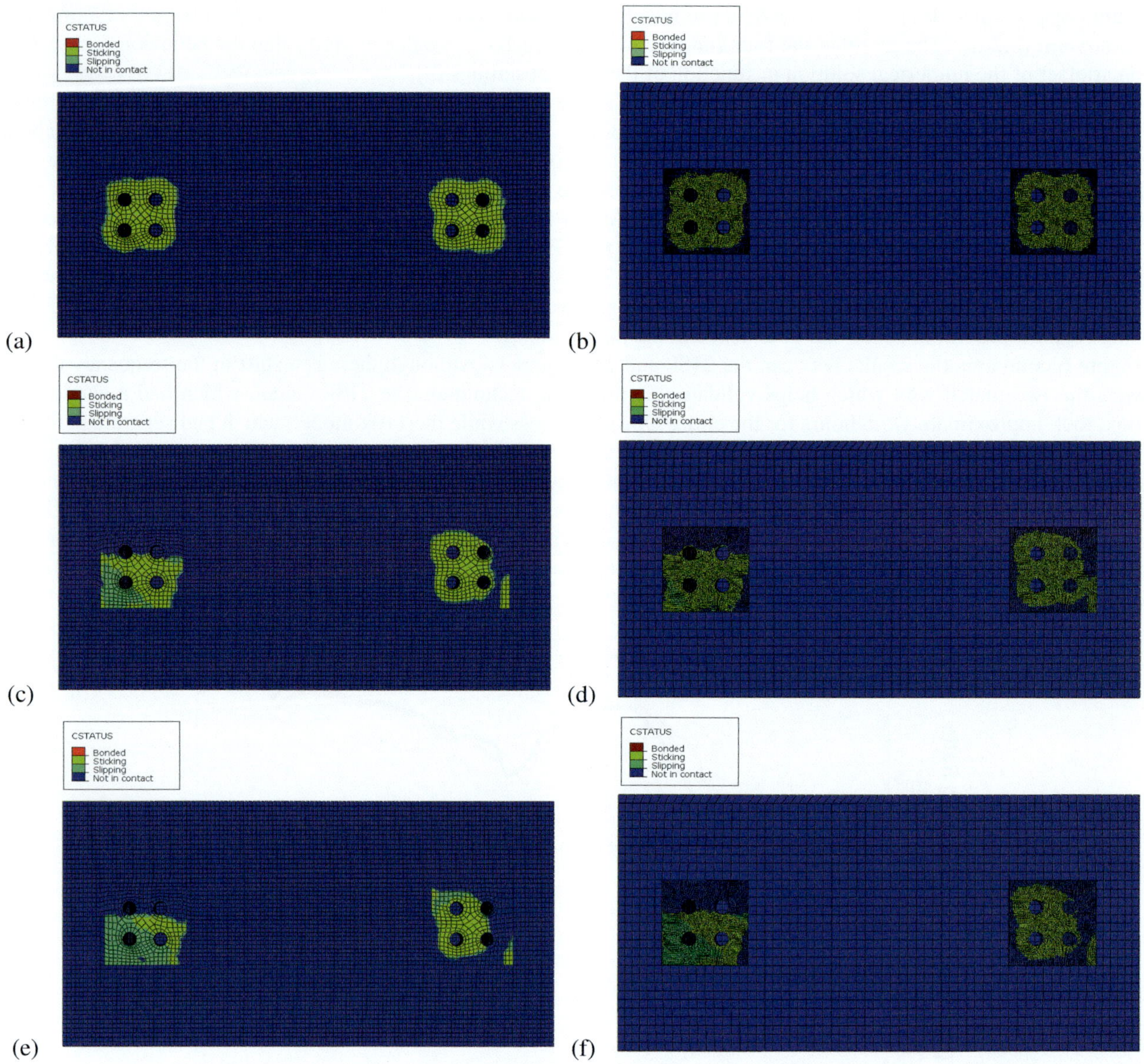

Fig. 7 The slip or stick status of the contact region on the box. (a), (c) and (e) show the 218k model, which has $\approx$ 5,600 nodes in the contact region while (b), (d) and (f) show the 190k model, which has $\approx$ 44,600 nodes on the contact region. (a) and (b) show the status at the beginning ($\approx$ 1g), (c) and (d) the middle ($\approx$ 500g), and (e) and (f) at the end ($\approx$ 4000g) of the QSMA analysis.

mode. One could instead use the approach developed by Shetty et al. [13], but time did now allow the authors to pursue this. instead it was simply noted that, based on the QSMA results, this mode might be expected to exhibit a bilinear-type nonlinearity.

Physical Testing

The BARC structure was manufactured per the drawings and instructions available on the Sandia website, which is in the process of being transitioned to another site hosted by SEM. A few of the washers had to be ground slightly so they wouldn't interfere when everything was assembled, but otherwise the structure was built per the drawings and reasonable tolerances seemed to have been achieved.

Linear Testing and Data Analysis

The BARC structure was hung from a metal frame by looping rope through the box component, then connecting the rope to 0.125 inch bungee cord, which then attached to the frame. Accelerometers were attached to several points, denoted $131Z+$, $133Z+$, $135Z+$, $231X-$, $231Y-$, $231Z+$, and $237Y-$ and shown in Figure 8. Using a tiny PCB impact hammer, each point was hit 3 times and averaged to obtain the frequency response functions (FRF) between the hammer and all accelerometers. The CMIF for the set of FRFs that was obtained is shown in Figure 9, and confirms that each mode appears at a distinct frequency in the spectrum.

(a) 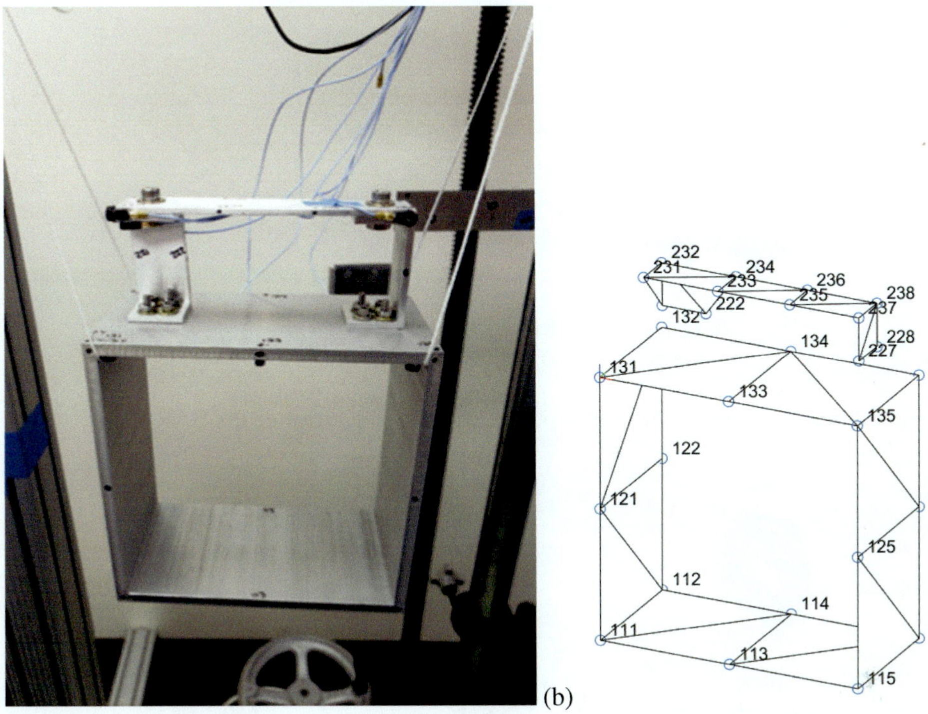 (b)

Fig. 8 (a) Photo of test setup for the BARC structure and (b) Measurement Grid

The set of FRFs was curve fit using the Algorithm of Mode Isolation and the modes below 2kHz were extracted. These are shown in Figure 10.

Nonlinear Testing and Data Analysis

A series of impacts was performed and the measurements processed to detect and characterize the nonlinearity in each of the first six modes. Based on the mode shapes shown in Figure 10, points for testing were chosen based on where the greatest displacement of the removable component was. A medium-sized hammer (PCB 086-C0x) was used for all modes except Mode 3. For that mode both a medium and large (PCB O86-D0x) hammers were used. Hence, the excitation location should have been that which best excites each mode. Data was processed in MATLAB using the Hilbert Transform as outlined in [14], which estimates the natural frequency and damping ratio as a function of vibration amplitude. The curves from various impacts were then combined into a single graph to check consistency.

Mode 1 responds in both the X and Z directions, and so impacts were applied in both directions. Figure 11 shows natural frequency and damping that were obtained from each of various impacts. The legend gives the impact number, maximum force, and the location and direction of the impact ($237X-$,$238X-$ or $231Z-$). The response was measured at point $231X-$ for X-direction hits and $231Z-$ for Z-direction hits. The amplitude on the x-axis is that of the acceleration at that point, which should be close to the maximum acceleration for the mode in question. Both sets of measurements show minimal change in the damping ratio, as the impact with the greatest damping ratio range (impact 28) only has a net change in damping of 0.08%. The natural frequency also shows a small change with amplitude. The results are similar for the Z-direction in Figure 11(b), although these results seem to show a sudden decrease in the natural frequency at higher amplitudes. The Z-direction measurements were taken on two different days, and it is interesting to note that the linear

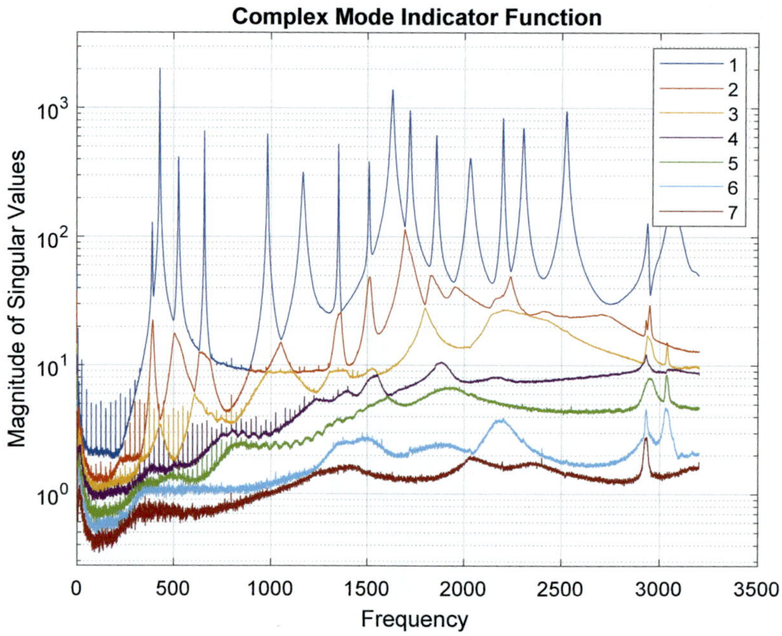

Fig. 9 Complex Mode Indicator Function for FRFs obtained in the linear test of the BARC

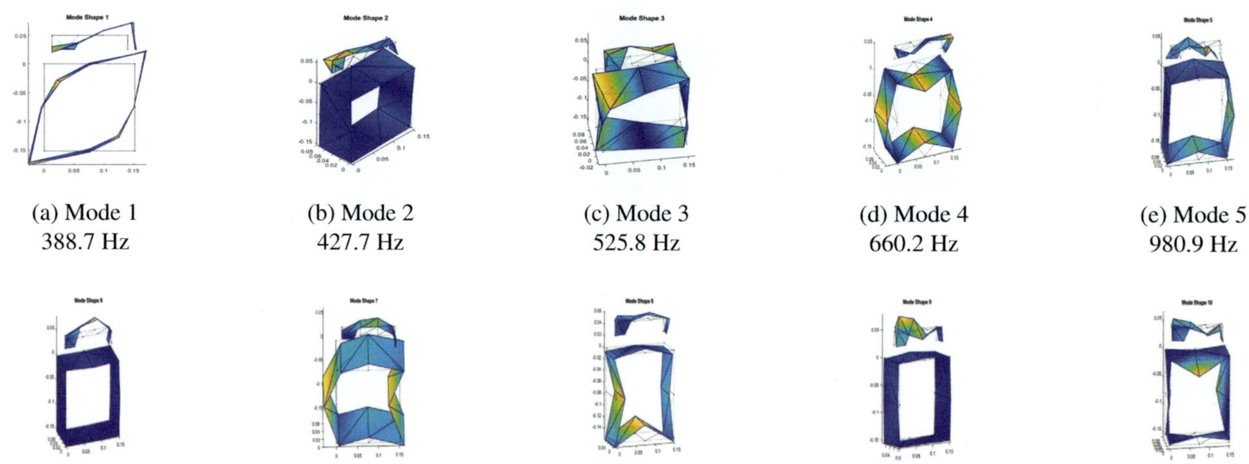

(a) Mode 1
388.7 Hz

(b) Mode 2
427.7 Hz

(c) Mode 3
525.8 Hz

(d) Mode 4
660.2 Hz

(e) Mode 5
980.9 Hz

(f) Mode 6
1166.8 Hz

(g) Mode 7
1350 Hz

(h) Mode 8
1508.2 Hz

(i) Mode 9
1631.6

(j) Mode 10
1721.1 Hz

Fig. 10 First ten elastic modes of the BARC and their natural frequencies in Hz

natural frequency appears to have shifted by about 0.2 Hz from one day to the next. In summary, Mode 1 a slight softening (stiffness) nonlinearity, yet it is small relative to experimental uncertainty.

Data for Mode 2 was taken on two separate days, 6/4/2024 and 6/7/2024, and as was the case for Mode 1, the linear natural frequency was slightly higher on the first day. The shift in the natural frequency is once again fairly small, only about 1.5Hz or 0.3%, all data does show the same trend. The damping increases by a factor of three or four. Both of these observations agree perfectly with the typical nonlinearity that is exhibited by bolted joints, as seen in [14, 15].

Data for Mode 3 was all collected on the same day, but using both the medium hammer (denoted *mh*, purple-yellow lines) and the big hammer (denoted *bh*, green and blue lines) in Figure 13. This mode was active along the top of the removable component near the center, so points 233-236 were excited in the negative Z direction. The big hammer didn't seem to be

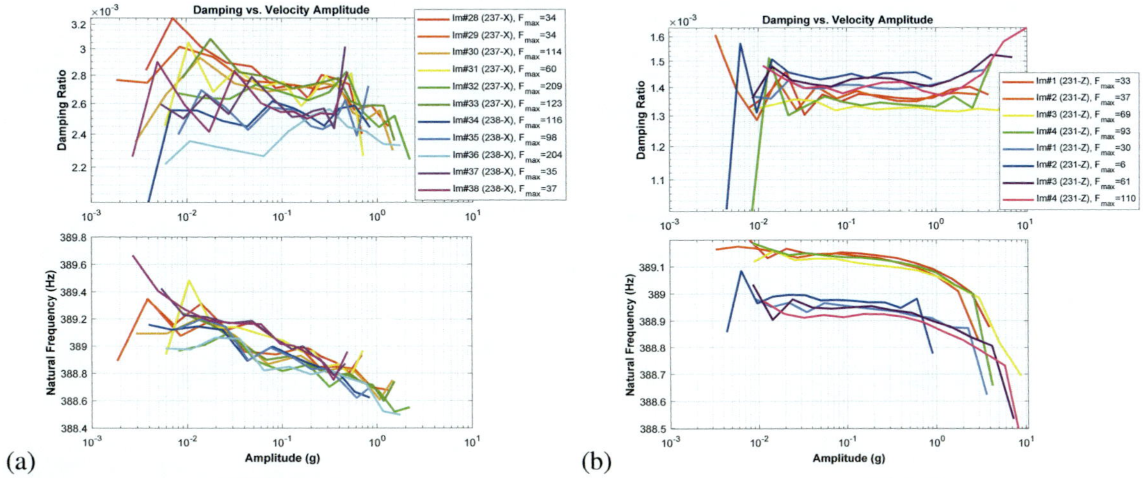

Fig. 11 Mode 1 Natural Frequency and Damping vs. Amplitude. The legend gives the impact number, maximum force, and the location and direction of the impact ($237X-$,$238X-$ or $231Z-$). Amplitude is that measured at points $231X-$ and $231Z-$ respectively.

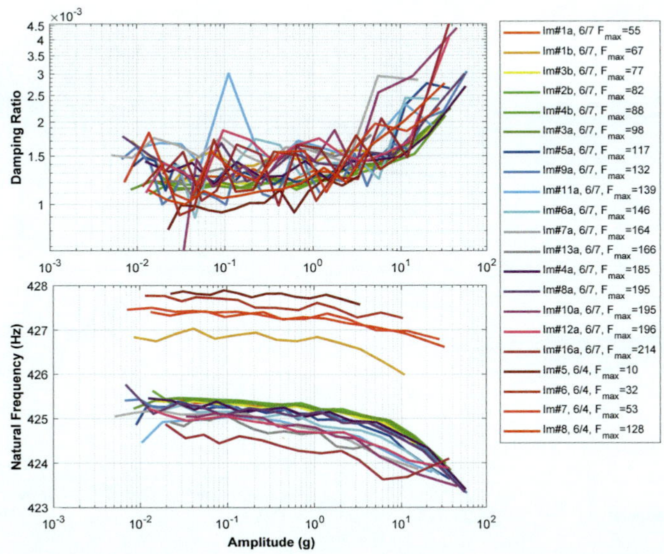

Fig. 12 Mode 2 Natural Frequency and Damping vs. Amplitude. All impacts were applied at $232Y-$ and amplitude is that measured at points $231Y-$.

able to excite any nonlinearity in this mode, as the measurements showed little to no change in the damping and natural frequency. The tip on that hammer may have been too soft to excite this mode effectively. The medium hammer did appear to produce changes at high amplitudes, the overall changes in the damping ratio (5e-4) and natural frequency (0.6 Hz or 0.2%) were fairly small.

Mode 4 was tested on 6/14/2024 (red-yellow lines) and 6/18/2024 (green-purple lines) at points 233 and 234 in the –Z direction. As with previous modes the linear natural frequencies were at slightly different values based on the day the data was taken, The damping shows no significant trend and while the frequency did decrease consistently on the second day of testing, the change was small (0.4 Hz or 0.06%).

Modes 5 and 6, shown in Figure 14, were somewhat difficult to process using the Hilbert Transform because they are somewhat close together and decayed quickly relative to the other modes. Hence, the results were somewhat noisy. Even then, the variation in these modes is small suggesting that they are behaving linearly.

To confirm that these modes were truly linear, the frequency response (FRF) was computed by dividing the FFT of the response by the FFT of the input force. The result, in Figure 15, shows that there is not a significant change in the stiffness

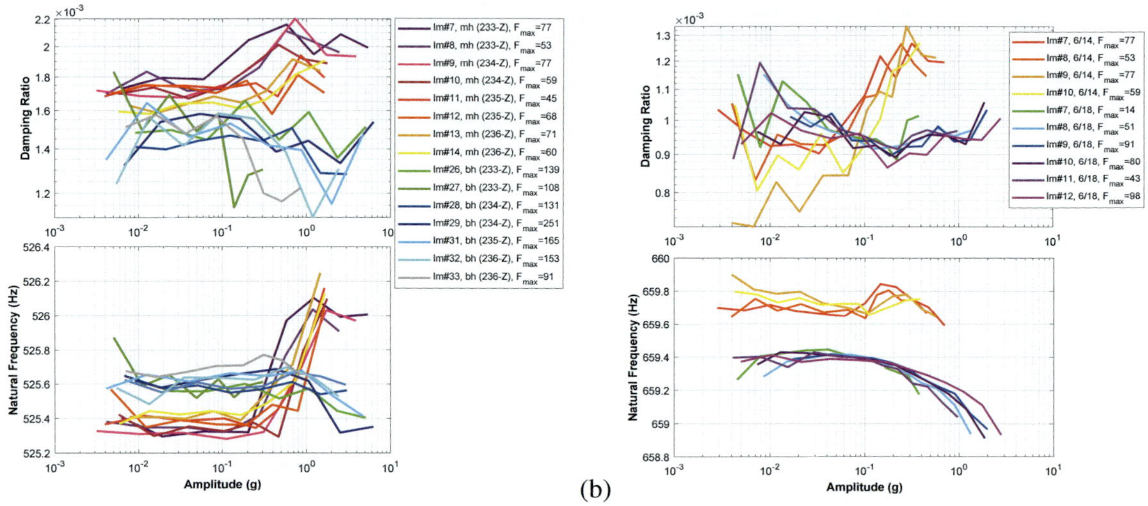

(a) (b)

Fig. 13 (a) Mode 3 and (b) Mode 4 Natural Frequency and Damping vs. Amplitude. For Mode 4, impacts 9 and 10 were at $234Z-$ and all others were at $233Z-$. Amplitude is that measured at point $231Z-$.

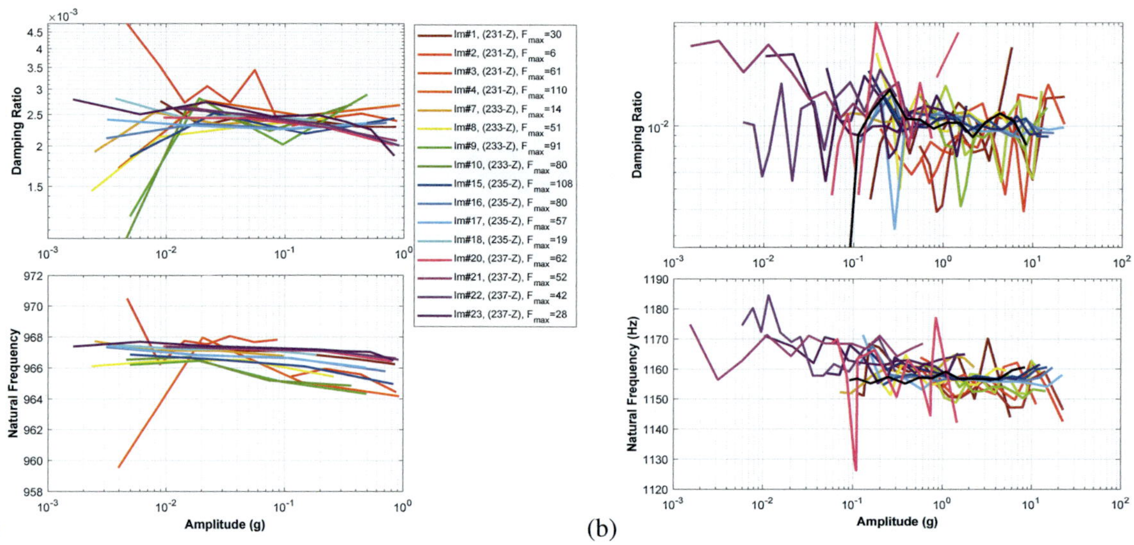

(a) (b)

Fig. 14 (a) Mode 5 and (b) Mode 6 Natural Frequency and Damping vs. Amplitude. Amplitude is that measured at point $231Z-$.

or damping of these modes for the various impacts. Mode 5 does show an amplification of about 50% for Impact 17, as compared to the other tests, but the trend is not consistent for Impacts 18 and 19, which also had lower amplitudes.

Comparing Test Results and Predictions

Linear Mode Shapes and Natural Frequencies

The mode shapes, shown previously in Figures 4 and 10 show a strong visual correspondence between the FEM and test. These are repeated in Figures 17 and 18 in the Appendix to allow an easier comparison. The natural frequencies are compared in Table 1. They differed by at most 1.7% except for Mode 6, which differed by 8.0%. This is an opening-closing mode; the other modes don't involve this type of motion. Most of the natural frequencies agreed to within 1-2%.

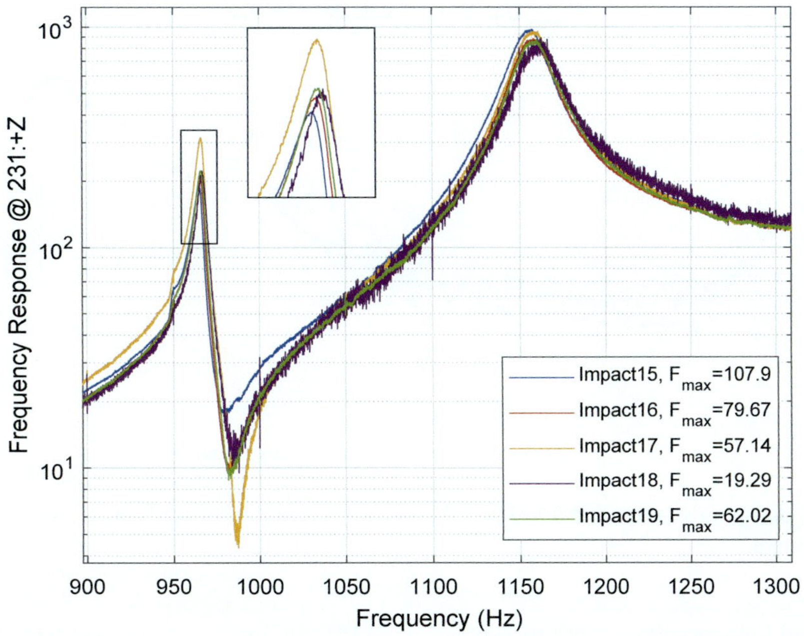

Fig. 15 Frequency Response Functions (FRFs) computed from several high amplitude impacts, to quantify nonlinearity in Modes 5 and 6.

Table 1 Comparison of Natural Frequencies between FEM and Test

Mode	FEM	Test	% Diff
1	391.77	388.7	0.79
2	426.02	427.7	-0.39
3	523.77	525.8	-0.39
4	664.21	660.2	0.61
5	977.38	980.9	-0.36
6	1072.9	1166.8	-8.0
7	1359.7	1350	0.72
8	1516	1508.2	0.52
9	1655	1629.3	1.6
10	1699	1721.1	-1.3

Nonlinearity

As was expected based on the FE analysis described in Section , Mode 2 showed the strongest nonlinearity in the test data. Furthermore, that mode exhibited a classical bolted-joint type nonlinearity, so one would expect QSMA to be effective at predicting its behavior. Figure 5 and 12 both show an increase in damping with amplitude as well as a decrease in frequency with amplitude. The damping also increases by a similar amount in both cases (about half an order of magnitude for a two-order of magnitude increase in vibration amplitude). These results are overlaid in Figure 16. To facilitate the comparison, material damping of $\zeta = 0.001$ was added to the FEM predictions, as this was clearly manifest in the test data. Note that the horizontal axis for the FEM data is the peak modal amplitude, or the product of the modal acceleration and the mode shape at the point where the displacement is largest. For the test data the horizontal axis is the acceleration amplitude at point $231Y-$, which appears to be the point of maximum displacement.

The FEM and Test results agree qualitatively, and have similar orders of magnitude. The anomaly in the damping that was observed in the FEM predictions around 3-6g happens to be the same region in which the test data shows the strongest nonlinearity; hence it would be wise to refine the model to make sure that the contact is well resolved for this amplitude. In contrast, if one extrapolates the power law behavior that the FEM exhibits from 200-2000g, that seems to agree with the lower bound of the test data. The lower bound is typically of interest in measurements such as this, as it

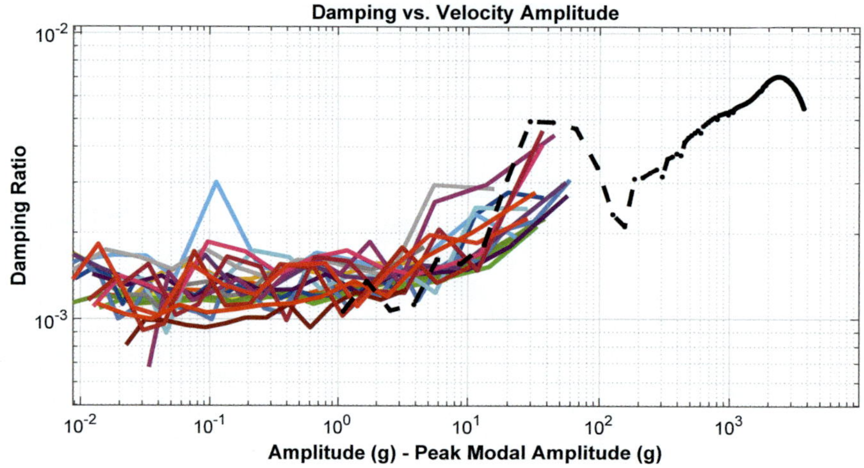

Fig. 16 Comparison of damping vs vibration amplitude between FEM and test for Mode 2: (black dashed) FEM Prediction, (colored lines) Measurements

would have the minimum influence from other modes (i.e. see [15]) and hence agrees best with the assumptions made in QSMA.

It is surprising that the results agree this well. In [10] the authors found that one had to model the interface very carefully, including any lack of flatness, to accurately capture damping.

The FEM analysis also noted that Mode 6 was likely to behave nonlinearly, with an opening-closing type nonlinearity. The measurements did not show significant nonlinearity in Mode 6. However, it should be noted that bilinear-type nonlinearities may not exhibit a significant change in the resonance frequency with vibration amplitude [16]. Further analysis would be required to see if the test data confirms any bilinear behavior of Mode 6.

Conclusion

The Box Assembly with Removable Component (BARC) was modeled using Abaqus FEA software in order to predict the linear mode shapes and natural frequencies. Then QSMA was performed on Mode 2, which was suspected to have nonlinear behavior, in order to predict the amplitude dependant damping ratio and natural frequency. These predictions from the FEM were complete before physical testing on the BARC structure was performed, thus giving blind predictions of the linear and nonlinear behavior based only on the nominal design of the components.

Once the FEM predictions were well along, the BARC was manufactured and linear and nonlinear testing was performed on the physical BARC structure. Modal analysis was performed in MATLAB to extract the linear mode shapes and natural frequencies. The mode shapes were found to visually match those of the FEM and the linearized natural frequencies from the FEM matched the tests well with typical errors of less than 1% and a maximum of 1.7% error, except for mode 6, which had an error of 8.0%. Mode 6 involves opening and closing of the joint while the other modes do not have this behavior. The first six modes were examined for nonlinearity and Mode 2 was found to exhibit a typical joint-type nonlinear damping behavior while the other modes had minimal nonlinearity. The nonlinearity predicted by applying QSMA on Mode 2 agreed quite well with the measurements.

This study demonstrates the feasibility of using finite element analysis and QSMA to predict the linear and nonlinear behavior of a component such as the BARC. The linear and nonlinear FEM predictions agreed very well with measurements in this study, but, as was cautioned earlier, research on other structures with joints [10] suggests that prediction may often be much more challenging. Still, these results are encouraging and motivate further use of these predictive tools.

Appendix

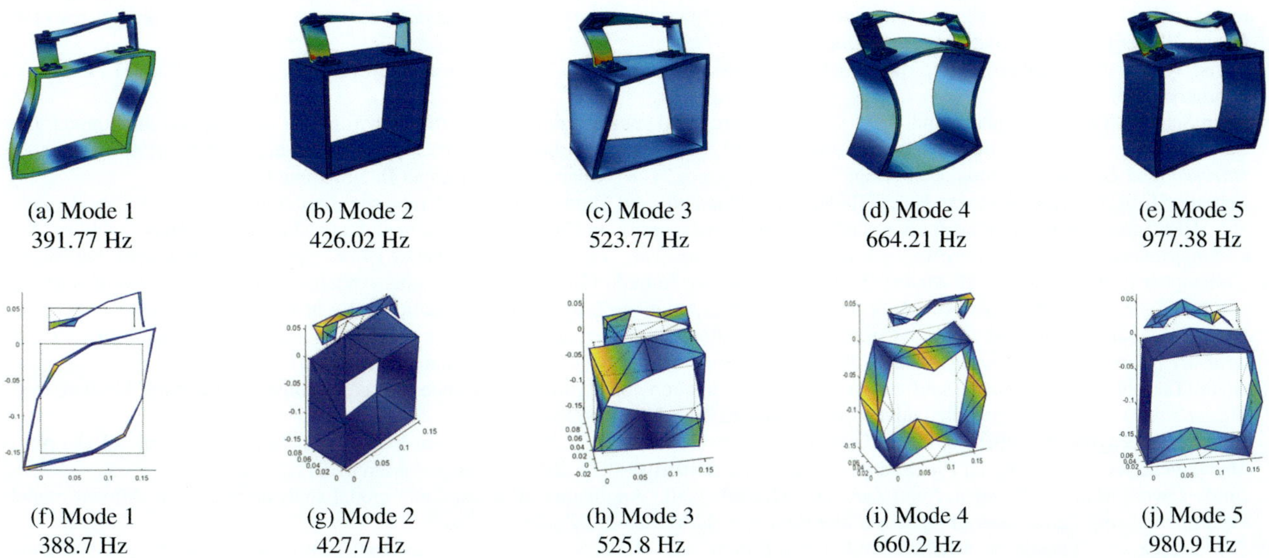

(a) Mode 1 391.77 Hz
(b) Mode 2 426.02 Hz
(c) Mode 3 523.77 Hz
(d) Mode 4 664.21 Hz
(e) Mode 5 977.38 Hz
(f) Mode 1 388.7 Hz
(g) Mode 2 427.7 Hz
(h) Mode 3 525.8 Hz
(i) Mode 4 660.2 Hz
(j) Mode 5 980.9 Hz

Fig. 17 Modes 1-5 with the FEM being on top and the mode shapes from tests on bottom.

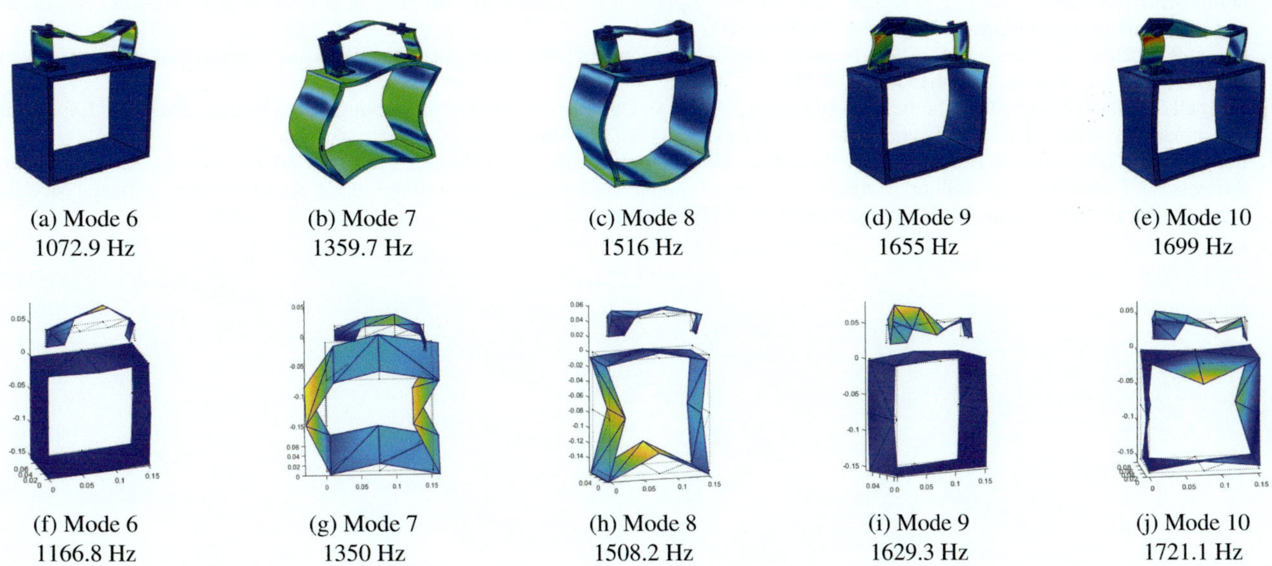

(a) Mode 6 1072.9 Hz
(b) Mode 7 1359.7 Hz
(c) Mode 8 1516 Hz
(d) Mode 9 1655 Hz
(e) Mode 10 1699 Hz
(f) Mode 6 1166.8 Hz
(g) Mode 7 1350 Hz
(h) Mode 8 1508.2 Hz
(i) Mode 9 1629.3 Hz
(j) Mode 10 1721.1 Hz

Fig. 18 Modes 6-10 with the FEM being on top and the mode shapes from tests on bottom.

References

1. David E. Soine, Richard J. Jones, Julie M. Harvie, Troy J. Skousen, and Tyler F. Schoenherr. Designing hardware for the boundary condition round robin challenge. In Michael Mains and Brandon J. Dilworth, editors, *Topics in Modal Analysis & Testing, Volume 9*, pages 119–126, Cham, 2019. Springer International Publishing.
2. Alvis, Tyler and Schoenherr, Tyler. Best Practices for Modeling Bolted Joints: Calibrating the BARC System. Orlando, FL, United states, February 2024.
3. Ryan Schultz, Tyler Schoenherr, and Brian Owens. A proposed standard random vibration environment for barc and the boundary condition challenge. In Chad Walber, Matthew Stefanski, and Steve Seidlitz, editors, *Sensors and Instrumentation, Aircraft/Aerospace, Energy Harvesting & Dynamic Environments Testing, Volume 7*, pages 77–84, Cham, 2022. Springer International Publishing.
4. Sebastian Chirinos, Aneesh Pawar, Haley Tholen, Scott Ouellette, and Thomas Roberts. Parametric simulations of the BARC model in SDOF and MIMO configurations for estimating service environment severity. In Chad Walber, Matthew Stefanski, and Julie Harvie, editors, *Sensors and Instrumentation, Aircraft/Aerospace and Dynamic Environments Testing, Volume 7*, pages 19–40. Springer International Publishing.
5. Christopher L. Padilla, Jonah M. Madrid, Ezekiel C. Granillo, Antonio Flores, and Abdessattar Abdelkefi. *Dynamics and nonlinear characterization of BARC systems with varying central cut widths*. American Institute of Aeronautics and Astronautics, Inc, 2024.
6. Ezekiel C. Granillo, Jonah M. Madrid, Christopher L. Padilla, Jorge Perez, and Abdessattar Abdelkefi. *Insights on the scalability of the BARC structure on its dynamical characteristics*. American Institute of Aeronautics and Astronautics, Inc, 2024.
7. Adam Takeshita. Experimental and computational investigations on fixture interference and accelerometer placement for barc systems. Master's thesis, New Mexico State University, Las Cruces, NM, USA, November 2022.
8. Robert M. Lacayo and Matthew S. Allen. Updating Structural Models Containing Nonlinear Iwan Joints Using Quasi-Static Modal Analysis. *Mechanical Systems and Signal Processing*, 118(1 March 2019):133–157, 2019. Number: 1 March 2019.
9. Emily Jewell, Matthew S. Allen, Iman Zare, and Mitchell Wall. Application of quasi-static modal analysis to a finite element model and experimental correlation. *Journal of Sound and Vibration*, 479(0022-460X):115376, 2020.
10. Seyed Iman Zare Estakhraji, Mitchell Wall, Jacob Capito, and Matthew S. Allen. A thorough comparison between measurements and predictions of the amplitude dependent natural frequencies and damping of a bolted structure. *Journal of Sound and Vibration*, 544:117397, February 2023.
11. Mitchell Wall, Matthew S. Allen, and Robert J. Kuether. Observations of modal coupling due to bolted joints in an experimental benchmark structure. *Mechanical Systems and Signal Processing*, 162(0888-3270):107968, 2022.
12. Daniel Peter Rohe. Modal data for the barc challenge problem test report. Technical Report SAND-2018-0640R; 660094, Sandia National Lab, 2018.
13. Drithi Shetty, Mathew S. Allen, and Kyusic Park. A New Approach to Model a System with Both Friction and Geometric Nonlinearity. *Journal of Sound and Vibration*, 522(26 May 2023):117631, 2023.
14. Daniel R. Roettgen and Matthew S. Allen. Nonlinear characterization of a bolted, industrial structure using a modal framework. *Mechanical Systems and Signal Processing*, 84:152–170, 2017.
15. Mitchell P. J. Wall, Allen Allen, Matthew S, and Robert J. Kuether. Observations of Modal Coupling due to Bolted Joints in an Experimental Benchmark Structure. *Mechanical Systems and Signal Processing*, 162:107968, 2022.
16. Benjamin R. Pacini, Wil A. Holzmann, and Randall L. Mayes. Performance of Nonlinear Modal Model in Predicting Complex Bilinear Stiffness. In Gaetan Kerschen, editor, *Nonlinear Dynamics, Volume 1*, Conference Proceedings of the Society for Experimental Mechanics Series, pages 101–112, Cham, 2019. Springer International Publishing.

Chapter 2
www.FreeDyn.at – A free, GUI based multi-body dynamics software for stand-alone use and with C API for linking it to Python, Matlab or custom code

Wolfgang Witteveen and Stefan Oberpeilsteiner

Abstract FreeDyn is a free multibody simulation (MBS) package developed by the Upper Austrian University of Applied Sciences. The precompiled Windows binaries can be downloaded from www.freedyn.at. It features a graphical user interface (GUI) and a C++ solver which is a modified version of the HHT time integration method. FreeDyn can be used as standalone software via its GUI or can be linked to other software such as Python, MATLAB, Scilab or custom codes through a C-Interface.

The new Python interface makes FreeDyn particularly attractive to universities and research groups. Researchers can outsource multi-body simulations to FreeDyn while addressing unique research questions using Python to call FreeDyn functions. FreeDyn supports rigid and flexible bodies, force elements and constraints. Users can define state and time-dependent force and constraint elements using mathematical expressions. Additionally, advanced force elements may be linked to FreeDyn using a user written subroutine (DLL).

The extended abstract contains a brief review of the underlaying theoretical concepts, a short summary of all modelling elements and of all callable FreeDyn functions. Furthermore, a simple example will be given in detail to demonstrate the binding of FreeDyn to Python.

Keywords Multibody Dynamics · Free Software · Python

Introduction and Theoretical Background

FreeDyn [1] is a freely available Windows® software package for the pre- and post-processing and numerical time integration of multibody dynamic systems. A user interface (UI) supports the intuitive modelling of complex models via typical graphic oriented methods. The final model can be exported into a solver for the numerical time integration. The results can be read in again, animated and plotted. The automatically generated equations of motion take on the form

$$\begin{aligned} &\mathbf{M}(\mathbf{q})\ddot{\mathbf{q}} + \mathbf{C_q}^{\mathbf{T}}(\mathbf{q})\lambda = \mathbf{Q}(\mathbf{q}, \dot{\mathbf{q}}, \mathbf{t}) \\ &\mathbf{C}(\mathbf{q}, \mathbf{t}) = \mathbf{0} \end{aligned} \tag{1}$$

where the state vector $\mathbf{q}$ contains the degrees of freedom (dof) of the system. It consists of n sub-vectors containing the dof of the n modeled bodies. The translation of a rigid body is characterized by the 3 coordinates of the center of mass. For the parameterization of the rotation, Euler parameters are chosen. The matrix $\mathbf{M}$ is the state dependent mass matrix with n block diagonal entries according to the mass matrices of the individual bodies. The vector $\mathbf{Q}$ contains the generalized forces and the quadratic velocity vector. The constraint equations are collected in the vector $\mathbf{C}$ and depend on the state vector $\mathbf{q}$ and the time t. The constraint forces are computed via the constraint Jacobian $\mathbf{C}_q{}^T$ and the vector of the Lagrange multipliers $\boldsymbol{\lambda}$. Readers who are more interested can find more details in the following list:

- On the numerical time integration via an HHT solver see [2].

Wolfgang Witteveen
University of Applied Sciences Upper Austria, Degree Program Mechanical Engineering, 4600 Wels, Austria
e-mail: wolfgang.witteveen@fh-wels.at

Stefan Oberpeilsteiner
principia MBS GmbH, 4113 St. Martin, Austria, www.principia-mbs.com
e-mail: stefan@principia-mbs.com

Ludovic Renson et al. (eds.), *Nonlinear Structures & Systems, Vol. 1*, Conference Proceedings of the Society for Experimental Mechanics Series, https://doi.org/10.13052/9788743801467-2

- On known problems concerning numerical damping in connection with Euler parameters see [3].
- On flexible bodies see [4] and [5].
- Additional publications on different theoretical and application-oriented topics see [6], [7] and [8].

Available Modelling Elements

Bodies:

- Rigid bodies can either be selected from a library of simple geometries (Sphere, Cone, Cylinder, Box) or imported via stl or obj file.
- The representation of flexible bodies is based on the Floating Frame of Reference Formulation (FFRF). A text file is required for the import. If an FE solver does not support the implemented format, a corresponding parser can be written.

Data Objects:
Data objects can be used to specify parameters and characteristic curves that can be referenced by other FreeDyn elements. This allows the parametrization of a model.

- A parameter consists of a name and a value. It can be referenced from other FreeDyn elements via its name which will be replaced by the value when numerical time integration starts.
- A spline can be used to describe a relationship between two quantities. It is defined by certain numbers of interpolation points. Once defined, the spline can be used with special access functions in force elements and motions.

Measures:

- Translational measures can be used to measure the displacement or velocity between two points of two bodies. In case of a measure between a body and ground the acceleration can be measured as well.
- Rotational measures can be used to measure the rotational displacement or rotational velocity between two vectors of two bodies.

Constraints:
Constraints restrict the motion possibilities of two bodies relative to each other. One of the two bodies may be ground.

- A generic constraint restricts one motion along or around one axis. By combining several such generic constraints, arbitrary or well-known constraints such as "fixed" or "spherical" can be realized.
- Motion constraints can be used to impose a certain motion on a body. The type and characteristics of the motion can be defined with measures, common mathematical expressions and the former mentioned splines.
- A gear constraint can be used to link the rotations of two bodies via a gear ratio.

Forces:

- Generic forces/torques can act between two bodies or between one body and ground. Measures, data objects and mathematical expressions are available for a wide range of modelling possibilities of state and time dependent forces and torques.
- With the external subroutine force element, arbitrary self-written forces can be linked to FreeDyn via a dll. With this interface, it would also be conceivable, for example, to integrate force elements that are written in Python.
- The tire force element provides a Pacejka tire model which acts between a rotating body and a street profile.
- Force contact offers the possibility to define a contact model between two rigid bodies based on the work and code published by G. Hippmann, see [7].

Stand-Alone Software or C-Api

Figure 1 illustrates how FreeDyn can be used. The left-hand side indicates the data flow for the use as a stand-alone tool. The user mainly interacts with the graphical UI. Once the model has been built, the numerical time integration can be started via a corresponding dialog box and the results are automatically fed back into the UI.

The right-hand side of Figure 1 illustrates the use of the C-API. The graphical UI can still be used to build up a model and investigate the results. Once the model has been exported (= model file) it can be referenced via functions calls of the C-API. The FreeDyn model can then be used and manipulated inside self-written program code like Python, Matlab, Scilab or custom code. Furthermore, it can be used as a block inside a Matlab/Simulink or Scilab/XCos simulation. For each current time integration point, all kind of information can be requested from the solver. These are for example the displacements, velocities, accelerations, measures, constraint forces, Lagrange multipliers, constraint Jacobian, Jacobian of the full system and other quantities. All this data can be used of optimization tasks or whatever is needed. The interface can also be used to modify model parameters. These are e.g. integration parameters, model states or forces. To give an impression of the possibilities offered by the C-API, some of the available commands are listed and explained in more detail:

- The function call `createFreeDynModel(...)` creates a FreeDyn model based on a model file name.
- `setModelAsActiv(...)` gives the possibility to select a specific FreeDyn model. This is of particular importance when multiple models have been created.
- `getModelInfos(...)` returns information like the number of degrees of freedom, the number of Lagrange multipliers and others.
- With `getMeasureValueWithMeasureName(...)` the content of a measure can be requested.
- The function call `solveTimeInterval(...)` allows to solve for a time interval. This interval can be very long, but also be very short, such as one single integration time step.
- `getPhysicalDofRelatedVector(...)` provides all kind of internal vectors like the quadratic velocity forces, the acceleration inertia forces, the sum of external applied forces, and others more.
- `getModelRelatedMatrix(...)` provides all kind of internal matrixes like the state depended mass matrix, the derivation of the external forces with respect to the generalized coordinates, the constraint Jacobian and others more.

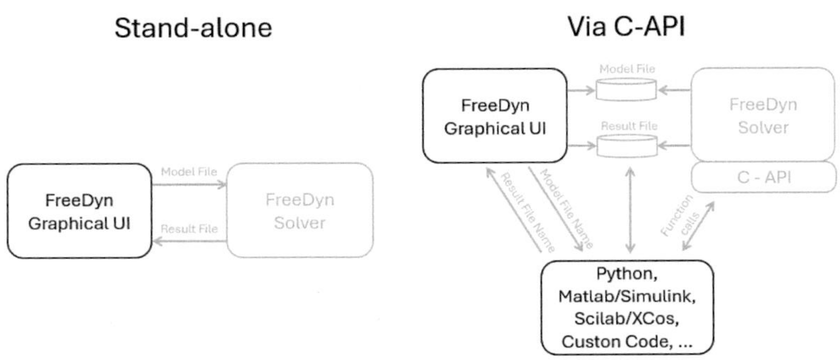

Fig. 1 FreeDyn use as stand-alone tool or via C-API

Python Example

Part of the latest release is the file *fdApi.py*. It can be included in Python and contains an interface to FreeDyn which can be controlled and used by the commands defined in this file. The functionality of this interface is explained using a simple demonstration example. The excitation force of a single-mass oscillator f(t) is searched via a so called "virtual iteration" in such a way that the mass follows a given motion x(t). The virtual iteration is a proven method of inverse dynamics and can be found in detail in section 3 of [13]. The method works in the frequency domain and requires the frequency response (FRF) functions between all outputs and inputs. In an iterative process, the input is changed in the frequency domain until the deviation of the output from the target curve is small enough. The input and output variables are transformed back and forth between the time and frequency domain, as the update based on the error takes place in the frequency domain, but the simulation takes place in the time domain. The single-mass oscillator was modeled in FreeDyn and the model file was stored accordingly. We assume that the FRF between the input f(t) and the output x(t) is already available. In the following the Python code is given for this example. The commands related to the FreeDyn API are highlighted with yellow color. Fig. 2 holds the result. It can be seen that the mass follows the specified motion.

```python
import numpy as np
import math
import sys
import os
import matplotlib.pyplot as plt
from ctypes import *
import utils
#-----------------------------------------------------------------------------------------
# Initialize FreeDyn API
# Add FreeDyn API to the system path
sys.path.insert(0, os.path.abspath(os.path.join(os.path.dirname(__file__), '..', '..', 'fdApi')))
import fdApi
fdApi.init()
#-----------------------------------------------------------------------------------------
# Initialize FreeDyn model
# Define filename+path of FreeDynModel
fds_file_path = os.path.abspath(os.path.join(os.path.dirname(__file__), "singleMassOscillator.fds"))
# Create and set model as active
model_index = fdApi.createModel(fds_file_path, 'no')
if model_index < 0:
    sys.exit("Error: Failed to create model")
fdApi.setModelAsActive(model_index)
states = fdApi.generateStateVectors()
#define name of variable to be modified (excitation force)
spline_label = "forceSpline"
#-----------------------------------------------------------------------------------------
# Prepare time vector according to model
v_time = utils.getTimeVector()
n_time_steps = v_time.size
#-----------------------------------------------------------------------------------------
# Load transferfunction and target Y-coordinate of single mass oscillator
H = utils.getTransferFunction(v_time)
Y_target = utils.getTargetSignal(v_time)
y_target = np.fft.irfft(Y_target, n_time_steps)
#-----------------------------------------------------------------------------------------
# Initialize variables
# compute first shot using transferfunction and target of cg_y
U = np.zeros_like(Y_target, dtype=np.complex128)  # Change dtype to complex128
for i, Yi in enumerate(Y_target):
    U[i] = Y_target[i] / H[i]
# Iteration parameters
it = 1; i_max = 1000; alpha = 1.0; eps = 0.001; e_old = 0.0; alpha_min = 1e-10
#-----------------------------------------------------------------------------------------
# Iteration loop
while True:
    # Update force on FreeDyn model
    u = np.fft.irfft(U,n_time_steps)
    spline_data = np.vstack((v_time, u)).T
    fdApi.modifySpline(spline_label, spline_data)
    # Simulate the model again
    if fdApi.solveEoM() < 0:
        sys.exit("Error: Simulation failed after modifying spline")
    # Copy generalized coordinates
    y = np.zeros(n_time_steps)
    for i in range(n_time_steps):
        fdApi.getStatesAtTimeIndex(i, states["Q"])
        y[i] = states["Q"][1, 0]
    Y = np.fft.rfft(y)
    # Compute error
    error = np.linalg.norm(y - y_target)
    utils.updatePlots(y,u, it, error)
    if error < eps:
        print("Converged: Error threshold reached")
        break
    elif it >= i_max:
        print("Failed to converge: Maximum iterations reached")
        break
    elif alpha < alpha_min:
        print("Failed to converge: Alpha below minimum")
        break
    # Update U and H
    if it > 1 and e_old < error:
```

```
        # Reduce alpha if the new result is not better
        U = np.copy(U_old)
        alpha /= 2.0
    else:
        U_old = np.copy(U)

    for i, Ui in enumerate(U):
        U[i] += alpha * (Y_target[i] - Y[i]) / H[i]
    # Store previous error value
    e_old = error
    #increment iteration counter
    it += 1
# Clean up and close
fdApi.deleteModel(model_index)
plt.ioff()
plt.show()
```

Additonal Examples

To give interested readers an even better idea of FreeDyn, some additional projects are listed below with corresponding references.

- In [8] an iterative coupling procedure of experimental and numerical subsystems to one overall system is introduced. The numerical subsystem is an offroad vehicle, which is modelled with FreeDyn, see the upper left picture in Fig. 3.
- In [10] FreeDyn is used for the development of motorcycle rider assistance systems. The motorcycle is modelled with FreeDyn and used as block in a Matlab/Simulink model. The upper right picture of Fig. 3 holds a screenshot of the Matlab/Simulink model.
- [11] reports on an optimization problem in which energy-optimal control of a robot motion are sought. A picture of the FreeDyn robot model is given in the lower right corner of Fig. 3.
- In [12], rolling processes of tooth flanks (lower left image of Fig. 3) are investigated under consideration of measured surface roughness.

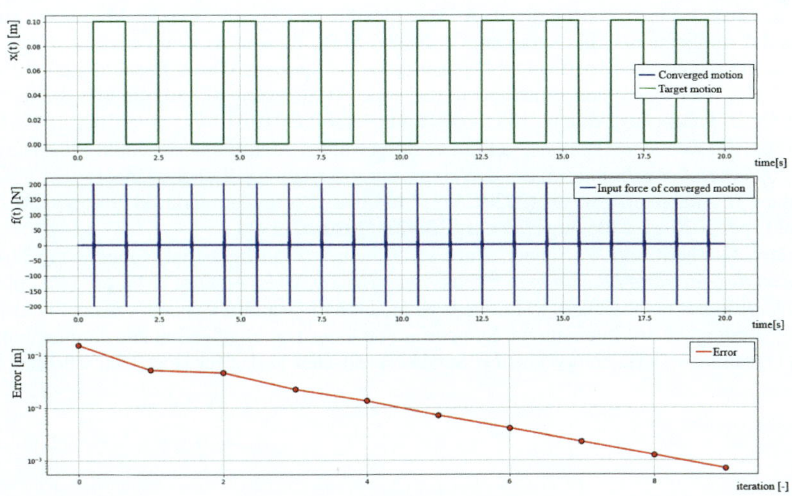

Fig. 2 Result of the virtual iteration.

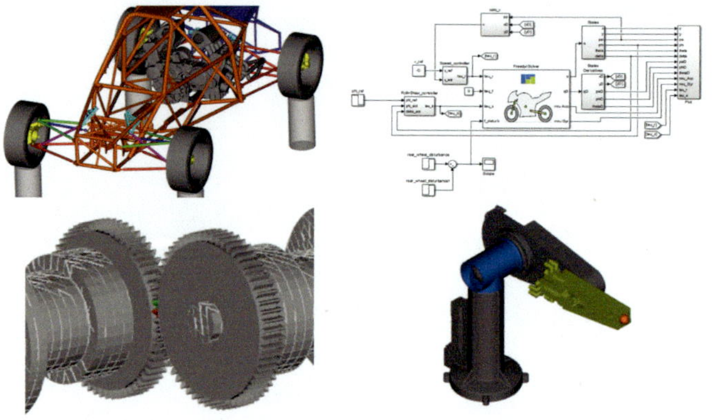

Fig. 3 FreeDyn examples.

References

1. www.freedyn.at
2. Negrut, D., Rampalli, R., Ottarsson, G., Sajdak, A., 2007, "On an Implementation of the Hilbert-Hughes-Taylor Method in the Context of Index 3 Differential-Algebraic Equations of Multibody Dynamics", Journal of Computational and Nonlinear Dynamics, 2, pp. 73–85, DOI:10.111./1.2389231
3. Sherif, K., Nachbagauer, K., Steiner, W., Lauß, T.: A modified HHT method for the numerical simulation of rigid body rotations with Euler parameters - Multibody System Dynamics, Vol. 46, No. 2, 2019
4. Sherif, K., Nachbagauer K., 2014, "A detailed derivation of the velocity-dependent inertia forces in the floating frame of reference formulation", Journal for Computational and Nonlinear Dynamics, doi:10.1115/1.4026083 (8 pages), 2014.
5. Witteveen, W., Pichler, F., 2019, "On the relevance of inertia related terms in the equations of motion of a flexible body in the floating frame of reference formulation", Multibody System Dynamics. 46, 1, S. 77–105
6. Sherif, K., Nachbagauer, K., Steiner, W., 2015, "On the rotational equations of motion in rigid body dynamics when using Euler parameters", Nonlinear Dyn 81, 343–352, https://doi.org/10.1007/s11071-015-1995-3
7. Witteveen, W., Thomas, L., Stefan, O., 2021, "FreeDyn: A Free, Flexible and State of the Art Multibody Simulation Package for Education, Research and Industrial Applications", In: Kerschen, G., Brake, M.R., Renson, L. (eds) Nonlinear Structures & Systems, Volume 1. Conference Proceedings of the Society for Experimental Mechanics Series. Springer, Cham. https://doi.org/10.1007/978-3-030-47626-7_8
8. Witteveen, W., Koller, L, 2023, "FRF-based non-simultaneous real-time hybrid testing of multiple subsystems with moderate nonlinearities", Mechanical Systems and Signal Processing, Volume 188, 109944, https://doi.org/10.1016/j.ymssp.2022.109944.
9. Hippmann G., Modellierung von Kontakten komplex geformter Körper in der Mehrkörperdynamik, PhD Thesis, Institut fur Mechanik und Mechatronik, TU Wien, 2004 (http://pcm.hippmann.org/doc/hippmann_dissertation.pdf,cited09.2024)
10. Haas, S., Dück, M., Winkler, A., Grabmair, G. et al., (2020) "Free Multibody Cosimulation Based Prototyping of Motorcycle Rider Assistance Systems", SAE Technical Paper 2020-32-2306,https://doi.org/10.4271/2020-32-2306.
11. T. Lauß, S. Oberpeilsteiner, K. Sherif and W. Steiner, "Inverse Dynamics of an Industrial Robot Using Motion Constraints," *2019 20th International Conference on Research and Education in Mechatronics (REM)*, Wels, Austria, 2019, pp. 1–7. doi:10.1109/REM.2019.8744124
12. Grünberger, J., Steiner, W., Witteveen, W. (2024), „Multibody Simulation of Gearbox Dynamics using 3D Contact Modeling and measured Tooth Geometries.", In Proceedings of 3rd International Symposium on Industrial Engineering and Automation Article 112
13. S. Reichl, 2011, Inverse Dynamics and Trajectory Tracing, Sündwestdeutscher Verlag für Hochschulschriften, ISBN 9783838121697

Chapter 3
Nonlinear Frequency Response Surfaces, Part I: Theoretical Formulation

D. Di Maio, Jip van Tiggelen, and Marcel H.M. Ellenbroek

Abstract The Frequency Response Functions (FRF) are the most widely used functions to characterize the dynamic behaviour of structures. The natural frequencies and damping behaviour can be easily and quickly detected from a Bode diagram. FRFs are used to calculate the modal properties by modal analysis methods, and in the last step, a frequency response model is used to synthesize the response functions to verify the quality of the identified modal properties identified. The circularity between 1) measurement, 2) identification, 3) regeneration and 4) comparison is ensured on the assumption that transfer functions are measured under strict conditions, such as linear vibrations. All mechanical systems are intrinsically nonlinear. Some sources of nonlinearity might be excited and revealed, and others not for various reasons. Anyhow, it is unavoidable to measure nonlinear vibrations when vibration tests are executed at various levels of excitation forces, and both linear and nonlinear vibrations are processed to obtain linear and nonlinear FRFs. The linear FRFs can be processed using the existing linear identification methods. The nonlinear FRFs are either stored in archive, or blandly processed to evaluate the level and the type of nonlinearity, such as hardening or softening behaviour.

This research aims to (i) formulate a new frequency response model and (ii) formulate a new identification method for extracting amplitude-dependent modal parameters. The first objective will demonstrate that a nonlinear frequency response function lays on a Force Plane which intercepts an FRF surface generated by equivalent linear FRFs. The second objective will demonstrate that a linear modal analysis method, called line-fit and based on the Dobson formulation, allows extracting amplitude-dependent modal parameters from nonlinear FRFs. By combining the results achieved by the two research objectives, one can regenerate and compare measured and synthetic nonlinear transfer functions.

Keywords Forced vibrations · system assembly · the nonlinearity of joints · Nonlinear Frequency Response Surface

Introduction

In the aerospace and space industry, vibration qualification tests serve an essential purpose - to evaluate the behaviour of a structure in conditions that it would typically experience during launch or flight. However, the current approach of performing qualification tests on fully assembled systems and sub-systems has its limitations. By the time the dynamic behaviour is tested, it may be too late to make any design changes that would have been required, which can result in costly redesigns and delays. Thus, it is crucial to develop a more proactive approach to assessing the vibration response of systems earlier on during the design and make.

While modelling the individual components in a system can be relatively straightforward, accurately modelling their assembly, particularly the connections between the components, is complicated due to the nonlinear transfer behaviour at the joints between components [1]. As such, predictive modelling of the forced vibrations of assembled structures has been challenging for many years [2, 3]. Reducing the computational efforts required for solving the dynamics response of nonlinear systems by lumped parameter joint models, such as the Iwan model [4, 5] or assuming a polynomial function. These could be considered phenomenological models and not fundamental models describing the physics, thus still requiring expert

D. Di Maio · Jip van Tiggelen · Marcel H.M. Ellenbroek
University of Twente, Drienerlolaan 5, 7522 NB Enschede, The Netherlands
e-mail: d.dimaio@utwente.nl, j.vantiggelen@utwente.nl

Marcel H.M. Ellenbroek
Airbus Defence and Space Netherlands, Mendelweg 30, 2333 CS Leiden, The Netherlands
e-mail: m.h.m.ellenbroek@utwente.nl

Ludovic Renson et al. (eds.), *Nonlinear Structures & Systems, Vol. 1*, Conference Proceedings of the Society for Experimental Mechanics Series, https://doi.org/10.13052/9788743801467-3

knowledge or calibration using test data. On the other hand, more efficient solution methods were developed for obtaining the forced vibration behaviour of nonlinear systems (as compared to i.e. direct time integration), such as harmonic balance method [6], continuation-based methods [7] and quasi-static approaches [8, 9]. Yet, prediction of the nonlinear dynamic behaviour of joints remains a challenge (in an industrial context).

This short context will help the reader understand that one of the main shortfalls can be seen in the characterization of the vibration response, which becomes nonlinear for several reasons, where interfaces play a large part in such a challenge. Although academia has developed several testing and analysis techniques for characterizing nonlinear vibrations during the past years, industry integrated a tiny fraction of that knowledge [10]. The main reason is the implementation of the proposed methods into industrial standards, which still relies on the measurement of Frequency Response Functions (FRF) for characterizing the dynamics of structures. Engineers find FRFs very useful because they can relate the response at any point on the structure to the input force exerted elsewhere. This means that an engineer can evaluate an appropriate response to design requirements. Unfortunately, the nonlinear vibrations disrupt the established procedures. Therefore, this paper aims to offer an approach that is simple and easy to integrate into the industry standards, thus providing invaluable information to engineers. This paper is divided into two parts, one more theoretically focussed and one more application-driven. The latter was authored by Jip van Tigglen and is listed in the same conference proceedings.

Theoretical Formulations of the Nonlinear Frequency Response Surface

This section will introduce the theoretical formulation of a nonlinear frequency response surface. It is a major breakthrough for engineering science and the practice of experimental modal testing. It is a custom to perform modal testing on structures before any other vibration test is executed. It is important to identify the natural frequencies, damping values, and mode shapes, all three of which form the modal parameters that can be used to characterize a structure's dynamic. Furthermore, these modal parameters validate the experimental model of the mechanical system. It is common practice to perform the modal testing using any desired excitation force and process the test data with modal analysis tools. The final verification of the data analysis consists of synthesizing the FRFs and overlaying them on the experimental ones to check if the modal analysis evaluated the modal parameters correctly. The process described so far stops when a test engineer acquires nonlinear test data, meaning that FRFs are distorted because of sources of nonlinearity. Hence, the modal analysis stops working correctly. The following sections will show how carrying out the same modal identification process is possible but using a different analysis method. Most of the research is already available in [11], so this paper offers a concise summary explaining the fundamental steps.

Linear frequency response surface (LFRS)

The relationship between force and displacement for a single degree of freedom system is linear when there is a single coefficient describing the system's stiffness. Many engineers are accustomed to seeing this relationship made a constant stiffness line, which intercepts the mass line and the system's natural frequency. The frequency response is linear for any vibration amplitude. Figure 1 shows a hypothetical waterfall of linear responses forming a Linear Frequency Response Surface. As said the stiffness of the system is not amplitude dependent; therefore, every FRF will start from the same stiffness amplitude. It is also simple to understand that any constant amplitude force plane, created by expanding over the frequency range of the red dashed line, cuts across the LFRS and will produce a linear FRF that lies on that force surface. Although this description reads very clearly, a linear system will produce linear responses, and the concept of frequency response surface will serve to understand the transition from linear to nonlinear vibrations.

Nonlinear Frequency Response Surface (NFRS)

This sub-section will discuss the transition form linear to nonlinear vibrations. It is worth mentioning that, hereafter, the nonlinear vibrations will be assumed to be at a steady state, for which the FRF is the ratio between the input and output at the fundamental drive frequency of the excitation. Hence, any high-order harmonics is neglected. Figure 2 shows the static receptance for a single DOF system whose stiffness relationship is amplitude-dependent. This dependency implies that any FRF will start from a different linear equivalent stiffness. The figure shows the system's linear and nonlinear static receptance, indicating receptance as the ratio between force and displacement.

Having generated a nonlinear static receptance using a cubic stiffness nonlinearity, the following step is to generate the Nonlinear Frequency Response Surface. Therefore, one can still use a linear frequency response model, starting from an equivalent linear stiffness amplitude, which is likewise carried out for the linear frequency response surface. One must catch this idea's major novelty, which suggests that a linear response model is still a valid selection for generating a nonlinear

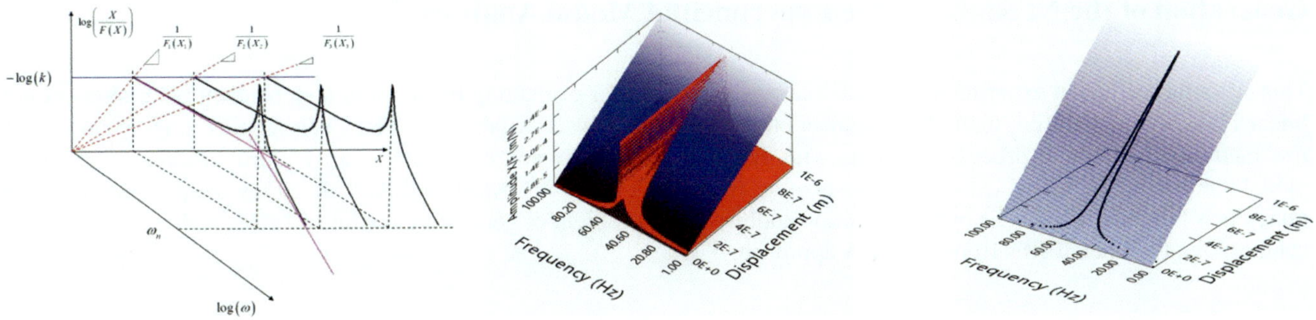

Fig. 1 Linear frequency response surface (LFRS)

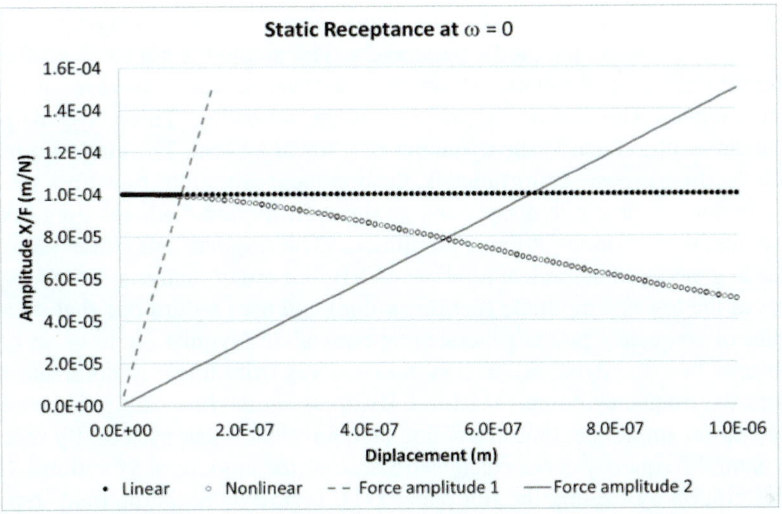

Fig. 2 Linear and nonlinear static receptance

response solution. Figure 3 shows the 3D NFRS in red and the constant amplitude force surface in blue. The nonlinear FRF is the unique solution between the interception of the NFRS and the desired force plane, the inclination of which can be changed. The concluding remarks of this section are straightforward. One can understand that the same NFRS can be generated by accessing modal amplitude-dependent parameters. The response model equation (1) can be used to generate that NFRS once the amplitude-dependent parameters are identified. The identification process will be briefly explained on the following section and the application to an experimental case will be provided as an example.

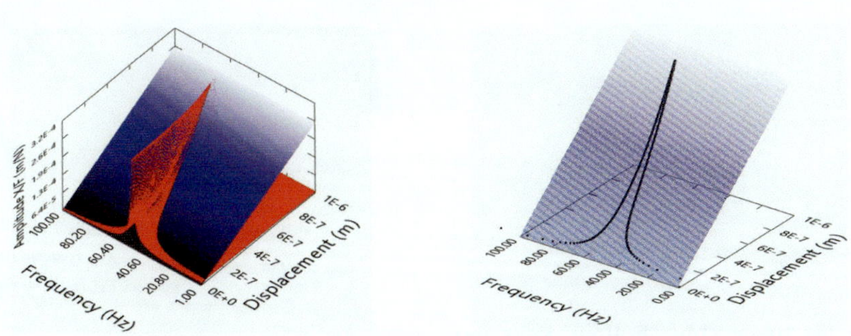

Fig. 3 Nonlinear Frequency Response Surface (NFRS) (red) and constant amplitude force plane (blue)

Generation of the NFRS Using the Experimental Modal Analysis

This section presents an experimental modal analysis technique for evaluating the amplitude-dependent parameters of nonlinear FRFs. As introduced, many test engineers find the data analysis of nonlinear FRFs very challenging because of the few methods available to process FRF data with distortion caused by source of nonlinearity. Furthermore, assuming that modal parameters are evaluated, the synthesis of an FRF is not straightforward, and thus, the assessment of the modal parameters is left unchecked. The previous section addressed the problem of generating nonlinear FRFs, which can be used for synthesizing nonlinear FRFs from the EMA approach.

EMA using the modified-Dobson method

The experimental identification is carried out on FRFs that are measured under steady-state conditions and for reasonably well-separated modes to avoid mode coupling. The Dobson method [11] is a modal analysis technique that allows using one of the properties of the inverse of the FRF. It works by line-fitting the real and imaginary parts of the inverse of the FRF and using the line-fitting coefficients to extract the modal parameters. The method is not used as often for linear modal analysis because of the more powerful techniques developed during the past two decades. However, it proves the best identification method for nonlinear FRF analysis. The underlying principle is the following. Three Nyquist points identify a circle, the properties of which can be curve-fit to identify the dynamics of a linear system. The circle -fit method was one of the most used techniques in the past for this purpose. Unfortunately, the nonlinear vibrations transform the circle into something else, for the stiffness nonlinearity part of the circle disappears, as shown in Figure 4 see the grey hollow dots. The circle stops at a certain amplitude level and then resumes at lower amplitudes. The response amplitude jump causes this because of the stiffness nonlinearity used to generate a numerical nonlinear FRF. The smart analysis, which exploits the Dobson method, fixes two frequency points at the lowest amplitude (the linear one) and uses a third one that sweeps all the other frequency points. Hence, every triplet of frequency points generates an equivalent Nyquist circle or an equivalent linear dynamical system. The goal is to observe how the dynamics of a system evolves from linear to nonlinear vibrations. Hence, one can generate as many triplets as the frequency points measured. By applying the line-fitting to the triplets, one can evaluate the modal properties at that vibration amplitude, thus evaluating an equivalent linear system. By repeating the analysis process, one can observe how the natural frequency curve changes because of the amount of vibrations. The reader is encouraged to expand this quite brief description by reading the full paper [11]. Figure 5 shows the final step of the analysis, where the natural frequency curve is evaluated at every amplitude level of the nonlinear FRF. The black dots of the natural frequency curve indicate the analysis carried out using the frequency points from 45 to approx. 51 Hz, while the blue dots the other part of the FRF branch. Hence, equation (1) can be used for generating an NFRS.

$$\alpha(\omega, X) = \frac{{}_rA(X)}{\omega_r^2(X) - \omega^2 + i\eta_r(X)\omega_r^2(X)}$$

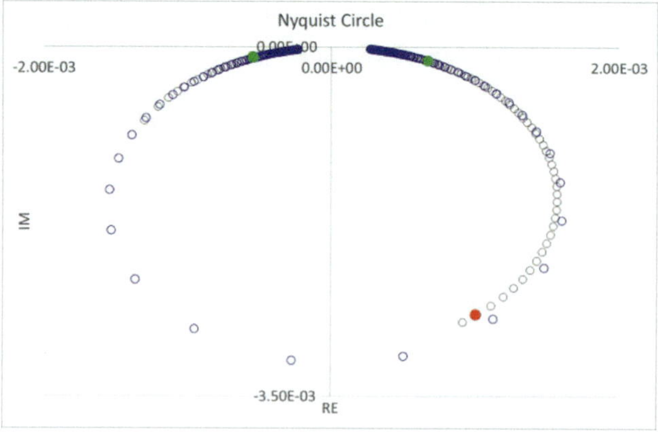

Fig. 4 Nyquist circles for linear (blue hollow dots) and nonlinear FRFs (grey hollow dots)

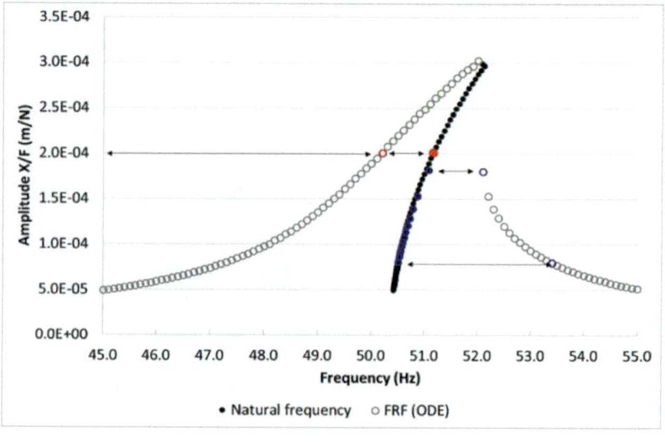

Fig. 5 Nonlinear FRF with the natural frequency curve

Application to an experimental blind case

Section 2 showed that a nonlinear FRF is the unique solution between an NFRS intercepting a force surface of any desired amplitude. The idea was demonstrated using a simple SDOF nonlinear system. However, engineers need to deal with real nonlinear FRFs and be able to evaluate properties that can be used to evaluate the system response at various amplitudes of vibration. This section will show an example of a blind case where the nonlinear FRF was provided without any further information of the test article or type of test. Besides, the FRF was a ratio between the base acceleration and the response measured at a point of the test article. Hence, it is a perfect case study to prove that the approach described so far is valid.

The process consists of:

1. Perform the modified-Dobson method to extract amplitude-dependent modal parameters.
2. Use equation (1) to generate a NFRS.
3. Decide the amplitude(s) of the force plane to extract the synthesize nonlinear FRF.
4. Overlay the measured and synthesized nonlinear FRFs.

Figure 6 shows the measured FRF in red. It was known that the FRF was generated using very slow sine sweep rate from low to high frequency and the FFT operator was used to generate the FRF. The modified-Dobson method was applied to the measured FRF (red) and the modal parameters were polynomial-fitted and used for the equation (1), which is valid for well-separated mode as it was assumed for this blind case. The NFRS was generated and a synthesized nonlinear FRF was extracted and compared to the experimental one. Figure 3.7 shows an overlay between the experimental (red) and synthesized (black) FRFs. Given the excellent overlay, any engineer would be interested in the response for a lower or higher amplitude of vibrations. Hence, by changing the force amplitude, two FRF can be evaluated for low and high vibration amplitudes, both in blue colour. Unfortunately, there is no possible comparison for the two FRFs in blue as experimental data were unavailable. The reference paper [11] shows an academic example where such a comparison was possible and successful. Therefore, it is assumed that this regeneration approach for low and high amplitudes can still be valid.

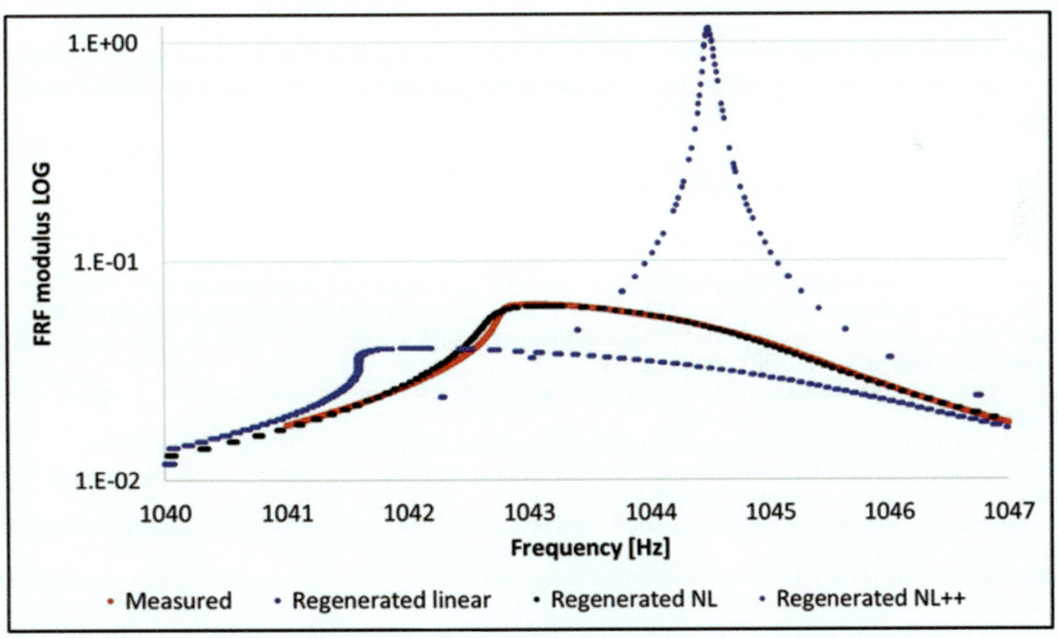

Fig. 6 Nonlinear FRFs measured (red) and regenerated in black and blue

Final remarks

This short paper summarizes the discovery reported in [11] by highlighting the fundamental steps for a test engineer to (i) measure, (ii) characterize, (iii) synthesize and (iv) compare the frequency nonlinear responses. These four steps are possible because a nonlinear frequency response curve can be built by using equation (1) as soon as the modal parameters of equation (1) are identified. Therefore, it is possible to evaluate for any desired level of excitation force a nonlinear frequency response which can be overlaid and compared to the experimental one. This final step is the most critical and valuable because the test engineer can evaluate if the identification process achieved the correct parameters. This is done is a matter of seconds by exploiting the fundamental equations of the linear modal analysis without using expensive and complex computational calculations.

Polynomials fit the modal amplitude-dependent parameters, and this process can intrinsically add some bias. Observing the response for much higher amplitudes than the maximum one measured is possible, but the risk is in the polynomial curves which might generate un-physical responses. Therefore, paying attention to simulation generated beyond the measured amplitudes of the responses is important.

A final remark is about the generation of the surface by using a force-displacement curve. Theoretically, any finite element method can be used to generate a fully reverse cycle of load by superimposing any desired deflection shape (at any resonance). It is a static analysis that allows the evaluation of a hysteresis curve. That curve can be used two ways, 1- to extract the static receptance and 2- to extract the damping from the hysteresis. Once these two parameters are extracted, an NFRS can be generated. It is straightforward to understand that any FE analyst can simulate a nonlinear frequency response given a stress model. Therefore, one can infer from the dynamic response (either linear or nonlinear) if the stress model needs better refinements, for instance, by increasing the damping by changing the material properties. Hence, the NFRS can become (i) a great design tool for analysts and (ii) a great experimental modal analysis tool for test engineers.

Conclusions

In conclusion, this paper provides some basic insights to the NFRS. It was shown that this surface can be generated by theoretical force-displacement full reverse cycle and by experimental identification of the nonlinear FRF. Both routes will meet simultaneously if a thorough experimental model validation is carried out. This paper and the primary reference [11] could not demonstrate it. Therefore, future work will be done to generate the NFRS from a theoretical route and validate the same surface by generating it using experimental measurements.

Acknowledgments Funded by the European Union under GA No. 101091409. Views and opinions expressed are however those of the author(s) only and do not necessarily reflect those of the European Union. Neither the European Union nor the granting authority can be held responsible for them.

References

1. Gaul, L., Lenz, J. (1997). Nonlinear Dynamics of Structures Assembled by Bolted Joints. Acta Mechanica 125, 169–181.
2. Segalman, D.J. (2006). Modelling Joint Friction in Structural Dynamics. Struct. Control Health Monit. 13(1), 430-453.
3. Brake, M.R.W. (2018). The Mechanics of Jointed Structures: Recent Research and Open Challenges for Developing Predictive Models for Structural Dynamics, Springer International Publishing.
4. Iwan, W. D. (1966). A Distributed-element Model for Hysteresis and its Steady-state Dynamic Response. Journal of Applied Mechanics, 33(4), 893–900.
5. Segalman, D. J. (2005). A Four-Parameter Iwan Model for Lap-Type Joints, ASME J. Appl. Mech., 72(5), 752–760.
6. Nakhla, M.S., Vlach, J. (1976). A Piecewise Harmonic Balance Technique for Determination of Periodic Response of Nonlinear Systems. IEEE Transactions on Circuits and Systems, 23(2):85–91.
7. Renson, L., Kerschen, G., Cochelin, B. (2016). Numerical Computation of Nonlinear Normal Modes in Mechanical Engineering. Journal of Sound and Vibration, 364, 177–206.
8. Festjens, H., Chevallier, G., Dion, J. L. (2013). A numerical tool for the design of assembled structures under dynamic loads. International Journal of Mechanical Sciences, 75, 170–177.
9. Allen, M. S., Lacayo, R. M., Brake, M. R. (2016). Quasi-static Modal Analysis based on Implicit Condensation for Structures with Nonlinear Joints. In: International conference on noise and vibration engineering, Leuven, Belgium.
10. Carrella A., Ewins D.J. (2011). Identifying and Quantifying Structural Nonlinearities in Engineering Applications from Measured Frequency Response Functions. Mech. Syst. Signal Process. 25(3), 1011–27.
11. Di Maio, D., 2024. A novel analysis method for calculating nonlinear Frequency Response Functions. arXiv preprint arXiv:2404.01972.

Funded by the European Union under GA No. 101091409. Views and opinions expressed are however those of the author(s) only and do not necessarily reflect those of the European Union. Neither the European Union nor the granting authority can be held responsible for them.

Chapter 4
Fatigue Damage Potential in a Nonlinear Dynamical System with Isolated Resonances

Brian Evan Saunders

Abstract Cyclical loading can cause fatigue damage in a structure, even at relatively low stress levels. If the structure is subject to nonlinear behaviors, then it may also experience complex phenomena such as super- or sub-harmonic resonances, chaotic motion, and isolated resonances, also known as isolas. Isolas in particular may be especially dangerous for a structure because they are dependent on initial conditions or perturbations to the structure, and they can drive the structure at amplitude outputs of an order of magnitude or higher than the expected amplitude levels. Although fatigue damage is a widely researched field, there is currently little work available on quantifying the effects of fatigue damage in a system that is experiencing isola motion. The goal of this work is to evaluate the fatigue damage potential of a harmonically driven, nonlinear dynamical system that is known to have isolas. The damage that can be accumulated while on an isola is compared to the damage accumulated at other points along the main frequency response branch of the system. Rainflow cycle counting is used to compute the fatigue damage potential, given a representative material with a range of S-N curve coefficients.

Keywords Fatigue, cyclical loading, nonlinear dynamics, isolated resonance, rainflow

Introduction

Nonlinearities give rise to a wide variety of strange phenomena in dynamical systems. Of particular interest in this work is the isolated resonance, which is a type of nonlinear behavior that forms closed solution curves separate from the main solution curve in a system's frequency response. Other names for this phenomenon include detached resonance, isolated response, detached response, or simply isola. Isolas have a tendency to occur at relatively high amplitudes and over narrow frequency ranges, and they can form near to or far from the system's primary resonances, depending on the system parameters. Isolas have been observed and studied since at least 1938, with Rauscher [1] analyzing a dynamical system whose frequency response contained solution curves that originated from "the delicate relation between damping and response in the present problem". Abramson [2] similarly observed isolas in a system with softening nonlinearity. As the excitation force was varied, two solution curves moved closer and eventually merged at a single point before detaching again into a classical softening resonance and a separate set of curves that "corresponds to motions of large amplitude and low frequency". Research on isolas picked up significantly in the 2010s, following the increased popularity of numerical techniques such as harmonic balance to simulate nonlinear systems. The inclusion of numerical continuation methods allows isolas to efficiently be traced out in the frequency space, and even with respect to multiple parameters. Mélot et al. [3] and Habib et al. [4] provide nice reviews on the historical development of isolas.

In many structural dynamics applications, isolas are considered dangerous for two main reasons: their susceptibility to perturbations, and their often relatively large amplitudes. As isolas are detached from the main frequency response curve, they are capable of forming far from resonances where the amplitude is otherwise expected to be small. A small bump, shock, or other perturbation sending a structure into unforeseen resonance is, understandably, quite an undesirable situation that places life and property at unacceptable risk. However, there has been seemingly little work to actually quantify the danger that an isola can pose. In this work, we evaluate the risk of isolas in a harmonically excited nonlinear dynamical system [5] using the Palmgren-Miner fatigue damage metric [6]. Our objective is to compare the damage accumulated due to motion on an isola to the damage accumulated at other co-existing solutions. The results of this work could help inform to

Brian Evan Saunders
Sandia National Laboratories, Albuquerque, NM 87123
e-mail: besaund@sandia.gov

Ludovic Renson et al. (eds.), *Nonlinear Structures & Systems, Vol. 1*, Conference Proceedings of the Society for Experimental Mechanics Series, https://doi.org/10.13052/9788743801467-4

what extent an isola motion can eat up the fatigue life of a structure, and how to better qualify a structure designed to operate in an environment that may be at risk of experiencing nonlinear behaviors such as isolas.

Methodology and System Model

Three simple and well-known methods are combined for this fatigue analysis: the log-linear (i.e., Basquin) model of the stress-cycles curve $\mathrm{N} = \mathrm{cS}^{-\beta}$, Palmgren-Miner linear damage $\mathrm{D} = \sum_{\mathrm{i}=1}^{\mathrm{k}} \frac{\mathrm{n_i}}{\mathrm{N_i}}$, and rainflow cycle counting [7]. An experimental apparatus of the dynamical system, along with parameter values, can be found in [3] and is modeled as follows:

$$\ddot{x} + 2\omega_n \zeta \dot{x} + \omega_n^2 x + \frac{\alpha}{m} x^3 + \frac{F_c(x)}{m} = \frac{p}{m} cos\left(\omega t\right) \quad , \tag{1a}$$

$$F_c = \begin{cases} K_c\left(x + j_1\right), & x < -j_1 \\ 0, & -j_1 \leq x \leq j_2 \\ K_c\left(x - j_2\right), & x > j_2 \end{cases} \tag{1b}$$

$$\sigma\left(t\right) = \frac{12Eb}{L^2} x\left(t\right) \tag{2}$$

Euler-Bernoulli beam theory, for a fixed-fixed beam with centered load, is assumed in order to approximate the maximum stress in the system in Equation (2) [8]. Figure 1 shows the nonlinear forced response curves for the system model, including superharmonic, subharmonic, and isola solutions.

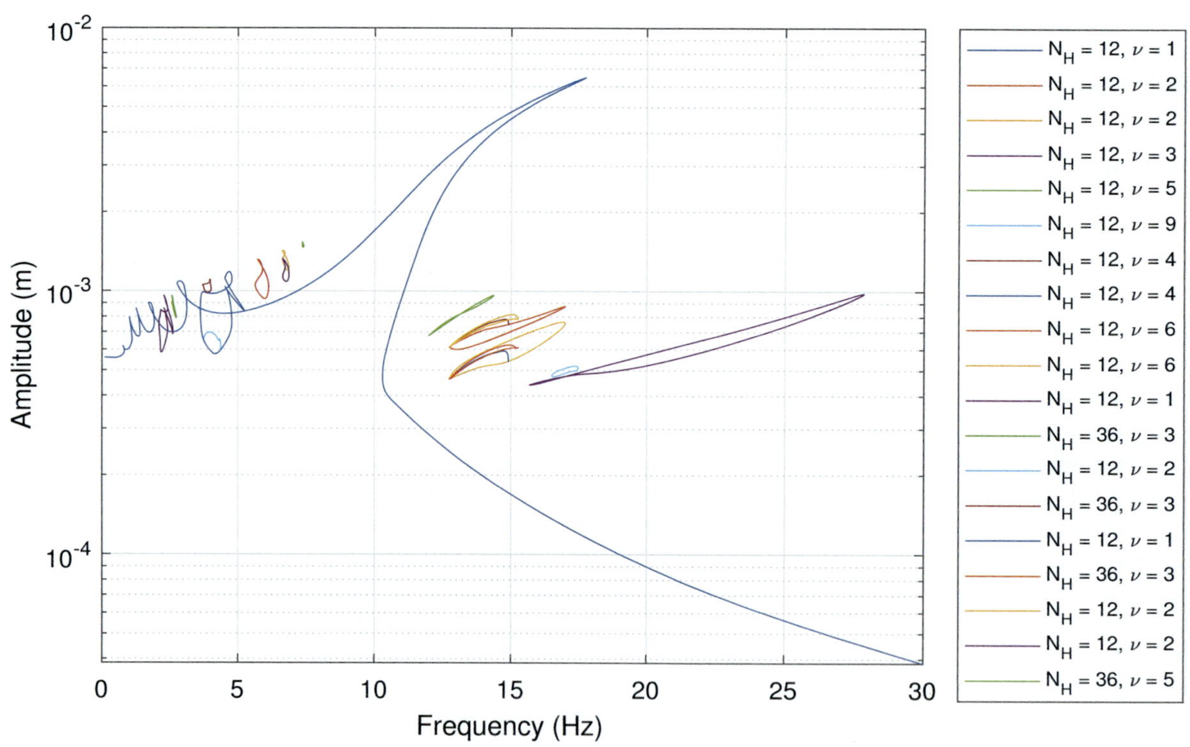

Fig. 1 Nonlinear forced response curves for the dynamical system described by Eq. (1). N_H is a solver convergence parameter, and ν is the subharmonic order of each solution curve

Fatigue Damage Analysis

Figure 2 shows the calculated fatigue damage for each of the solution curves in Figure 1. A fatigue damage coefficient of $\beta = 4.22$ was used in the following calculations. Complete analysis details and parameter values can be found in the final conference presentation. Notice that while the amplitudes span from 0.4 mm to 6 mm, the damage values span a much larger range from $3 * 10^{-13}$ to $4 * 10^{-4}$. Because the damage metric scales linearly with time duration, we can perform relative comparisons. At 20 Hz, for instance, the subharmonic isola of order 3 (in purple) has an amplitude approximately 6 times larger than the off-resonance amplitude that one would traditionally expect to encounter in this frequency region, but the fatigue damage for the same time duration is almost 2,400 times higher on the isola—a difference greater than three orders of magnitude. This difference means a 20-year life of operation at the off-resonance motion is reduced to a mere 3 days and 2 hours on the isola. For a material with fatigue coefficient $\beta = 6.67$, that lifetime is reduced to just 45 minutes.

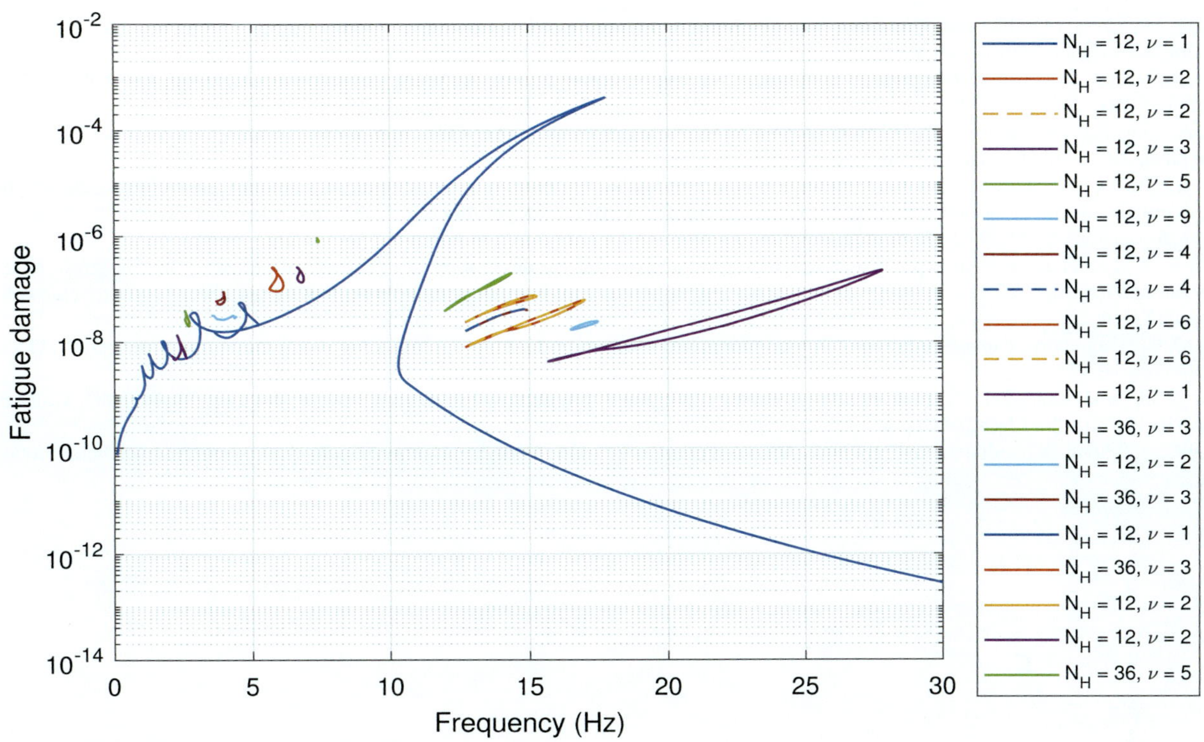

Fig. 2 Fatigue damage for the solution curves in Fig. 1 for a fatigue damage coefficient of $\beta = 4.22$. N_H is a solver convergence parameter, and ν is the subharmonic order of each solution curve

Conclusions

In this work, we assessed the risk of encountering an isola in a nonlinear dynamical system. Palmgren-Miner fatigue accumulation and rainflow stress cycle counting were applied to a structure treated as a loaded Euler-Bernoulli beam with a Basquin model of fatigue life. It was shown that the fatigue damage due to an isola motion can easily be orders of magnitude higher than an off-resonance motion, significantly eating into the fatigue life of a structure. The most straightforward defense against this type of damage is to treat an isola like a regular resonance: design the operating characteristics of the structure around the frequency ranges of any isolas; or design the structure to simply not have isolas in the operating frequency range. There are several avenues for future work on this topic.

Acknowledgements

This article has been authored by an employee of National Technology & Engineering Solutions of Sandia, LLC under Contract No. DE-NA0003525 with the U.S. Department of Energy (DOE). The employee owns all right, title and interest in and to the article and is solely responsible for its contents. The United States Government retains and the publisher, by accepting the article for publication, acknowledges that the United States Government retains a non-exclusive, paid-up, irrevocable, world-wide license to publish or reproduce the published form of this article or allow others to do so, for United States Government purposes. The DOE will provide public access to these results of federally sponsored research in accordance with the DOE Public Access Plan https://www.energy.gov/downloads/doe-public-access-plan. SAND2024-13418A.

References

1. Rauscher, M., "Steady Oscillations of Systems With Nonlinear and Unsymmetrical Elasticity," ASME. J. Appl. Mech., **5**(4), pp. A169–A177, 1938. Doi: https://doi.org/10.1115/1.4008875
2. Abramson, H. N., "Response Curves for a System With Softening Restoring Force," ASME. J. Appl. Mech., **22**(3), pp. 434–435, 1955. Doi: https://doi.org/10.1115/1.4011105
3. Mélot, A., Goy, E.D., and Renson, L., "Control of isolated response curves through optimization of codimension-1 singularities," Computers & Structures, 299, 107394, 2024. Doi: https://doi.org/10.1016/j.compstruc.2024.107394
4. Habib, G., Cirillo, G.I., and Kerschen, G., "Uncovering detached resonance curves in single-degree-of-freedom systems," Procedia Engineering, X International Conference on Structural Dynamics, EURODYN 2017, 199, pp. 649-656, 2017. Doi: https://doi.org/10.1016/j.proeng.2017.09.116
5. de Langre, E., Lebreton, G., "An experimental and numerical analysis of chaotic motion in vibration with impact," In: Pettigrew, M. (ed.) Flow-Induced Vibration, 328, pp. 317–325. American Society of Mechanical Engineers, New York, 1996. https://www.osti.gov/etdeweb/servlets/purl/427216
6. Miner, M.A., "Cumulative damage in fatigue," Journal of Applied Mechanics, 12(3), pp. A159-A164, 1945. Doi: https://doi.org/10.1115/1.4009458
7. Downing, S.D. and Socie, D.F., "Simple rainflow counting algorithms," International Journal of Fatigue, 4(1), pp. 31-40, 1982. Doi: https://doi.org/10.1016/0142-1123(82)90018-4
8. MechaniCalc, "Beam Deflection Tables," [online], Accessed October 3, 2024. Available at https://mechanicalc.com/reference/beam-deflection-tables

Chapter 5
Robust Optimization of Surface Topology for Jointed Connections

Aalokeparno Dhar, Justin H. Porter, and Matthew R. W. Brake

Abstract Bolted connections are ubiquitous in mechanical and aerospace engineering. However, these connections introduce significant uncertainties in vibration responses. Given sufficient information about a jointed connection, a recent physics-based rough contact modeling has provided significant improvements towards predicting nonlinear trends in frequency and damping. For jointed connections, increasing vibration amplitudes result in a decrease in the natural frequency and an increase in damping as the system transitions from sticking to slipping. While the physics-based model showed improvements towards predicting these trends, it required knowledge of the as-built mesoscale topology of the interface. Specifically, an approximately 100 micron gap in the center of the nominally flat interface resulted in an approximately 7% shift in the natural frequency compared to simulations with a flat interface. This uncertainty is due to manufacturing tolerances and cannot easily be eliminated for mass production of engineering structures. To address this challenge, the present work seeks to understand how the contact interface within a joint can be designed to reduce uncertainty due to uncontrolled machining deviations. Thus, the physics-based rough contact model that represents the best available predictions for jointed connections is employed to simulate the Brake-Reuß Beam. These simulations consider the effects of different interface designs while also analyzing the sensitivity of the responses to mesoscale topology variations. The extended periodic motion concept (EPMC) is utilized to obtain nonlinear modal properties. Sensitivities are calculated directly based on the solution points and verified with finite difference calculations. These sensitivities provide groundwork for optimizing designs to reduce uncertainty due to manufacturing deviations. Overall, this work provides insights into the effects of machining variations on jointed connections and how to improve designs to limit such effects.

Keywords Friction · Jointed Connection · Mesoscale Topology

Introduction

Bolted connections in a structure can introduce significant nonlinearities, which makes it challenging to analyze and predict the structure's behavior. In terms of the system's dynamic characteristics, the nonlinearity due to the bolted connections affects both the free vibration response's resonant frequencies and dissipative behavior. The damping factor and modal frequency can vary with the types of bolted interfaces, as seen across a number of experiments [1–7]. Lap joints can exhibit a 1-2% decrease in resonant frequency as the excitation amplitude is increased; however, over the same range the damping capacity can increase by an order of magnitude [1]. Flange joints, by contrast, can exhibit a much smaller increase, or even no increase, in damping as compared to lap joints [2, 3]. Dovetail joints can even exhibit a decrease in damping ratio as excitation amplitude increases (for some modes, and for other modes the damping ratio increases) [6, 7]. Across each of these types of joints, though, it is clear that the meso-scale topography and relative alignment of the joint surfaces can lead to much greater variability in the response of the structure than is due to preload or vibration amplitude [3, 7, 8].

The system characteristics of a jointed structure can be derived from free vibration responses, as demonstrated in [9–12]. In looking at different types of jointed connections, the participating modes significantly influence the system's characteristic damping frequency, while joint parameters, such as bolt torque, have been shown to affect modal interaction behavior [3, 13]. It is also found that changing the bolt torque can cause a different pressure distribution around the bolted joint [7, 14]. Upon dynamic excitation, the pressure distribution can exhibit highly nonlinear behavior. Variations in pressure distribution and the resulting system response can also stem from non-conformal interfaces, often caused by manufacturing uncertainties.

Aalokeparno Dhar · Justin H. Porter · Matthew R. W. Brake
Department of Mechanical Engineering, Rice University, 6100 Main St., Houston, TX 77005, USA
e-mail: ad205@rice.edu; jp88@rice.edu; brake@rice.edu

Ludovic Renson et al. (eds.), *Nonlinear Structures & Systems, Vol. 1*, Conference Proceedings of the Society for Experimental Mechanics Series, https://doi.org/10.13052/9788743801467-5

After applying prestress through bolt torque, gaps between connected surfaces can exist at macro, meso, or micro levels. Chen et al. [15] analyzed the effect of non-flat interfaces in a Brake-Reuß Beam (BRB) and found that end-gap configurations reduce the natural frequency at low amplitudes, while central gaps increase it. To mitigate such uncertainties, deliberate macro-level variations can be introduced at joints, such as adding padding around bolt holes. This approach creates a high-pressure field at predictable locations [1]. Although this design enhances system determinism, concentrated local stresses may not be a desirable for a design. On the micro-scale, contact between two surfaces is governed by variations in asperity heights, which primarily drive frictional contact. Pressure distribution tends to be more localized in interfaces with higher surface roughness than those with smoother surfaces.

Surface errors due to manufacturing can also occur at the meso-scale. Once the bolt torque is applied, the mesoscale profile at the connecting surfaces can create an interference contact at some locations and a gap in some other locations. However, the nonlinear response of various gap topologies can also be something the designer can use to increase the joint interface's robustness. In this paper, we hypothesize that one can design a topology which makes the damping parameter robust to the modal amplitude. That is, it will be beneficial to find a surface profile that is less sensitive to the manufacturing perturbation than a flat, conformal interface. Designing such an interface, however, requires a predictive modeling methodology of structure dynamics with bolted joints with various mesoscale gap topologies.

To analyze the effect of mesoscale variability due to manufacturing error and to design a robust surface topology, it is essential to have a detailed model of the system that includes enough information related to the joint interface. Numerous studies [16] have attempted to understand contact mechanics, how to model the joints and friction and study its dynamic behavior. Among different models, physics-based modelling [17] is applicable to a wide range of joints as long as underlying surface properties are captured well, and is more suitable for the current problem. This model takes care of the micro, meso, and macro contacts at a hierarchical dependency from smaller to a larger scale. The microslip modelling [8] is preferred over the traditional stick-slip models because the former is found to be more accurate while compared with experimental behaviors. The rough contact modeling framework makes use of the statistical representation of the micro-scale asperity heights obtained from the experimental measurements.

In this work, the Brake-Reuß Beam to study the effect of mesoscale gap topology under modal vibration. Modelling provides a brief description of the methodology and the contact model used. The Extended Periodic Motion Concept [18] is used to solve this model using the response amplitude as a continuation parameter. Different gap profiles have been fed to the model to check their implications for the frequency and damping response. The sensitivity of these topologies over perturbations has been studied to gain an understanding of the variability of the response. The paper concludes with thoughts on the more quantitative approaches to obtain a robust manufacturable surface profile.

Modeling

The Brake-Reuss Beam (BRB) is a an assembly of two half beams connected with a three bolted lap joint. The finite element modelling of BRB is initially done in Abaqus (Figure 1a), and model reduction is applied to the interface degrees of freedom plus 50 fixed interface modes using the procedure described in [8, 19]. The model is then further reduced using relative coordinates to a model with 588 Zero Thickness Elements (ZTEs) and 14 fixed interface modes [8]. For a faster model, the mesh is further reduced to 232 ZTEs using the procedure in [20] (and shown in Figure 1b). These ZTEs are modeled to capture the interfacial kinematics and asperity heights obtained from the statistical distribution and the mesoscale gaps or interferences. Friction inside each of the ZTEs is modeled with a physics-based rough contact model [17] that has been reimplemented in Python using automatic differentiation [21]. The normal contact between the asperities at each element is described using the ellipsoidal Hertzian contact law and principles of strain hardening, depending on the elastic and plastic regimes. The tangential contact is modeled using the tangent asperity model [17]. Experimental data as well as the plasticity models have been used to obtain a friction coefficient. A friction coefficient value of 0.55 is obtained for this system. A preload of 12.249kN is applied to the bolts. The latter implementation is adapted for the present models with simulation routines and examples available at [22]. The schematic of the overall procedure is shown in Figure 2.

The Extended Periodic Motion Concept (EPMC) [18] is the primary solution method adopted for this work, which is suitable for lightly damped systems. This method helps obtain a periodic response of a free dissipative system by including an artificial negative damping to the system. This way, the method is useful for obtaining the natural frequency as well as the mass proportionate damping factor, which are the two additional variables along with the harmonic coefficients of each DOFs. A nonlinear prestress analysis and eigenvalue analysis around the preload condition provide the initial guess for the frequency and displacement coefficients for each DOF under calculation. the EPMC calculation requires the Fourier coefficients of the nonlinear force terms associated with the model, which is obtained using the Alternating Frequency-Time (AFT) procedure.

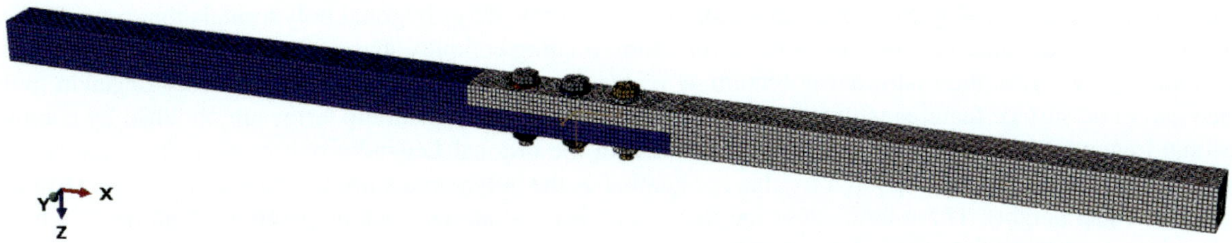

(a)

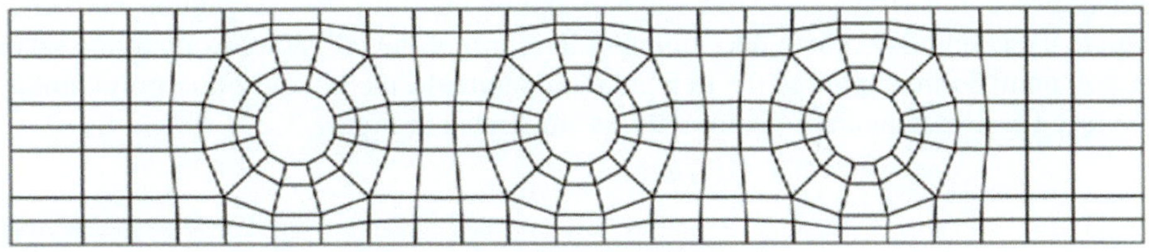

(b)

Fig. 1 Original mesh of BRB model in ABAQUS (a) and Reduced interface mesh of ZTEs (b)

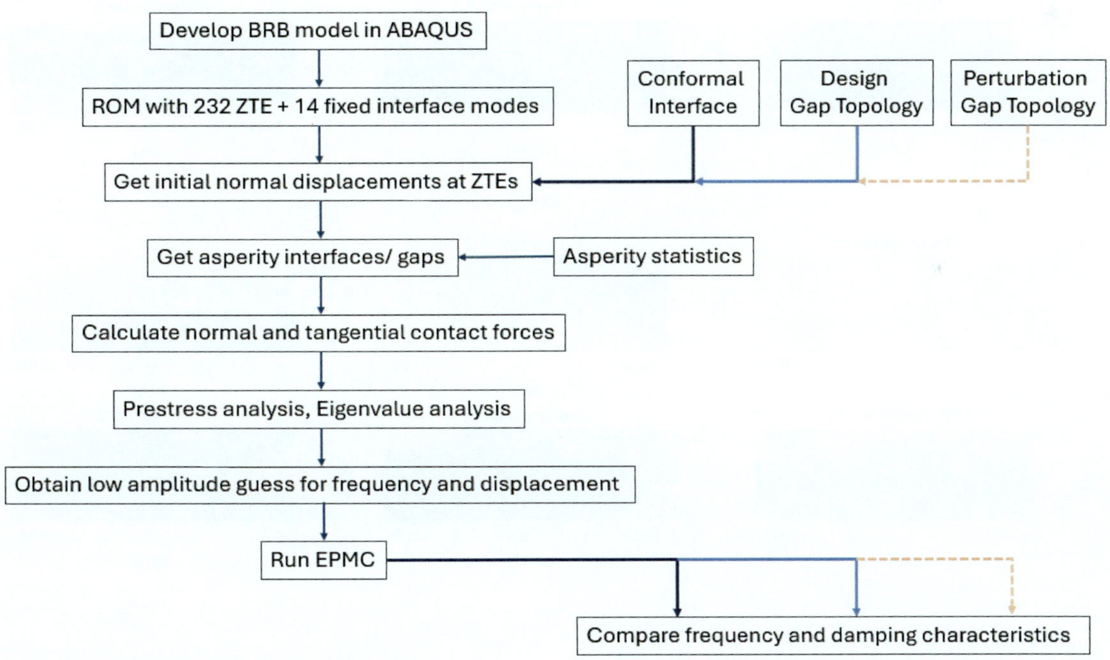

Fig. 2 Flowchart of the modelling approach.

The mesoscale topologies are applied within the ZTEs to offset the initial distances prior to contact of each ZTE. This approach allows for a single Abaqus model of a nominally flat interface to be used for all perturbations. It is expected that this approach is only valid for perturbations up to a certain amplitude; in the subsequent parameter studies, it is demonstrated that the perturbations considered in the present work are within the limit of validity for this modeling approach before a new Abaqus model of a larger perturbed surface is necessitated. This approximation can be validated by a detailed ABAQUS modelling of the perturbes surfaces in future work, where one can estimate the maximum height of the topology allowable by the ZTE.

The possible types of topologies are taken form the rectangular Legendre polynomials [23] and rectangular Zernike polynomials [24]. The Legendre polynomials, traditionally defined over a continuous interval, have been adapted to represent two-dimensional surface topologies over a rectangular domain. They are orthogonal polynomials that provide a systematic way to describe surface variations, offering a basis for approximating complex shapes. The zeroth order polynomial resembles flat topology, whereas the first-order polynomials represent y-tilt and y-tilt. The second-order Legendre polynomials introduce quadratic surface features with terms like x^2, xy, and y^2. These quadratic terms are obtained by combining one-dimensional Legendre polynomials in X and Y directions (in the original Legendre polynomials there are constant terms associated with x^2 and y^2 that are ignored for this application as the polynomials need to be scaled according to the domain boundaries and gap height). These terms describe more complex curvatures, such as parabolic features in either axis and coupled variations in both directions. Zernike polynomials, another class of orthogonal polynomials, are particularly well-suited for describing wavefront deformations and are often applied in optical systems for their radial symmetry. Though these polynomials are defined over a unit circle, there are ways to extend them for the rectangular domain. The transformation from a circular domain to a rectangular domain does not alter the form of the polynomials up to the second-order. While the zeroth and first-order polynomials are similar to the Legendre polynomials, the second-order polynomials include the shapes $x^2 + y^2$, xy, and $x^2 - y^2$. These rectangular polynomials are illustrated in Figure 3.

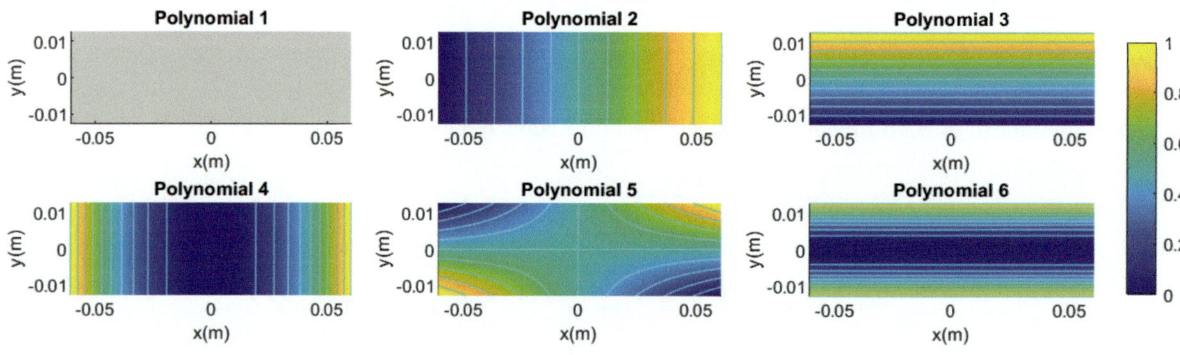

(a)

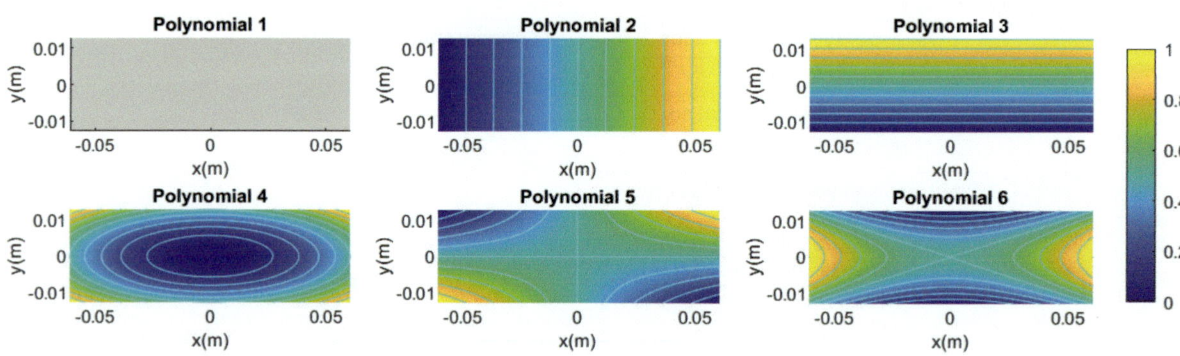

(b)

Fig. 3 First six shapes of rectangular Legendre (a) and Zernike (b) polynomials. The values are scaled with respect to the BRB interface dimension and normalized.

Over these designs, perturbation topologies have been added to determine the sensitivity of these above surface profiles. From the literature on manufacturing defects [25–27], various topologies have been observed. Based on that, five simple one-dimensional perturbations have been chosen for the preliminary analysis perturbation. The models are as follows: Perturbation 1: full sine wave with a maximum height of 10μm, Perturbation 2: full sine wave with a maximum height of 50μm, Perturbation 3: reverse of Perturbation 2, Perturbation 4:Perturbation 2 with the wavelength 2/3rd of the length of an interface, and Perturbation 5: a 50μm step at 2 cm away from the center of the interface. Perturbation 2 is a scaled version of Perturbation 1, and a comparison between these two perturbations can be useful to assess how much the scaling of the perturbation affects the response. The perturbations are shown in Figure 4.

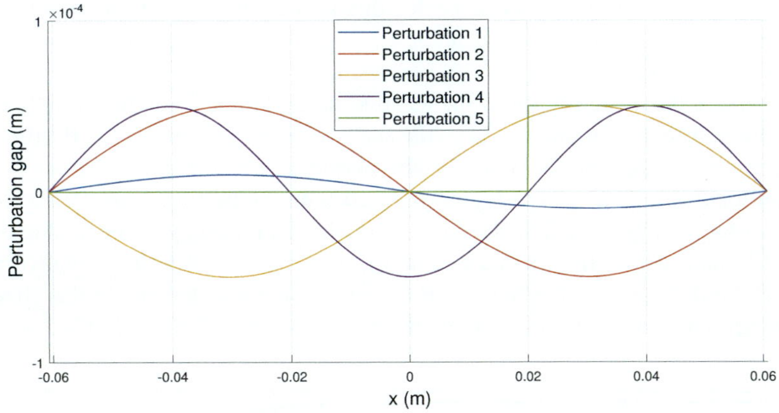

Fig. 4 The perturbation gaps used for the BRB interface

Results

The amplitude-dependent frequency and damping of the BRB is characterized with different mesoscale topologies by using the EPMC method. The initial goal is to understand the effect of different mesoscale topologies on the nonlinear dynamics of the system and to assess how the nonlinear dynamics change with the addition of perturbations. To investigate this, each topology effect is compared with respect to a flat, conformal interface. Then, a perturbation is added to each topology to study the effect of it. But how the conformal interface itself behaves under perturbation is studied first, and the results are shown in Figure 5. The 'flat' response is the baseline topology that contains no mesoscale curvature but a statistically described set of asperities (that are held constant across all simulations), and its frequency and damping characteristics are obtained using both 232 ZTE and 588 ZTE elements to assess model convergence. The 232 ZTE and 588 ZTE interface models yield identical responses, indicating that subsequent simulations can be conducted using just 232 ZTE. Also, it is found that the prestress and linearized eigenvalue analysis provides a useful initial guess for low-amplitude responses in the context of solving the EPMC problem. With the five different perturbations, a significant change in both frequency and

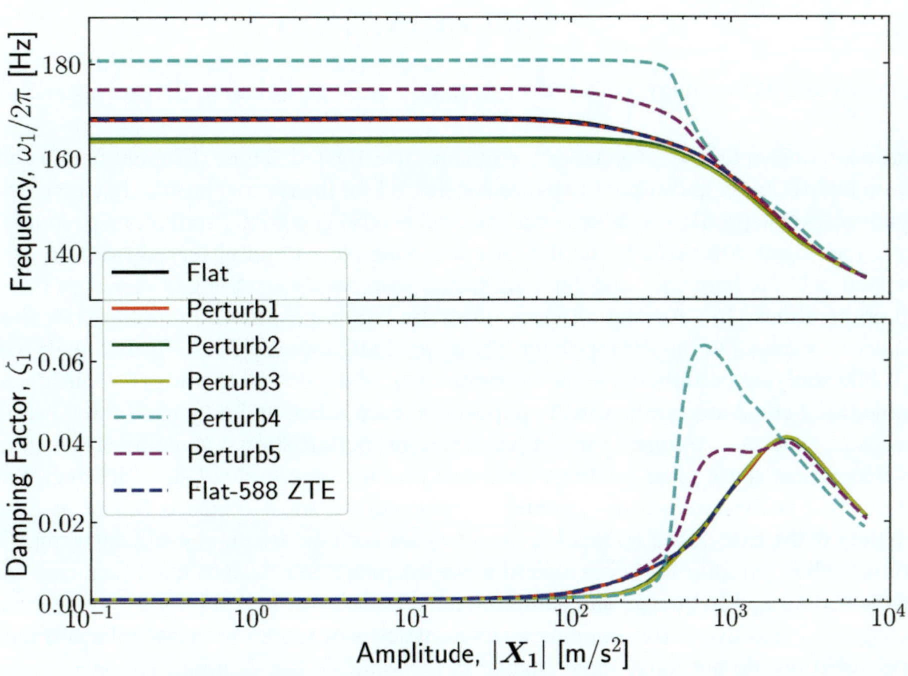

Fig. 5 EPMC results for the BRB with conformal interface and five different perturbations shown in Figure 4

damping is observed. These design perturbations chosen for this initial study are at the extreme of realistic perturbations, which indicates that this is a valid modeling approach for considering smaller perturbations as the design trends can be validated against experimental results [15].

In order to rigorously explore the design space of the meso-scale topology, the x-tilt, y-tilt and xy-tilt (addition of both) topology is taken into account. The maximum gap in x-tilt and y-tilt design is taken as 100μm. Later, a superposition of both these two makes the xy-yilt configuration. However, the frequency and damping characteristic curves are found to be precisely similar to the flat topology, as shown in Figure 6. Even with the perturbation (Perturbaton 2 in Figure 4) the the response is similar to the conformal topology. The deviation from the flat topology in the prestress frequency is in the fourth decimal in the case of x-tilt, and it is in the second decimal in the case of y-tilt. The relative position of the contacting surfaces after preloading is similar for both the flat and tilt cases, which resulted in similar characteristic curves. That is, upon assembly, the initial gap is able to be closed completely in the case of a planar tilt of the interface, as expected.

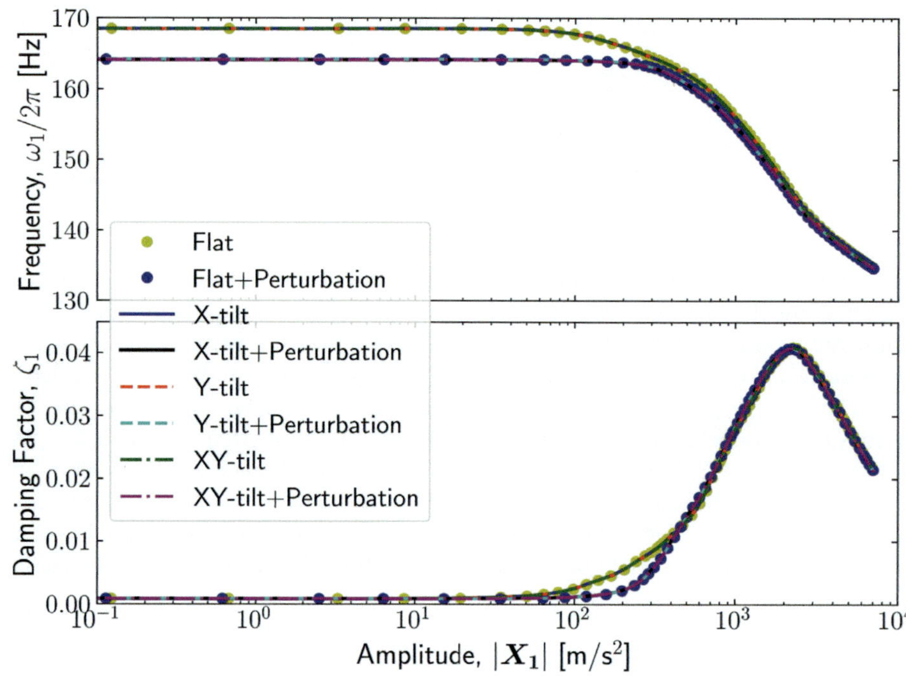

Fig. 6 Comparing linear and flat topology, with and without mesoscale perturbation (Perturbation 2)

Next, the quadratic topologies (x^2, y^2, xy, and $x^2 - y^2$) are investigated. Three different heights are considered for the profiles: 10μm, 100μm and 1000μm, and similar gaps are considered for the reverse profile. In case of the x^2 and y^2 profiles, the positive and negative/ reverse profile resembles edge gap and central gap configuration, respectively. Figure 7 shows the effect of the quadratic topologies compared to the flat interface. One clear trend observed from the x^2 topology is that the edge gap causes a reduction in the frequency and damping factor, whereas the central gap increases it. This is also confirmed by the experimental observations [15]. Among different quadratic topologies, the xy topology has shown minimal change in its characteristic curve compared to the flat topology. These quadratic topologies are further analyzed with perturbations, shown in Figure 8. While studying perturbations, the maximum gap of the design topology is considered to be 200μm.

Perturbation topologies 2 and 3 are symmetrically opposite to each other, and they have shown similar responses for all the topologies, which is as expected. Amongst various perturbations, perturbation 4 is the most sensitive one. This indicates that the shorter wavelength can cause more nonlinear interface profiles once preloaded, which increases the uncertainty in the system. After a detailed analysis of various profiles, xy showed the most robust behavior against perturbations. The sensitivity to perturbation in the case of the xy topology is lesser for both the frequency and damping. Thus, topologies with higher-order polynomials show promise for being able to make the interface robust to manufacturing variability.

If one compares the variation in frequency and damping, the relative error in damping is much larger than the frequency. So, making the topology less sensitive to the damping characteristics is of more importance. From Figure 8, it can be found that the first three perturbations do not make huge change in the value of the damping factor. However, the perturbations 4 (sine with shortened wavelength) and 5 (step) show significant changes in the damping value. The step perturbation increased the maximum damping factor over the response domain, and the increase was more than double in the case of flat

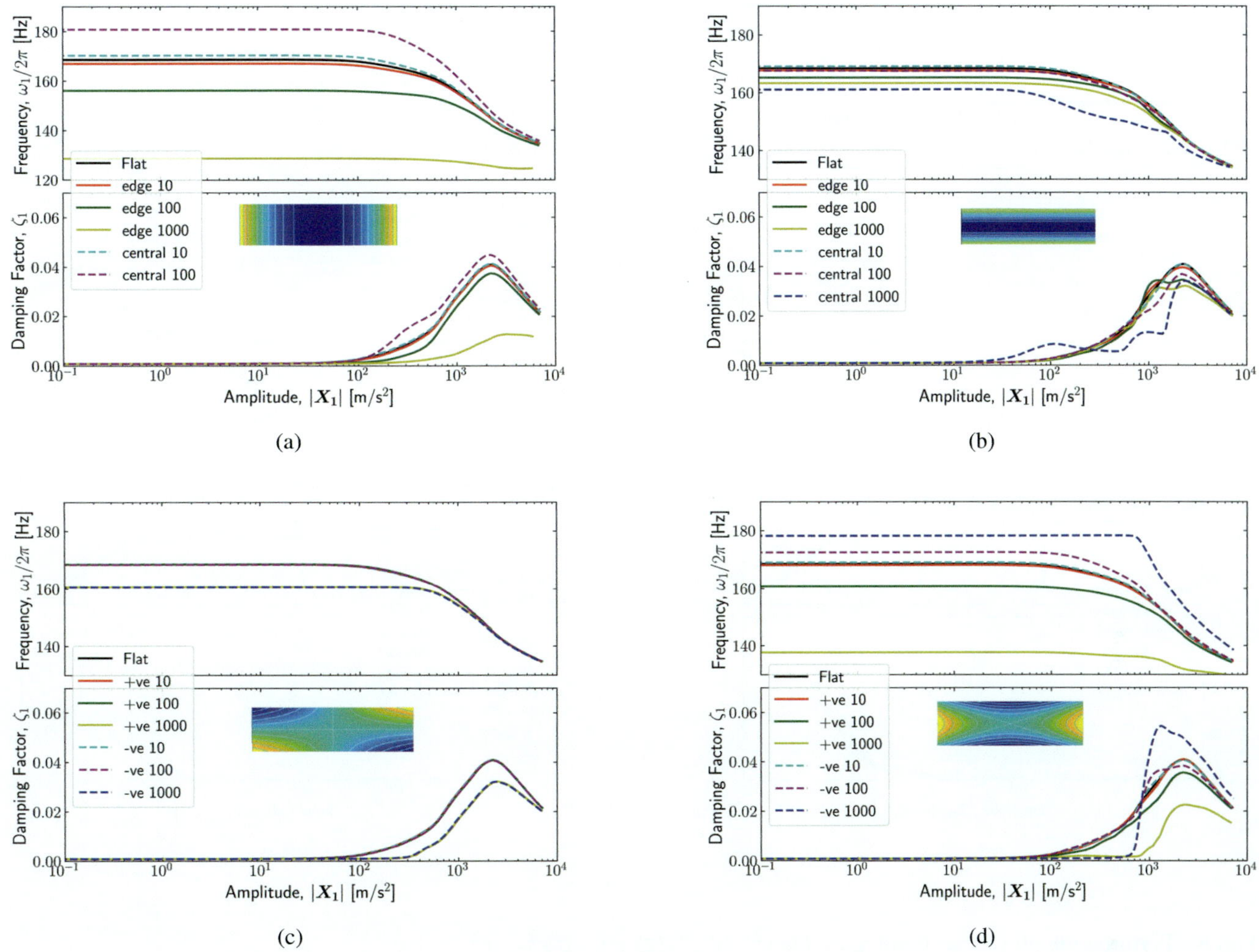

Fig. 7 EPMC results for BRB with quadratic topologies: (a) x^2, (b) y^2, (c) xy (d) $x^2 - y^2$. Each topology compared with flat and amplitude varied as 10μm, 100μm and 1000μm, for both positive and negative direction

and x-quadratic profiles. For the xy-topology, the growth is less, though significant. The increase in damping factor due to perturbation 4 is 55.8% for flat design (Figure 5), where it got reduced to 17.7% for xy design. A more detailed investigation with the surface profile might be able to mitigate the effect of these strong perturbations.

Conclusion and Future Work

This work represents a foundational step towards robust optimization of bolted connections. Specifically, it focuses on analyzing various interface topologies to identify an optimal design with minimal sensitivity to manufacturing errors. The BRB is employed as an example problem; it is found that by modifying the interface to have xy curvature, the design can be made robust to manufacturing variability. The EPMC method has proven effective for analyzing vibration characteristics at higher amplitude levels. Additionally, the use of ZTEs at the interface offers a highly effective modeling approach, allowing for detailed interfacial contact mechanics to be captured without the need for re-meshing with each surface topology change.

Future work will aim to experimentally validate these findings and explore the use of more sophisticated optimization algorithms. A key requirement will be the development of metrics to quantify the variability in behavior for each interface topology. One potential approach is to introduce a sensitivity metric to measure the variability of the damping factor ζ such as $||(\zeta_{\text{Design + Perturb}} - \zeta_{\text{Design}})/\zeta_{\text{Flat}}||_2$. Other than this response metric, another system parameter metric can also be defined as the ratio of gap variance in the perturbed system to that of the nominal design. Additionally, the design of such surfaces must

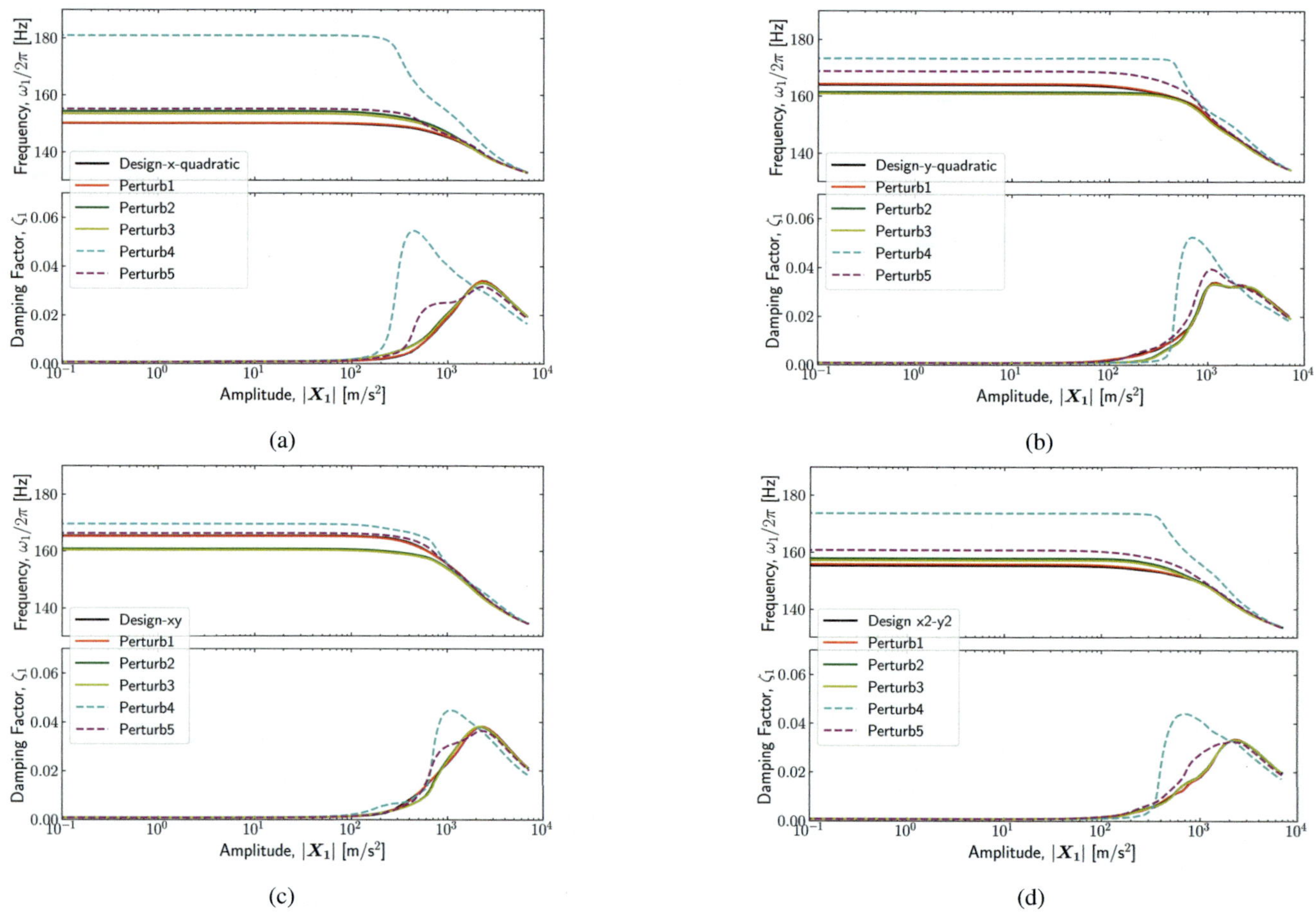

Fig. 8 Perturbation on various topologies: (a) x^2, (b) y^2, (c) xy, (d) $x^2 - y^2$

be informed by the capabilities of both conventional and non-conventional manufacturing processes [28, 29], as fabricating complex surface profiles will pose significant challenges.

Acknowledgments This material is based upon work supported by the U.S. Department of Energy, Office of Science, Office of Advanced Scientific Computing Research, Department of Energy Computational Science Graduate Fellowship under Award Number(s) DE-SC0021110. This work was supported in part by the Big-Data Private-Cloud Research Cyberinfrastructure MRI-award funded by NSF under grant CNS-1338099 and by Rice University's Center for Research Computing (CRC). The authors are thankful for the support of the National Science Foundation under Grant Number 1847130.

This report was prepared as an account of work sponsored by an agency of the United States Government. Neither the United States Government nor any agency thereof, nor any of their employees, makes any warranty, express or implied, or assumes any legal liability or responsibility for the accuracy, completeness, or usefulness of any information, apparatus, product, or process disclosed, or represents that its use would not infringe privately owned rights. Reference herein to any specific commercial product, process, or service by trade name, trademark, manufacturer, or otherwise does not necessarily constitute or imply its endorsement, recommendation, or favoring by the United States Government or any agency thereof. The views and opinions of authors expressed herein do not necessarily state or reflect those of the United States Government or any agency thereof.

References

[1] M. R. W. Brake, C. W. Schwingshackl, and P. Reuß. "Observations of variability and repeatability in jointed structures". In: *Mechanical Systems and Signal Processing* 129 (2019), pp. 282–307.

[2] T. W. Jerome, M. R. Shepherd, and S. A. Habric. “Acoustic excitation of a flanged joint”. In: *36th International Modal Analysis Conference (IMAC XXXVI)*. Orlando, FL, 2018.

[3] T. Jerome, M. Shepherd, and S. Hambric. “Variability in measured resonance frequencies and loss factors of a bolted panel structure”. In: *INTER-NOISE and NOISE-CON Congress and Conference Proceedings*. Auckland, New Zealand, 2023.

[4] B. J. Deaner, M. S. Allen, M. J. Starr, D. J. Segalman, and H. Sumali. “Application of Viscous and Iwan Modal Damping Models to Experimental Measurements From Bolted Structures”. In: *Journal of Vibration and Acoustics* 137.2 (2015), p. 021012. ISSN: 1048-9002. URL: https://doi.org/10.1115/1.4029074.

[5] S. I. Zare Estakhraji, M. Wall, J. Capito, and M. S. Allen. “A thorough comparison between measurements and predictions of the amplitude dependent natural frequencies and damping of a bolted structure”. In: *Journal of Sound and Vibration* 544 (2023), p. 117397. ISSN: 0022-460X. URL: https://www.sciencedirect.com/science/article/pii/S0022460X22005806.

[6] H. Dastani, D. Botto, and M. Glorioso. “Experimental and numerical investigation of contact parameters in a dovetail type of blade root joints”. In: *Applied Sciences* 11 (2021), p. 12008.

[7] J. Yuan, L. Salles, D. Nowell, and C.Schwingshackl. “Influence of mesoscale friction interface geometry on the nonlinear dynamic response of large assembled structures”. In: *Mechanical Systems and Signal Processing* 187 (2023), p. 109952.

[8] N. N. Balaji, W. Chen, and M. R. W. Brake. “Traction-based multi-scale nonlinear dynamic modeling of bolted joints: Formulation, application, and trends in micro-scale interface evolution”. In: *Mechanical Systems and Signal Processing* 139 (2020), p. 106615. ISSN: 0888-3270.

[9] M. W. Sracic, M. S. Allen, and H. Sumali. “Identifying the modal properties of nonlinear structures using measured free response time histories from a scanning laser Doppler vibrometer”. In: *30th International Modal Analysis Conference (IMAC XXX)*. Jacksonville, FL, 2012.

[10] J. P. Noël and G. Kerschen. “Nonlinear system identification in structural dynamics: 10 more years of progress”. In: *Mechanical Systems and Signal Processing* 83 (2017), pp. 2–35.

[11] M. Scheel, G. Kleyman, A. Tatar, M. R. W. Brake, S. Peter, J.-P. Noel, M. S. Allen, and M. Krack. “System identification of jointed structures: nonlinear modal testing vs. state-space model identification”. In: *36rd International Modal Analysis Conference (IMAC XXXVI)*. Orlando, FL, 2018.

[12] M. Jin, W. Chen, M. R. W. Brake, and H. Song. “Identification of instantaneous frequency and damping from transient decay data”. In: *Journal of Vibration and Acoustics* (2020), pp. 1–36.

[13] W. Chen, D. Jana, A. Singh, M. Jin, M. Cenedese, G. Kosova, M. R. W. Brake, C. W. Schwingshackl, S. Nagarajaiah, K. J. Moore, and J.-P. Noël. “Measurement and identification of the nonlinear dynamics of a jointed structure using full-field data, Part I: Measurement of nonlinear dynamics”. In: *Mechanical Systems and Signal Processing* 166 (2022), p. 108401. ISSN: 0888-3270.

[14] T. Dreher, M. R. W. Brake, B. Seeger, and M. Krack. “In situ, real-time measurements of contact pressure internal to jointed interfaces during dynamic excitation of an assembled structure”. In: *Mechanical Systems and Signal Processing* 160 (2021), p. 107859. ISSN: 0888-3270.

[15] W. Chen, M. Jin, I. G. Lawal, M. R. W. Brake, and H. Song. “Measurement of slip and separation in jointed structures with non-flat interfaces”. In: *Mechanical Systems and Signal Processing* 134 (2019), p. 106325.

[16] M. R. W. Brake, M. S. Allen, D. D. Quinn, D. Nowell, and D. R. Roettgen. “Testing and Modeling of Friction and Slip in Mechanical Interfaces: State of the Art and Perspectives for the Next Decade”. In: *Journal of Structural Dynamics* (To be published).

[17] J. H. Porter and M. R. W. Brake. “Towards a predictive, physics-based friction model for the dynamics of jointed structures”. In: *Mechanical Systems and Signal Processing* 192 (2023), p. 110210. ISSN: 0888-3270.

[18] M. Krack. “Nonlinear modal analysis of nonconservative systems: Extension of the periodic motion concept”. In: *Computers & Structures* 154 (2015), pp. 59–71. ISSN: 0045-7949.

[19] N. N. Balaji. *An ABAQUS-MATLAB tutorial for jointed systems*. 2024. URL: https://nidish96.github.io/Abaqus4Joints/.

[20] N. N. Balaji, T. Dreher, M. Krack, and M. R. W. Brake. “Reduced order modeling for the dynamics of jointed structures through hyper-reduced interface representation”. In: *Mechanical Systems and Signal Processing* 149 (2021), p. 107249. ISSN: 0888-3270.

[21] J. H. Porter and M. R. W. Brake. “Efficient model reduction and prediction of superharmonic resonances in frictional and hysteretic systems”. In: *Mechanical Systems and Signal Processing* arXiv:2405.15918 (Under Review). arXiv: 2405.15918.

[22] J. H. Porter and M. R. W. Brake. *TMDSimPy: Tribomechadynamics Simulations for Python*. URL: https://github.com/tmd-lab/tmdsimpy.

[23] Y. Li, J. Guo, and R. Liu. “Exploring wavefront detection in imaging systems with rectangular apertures using phase diversity”. In: *Sensors* 24.4 (2024). ISSN: 1424-8220. URL: https://www.mdpi.com/1424-8220/24/4/1191.

[24] K. Niu and C. Tian. “Zernike polynomials and their applications”. In: *Journal of Optics* 24.12 (2022), p. 123001. URL: https://dx.doi.org/10.1088/2040-8986/ac9e08.

[25] C. Zhao, J. Lv, S. Du, and Y. Deng. “Geometric error modeling and monitoring of the 3d surface by gaussian correlation model”. In: *2019 IEEE International Conference on Industrial Engineering and Engineering Management (IEEM)*. 2019, pp. 1022–1025.

[26] Y. Chen, H. Tang, Q. Tang, A. Zhang, D. Chen, and K. Li. “Machining error decomposition and compensation of complicated surfaces by emd method”. In: *Measurement* 116 (2018), pp. 341–349. ISSN: 0263-2241. URL: https://www.sciencedirect.com/science/article/pii/S0263224117307364.

[27] J. Wu, L. Qiao, and Z. Huang. “Deviation modeling of manufactured surfaces from a perspective of manufacturing errors”. In: *International Journal of Advanced Manufacturing Technology* 98 (2018), 1321–1337. URL: https://doi.org/10.1007/s00170-018-2305-2.

[28] C. Zuo, X. Zhou, Q. Liu, R. Wang, J. Lin, P. Xu, and X. Zhang. “Analytical Topography Simulation of Micro/Nano Textures Generated on Freeform Surfaces in Double-Frequency Elliptical Vibration Cutting Based Diamond Turning”. In: *Journal of Manufacturing Science and Engineering* 140.10 (2018), p. 101010. ISSN: 1087-1357. URL: https://doi.org/10.1115/1.4040616.

[29] D. Zhao and W. Guo. “Shape and Performance Controlled Advanced Design for Additive Manufacturing: A Review of Slicing and Path Planning”. In: *Journal of Manufacturing Science and Engineering* 142.1 (2019), p. 010801. ISSN: 1087-1357. URL: https://doi.org/10.1115/1.4045055.

Chapter 6
Nonlinear Vibration Control on Complex Structures via Model Predictive Control Exploiting Koopman Model

Yichang Shen and Ludovic Renson

Abstract Active vibration control of thin structures is often necessary to enhance or ensure performance, and it can also be used to characterize their dynamic behavior during experimental testing through techniques like control-based continuation. In this work, Model Predictive Control (MPC) is applied to control the nonlinear vibration of a thick (i.e. linear) cantilever beam with artificial nonlinearity and track periodic signals. The Koopman operator is used to linearise the dynamics of the system, thereby obtaining an effective real-time MPC implementation. Virtual experimental results show that using quadratic programming MPC with the Koopman model leads to efficient and accurate control performance.

Keywords Vibration control · Model predictive control · Koopman operator · Nonlinear vibration

Introduction

The constant drive to improve the performance of mechanical structures is leading to new designs that are increasingly lighter, more flexible, and inherently nonlinear. Active vibration control of thin structures is sometimes necessary to enhance or guarantee their performance. Active control can also be used to characterize the dynamic behavior of nonlinear structures during experimental tests using techniques like control-based continuation (CBC) [1]. Model Predictive Control (MPC) is an advanced control technique widely employed in industrial settings. MPC allows for the use of existing models to control multiple-input, and multiple-output systems and handle input-output interactions, making it appealing for complex systems. Moreover, MPC uses constraints to limit response amplitudes and reduce the risk of failure during large amplitude vibrations. Thus, it serves as a potential advanced feedback controller that may extend CBC to more complex, real-life applications.

To implement MPC effectively, it is important to develop an accurate yet computationally efficient model of the system's dynamics. The author's previous study [2] validates that, although exploiting nonlinear reduced-order models (NLROM) in (N)MPC algorithms leads to good control performance and low-dimensional representations of the system's dynamics, nonlinear models remain computationally expensive to solve numerically. To take advantage of powerful convex programming solvers, such as quadratic programming (QP), in this work, the Koopman operator has been utilized to convexify the originally nonconvex problem by generating a linear model from experimental data. Preliminary virtual experiments indicate that MPC, exploiting the Koopman model, is able to control nonlinear structural vibrations for tracking periodic signals.

Experimental Set-up of Benchmark Structure

The benchmark structure is a straight, thick cantilever beam with arbitrary artificial nonlinearity generated by an electrodynamic shaker. Key components of the setup include a real-time controller (dSPACE MicroLab Box), an electrodynamic shaker, a stinger, an impedance head attached to the beam for measuring the force and acceleration at the excitation point, and a laser sensor for measuring the beam's displacement. The assembly details and schematic view of the experimental setup are shown in Figure 1. A similar benchmark structure was previously investigated in [3]. The nonlinear system is investigated as a single-input, single-output system. The nonlinearity in the system was artificially implemented using the real-time

Yichang Shen · Ludovic Renson
Department of Mechanical Engineering, Imperial College London, London, SW7 2BX
e-mail: yichang.shen@imperial.ac.uk; l.renson@imperial.ac.uk

Ludovic Renson et al. (eds.), *Nonlinear Structures & Systems, Vol. 1*, Conference Proceedings of the Society for Experimental Mechanics Series, https://doi.org/10.13052/9788743801467-6

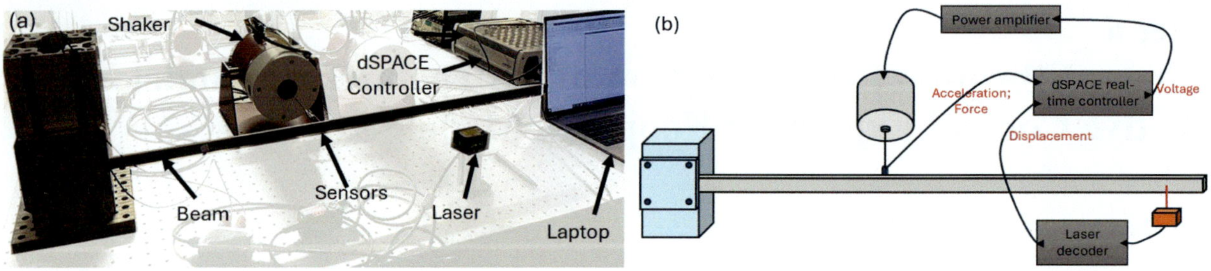

Fig. 1 Control performance of using the Carleman linearised model to generate model predictions. (a): Experimental set-up. (b): Schematic view of the experiments.

feedback control based on the measurement displacement. The force applied by the shaker to the structure was composed of the control output $f_{ex}(t)$ and virtual nonlinearity $f_{nl}(t)$. In this work, $f_{nl}(t) = K_c y^3$, where K_c is the nonlinear restoring force coefficient, and y is the beam's tip displacement measured by the laser.

Koopman Linear reconstruction and Control objective

The Koopman operator is an effective tool for providing a global linear representation of a nonlinear system [4]. This linearization allows for the application of well-established control techniques for linear systems. However, its infinite-dimensional nature presents considerable challenges for practical use. To address this issue, the method employed in this work is the EDMD method [5], which enables the approximation of the Koopman operator's eigenvalues, eigenfunctions, and modes from data. This data-driven approach allows us to obtain a linear reconstruction of the nonlinear system's dynamics from experimental data in the form of

$$\dot{\mathbf{z}}^{[\mathbf{k}]} = \bar{\mathbf{A}}\mathbf{z}^{[\mathbf{k}]} + \bar{\mathbf{B}}u. \tag{1}$$

The class of functions used for lifting $\mathbf{z}^{[\mathbf{k}]}$ used here includes polynomials up to third-order and first-order Fourier modes. In this model predictive control application, the objective is for the nonlinear structure vibration to follow a periodic reference signal X_{ref} composed of multiple sine functions. The cost function minimized by the MPC algorithm at time step k is expressed by

$$J_k = \sum_{i=0}^{n_p-1} \left((x_{k+i} - X_{ref}(t_k))^T \mathbf{Q}(x_{k+i} - X_{ref}(t_k)) + u_{k+i}^T \mathbf{R} u_{k+i}\right) + x_{k+n_p}^T \mathbf{P} x_{k+n_p}, \tag{2}$$

where x_{k+i} are the future system states predicted using the system model (Koopman) for a number of time steps, n_p, ahead of the current time t_k. The input u and the beam tip displacement are constrained to remain in a desired range to avoid damaging the experimental set-up.

Virtual Experimental results

Virtual tests were conducted in SIMULINK to validate the overall control design methodology. The virtual experiment model was tuned to match the physical setup, as shown in the SIMULINK model in Figure 2(a). The cantilever beam was represented as a single degree of freedom (1-DOF) system derived using MATLAB's n4sid.m system identification function based on experimental data obtained using a multiple sine excitation signal. The Koopman model employed in the MPC is measured by experimental data with the same input signal but with the virtual cubic nonlinearity included. The shaker is system-identified to the input voltage and output force transfer function. Fig. 2 (b) illustrates the performance of the model predictive control, demonstrating that the displacement simulated by the beam model closely aligns with the target signal. Figure. 2 (c) and (d) depict the artificially induced nonlinear component of the system dynamics and the optimal control output computed by MPC in real-time, respectively.

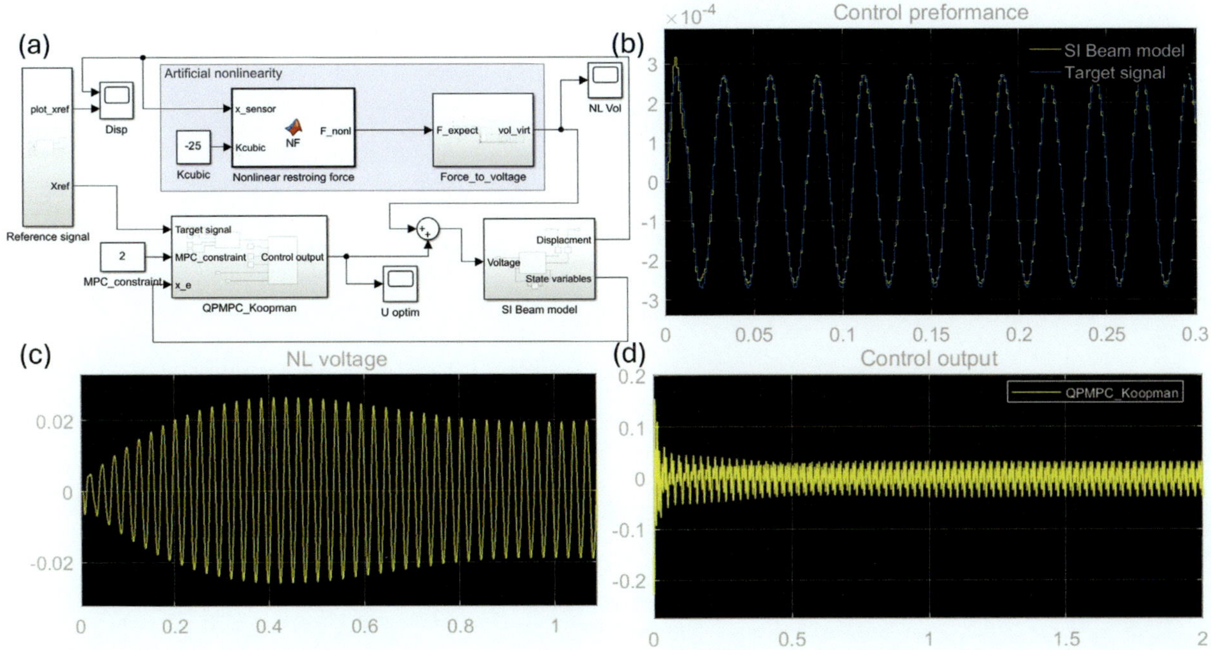

Fig. 2 Virtual experimental set-up and results. (a): SIMULINK model. (b): MPC Control performance. (c) Virtual nonlinear component. (d) Control output.

Conclusion

In this work, a cantilever beam with artificial nonlinearity is investigated, and model predictive control (MPC) exploiting the Koopman operator is applied to control the nonlinear structure vibrations to track a desired, multi-harmonic periodic signal. Results from virtual experiments demonstrate that using Koopman linearized models within MPC enables efficient and accurate control performance. Full-scale experimental testing is currently being prepared and will be conducted in the next phase of the research.

Acknowledgments The authors would like to thank HORIZON MSCA Postdoctoral Fellowship action and the UKRI Postdoc Guarantee Fellowship provided by the EPSRC, with Grant Ref: EP/X026027/1.

References

1. Renson, L., Sieber, J., Barton, D.A., Shaw, A., and Neild, S. "Numerical continuation in nonlinear experiments using local gaussian process regression". *Nonlinear Dynamics*, 98:2811–2826 (2019)
2. Shen, Y. and Renson, L. "Linearization and nonlinear model reduction for the model predictive control of nonlinear structure vibrations". *Proceedings of the 42th IMAC, A Conference and Exposition on Structural Dynamics 2024*.
3. Abeloos, G., Renson, L., Collette, C., and Kerschen, G. "Stepped and swept control-based continuation using adaptive filtering". *Nonlinear Dynamics*, 104:3793–3808 (2021).
4. Bevanda, P., Sosnowski, S., and Hirche, S. "Koopman operator dynamical models: Learning, analysis and control". *Annual Reviews in Control*, 52:197–212 (2021).
5. Williams, M.O., Kevrekidis, I.G., and Rowley, C.W. "A data-driven approximation of the koopman operator: Extending dynamic mode decomposition". *Journal of Nonlinear Science*, 25:1307–1346 (2015).

Chapter 7
Stochastic Dynamics and Surrogate Modeling for Nonlinear Aerospace Nozzle Systems with Partial Observations

Evangéline Capiez-Lernout, Olivier Ezvan, and Christian Soize

Abstract This research addresses the computational updating of a statistical surrogate model using a partial target dataset. A parameterized nonlinear stochastic computational model of a three-dimensional engine nozzle undergoing large displacements under a stochastic pressure jet is considered with a small, incomplete target dataset of normal displacements at the nozzle exit. The probabilistic PLoM algorithm is used and adapted to the target constraint in order to construct a surrogate computational model, enabling updates that allow for the response to better match the target response despite data limitations.

Keywords uncertainty quantification · probabilistic learning · nonlinear stochastic dynamics · nozzle · surrogate model

Introduction

Numerous works have highlighted the significance of nozzles in various engineering aspects, including flow analysis, jet noise source characterization and acoustics, material properties, design optimization, and the impact of uncertainty on nozzle computational models. More precisely, the vibrational behavior of nozzle structures is a critical concern in the aerospace engineering field. This research focuses on a computational framework for analyzing the updating process in the nonlinear stochastic dynamics of a nozzle, accounting for model uncertainties, partial observability, and limited data. A parameterized high-fidelity stochastic nonlinear uncertain computational model of the nozzle is meticulously designed, optimizing of the numerical parameters to balance accuracy and efficiency of the numerical solvers [1], yielding a small training dataset to be available. Then, the Probabilistic Learning on Manifolds (PLoM) algorithm adapted to an existing incomplete target data constraint [2, 3] and which relies on a purely probabilistic approach, is used in order to construct an updated surrogate computational model [4].

Background

The inherent complexity of this nozzle computational application emerges from multiple sources: (1) the high number of degrees-of-freedom present in the nozzle computational model whose mesh is represented in Figure 1, and which is a consequence of its structural complexity driven by its small thickness; (2) the geometric nonlinear effects in its dynamic behavior, induced by displacements of the structure that are comparable to the nozzle thickness; (3) the presence of model uncertainties in the representation of homogenized materials that arise from an uncontrolled parameter represented by random vector $\mathbf{U}$, (4) the influence of an external load represented by a stationary stochastic process. This latter one is modeled by a time-dependent random pressure at the nozzle wall $P(z, \theta, t; \mathbf{w})$, using cylindrical coordinates and dependent on the control parameter $\mathbf{w}$. The load intensity is assumed to be sufficiently high for triggering nonlinear displacements responses $\mathbf{Y}(t; \mathbf{w})$, (5) the under-observability and the limited data availability, complicating the resolution of the statistical inverse problem required for the updating, and (6) the high-computational cost associated with the construction of a single realization of the stationary random displacements responses, as this requires solving a set of nonlinear stochastic coupled differential equations in the time domain, using iterative algorithms combined to multiple time steps. The observations of interest $\mathbb{O}(\nu; \mathbf{w})$ are located at the nozzle exit and are then expressed as displacement amplitudes in the frequency domain. Such parameterized

Evangéline Capiez-Lernout · Olivier Ezvan · Christian Soize
MSME UMR 8208 CNRS, Université Gustave Eiffel, 77454 Marne-La-Vallée, France
e-mail: evangeline.capiez-lernout@univ-eiffel.fr; olivier.ezvan@univ-eiffel.fr; christian.soize@univ-eiffel.fr

Ludovic Renson et al. (eds.), *Nonlinear Structures & Systems, Vol. 1*, Conference Proceedings of the Society for Experimental Mechanics Series, https://doi.org/10.13052/9788743801467-7

Fig. 1 Finite element mesh of the nozzle.

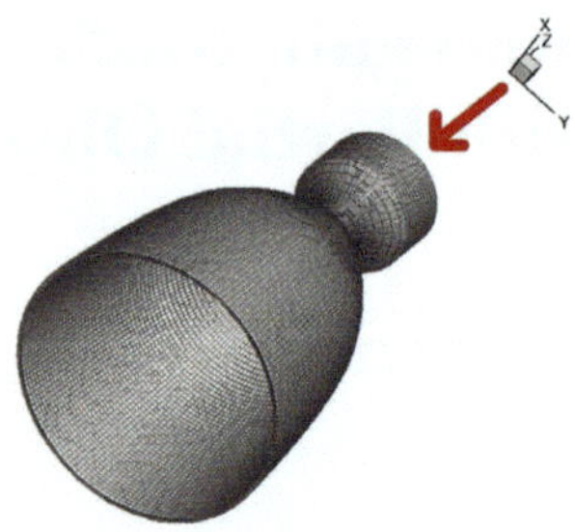

computational stochastic model is meticulously designed, allowing for computing accurate statistics of output observations but for which only a small training dataset is available.

The updating of the nonlinear uncertain stochastic computational model is carried out with respect to a given target dataset that is constituted of an incomplete data, corresponding for instance, to the case of a limited number of sensors located at the nozzle exit wall. In order to construct the statistical surrogate model of quantities of interest with respect to control parameter $\mathbf{w}$, using the PLoM algorithm under constraints, vector $\mathbf{w}$ is modeled by a vector-valued random variable $\mathbf{W}$. The statistical surrogate model is not represented by a computational model but by a learned probability measure. The quantities of interest are then derived from conditional statistics of the quantities of interest for given value of the control parameter, using the learned dataset for the updated stochastic system. To aid in the comprehension of this paper, the figure 2 presents a simplified flowchart of the used methodology.

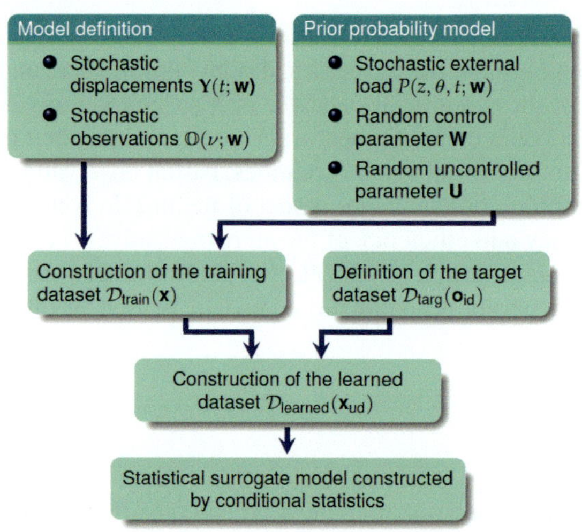

Fig. 2 Flowchart of the methodology.

Analysis of the Results

Figure 3 present the computational results corresponding to the responses as a function of the frequency for a given observation point. Each figure includes: (a) the mean response of the training dataset, both nonlinear and its linear counterpart (left upper figure), (b) a comparison between the mean response of the training dataset and the target (right upper figure), (c) the mean response of the learning dataset and the target (left lower figure), and (d) the mean response of the training dataset, the target, and the conditional confidence region given $\mathbf{w}_0$ at a 95% probability level. In (a), nonlinear effects are highlighted by shifts in resonance peaks and reduced amplitudes due to energy spreading across the excitation band. In (b), it is seen that the mean response of the training set does not match with the target, showing shifts in the second and third resonance peaks. In (c), it is shown that the constrained learning process captures the mean statistics of the target, even if limited observation data is available. Finally, in (d), three features are emphasized: (1) the mean target response is centered within the confidence region, (2) the confidence envelope peaks align well with the target's, and (3) the confidence region remains relatively narrow, reflecting the influence of uncontrolled parameters and excitation variability. A similar analysis is observed across various observation points, whether they correspond to identification points (used during the updating process) or validation points (not used), which validates the proposed methodology.

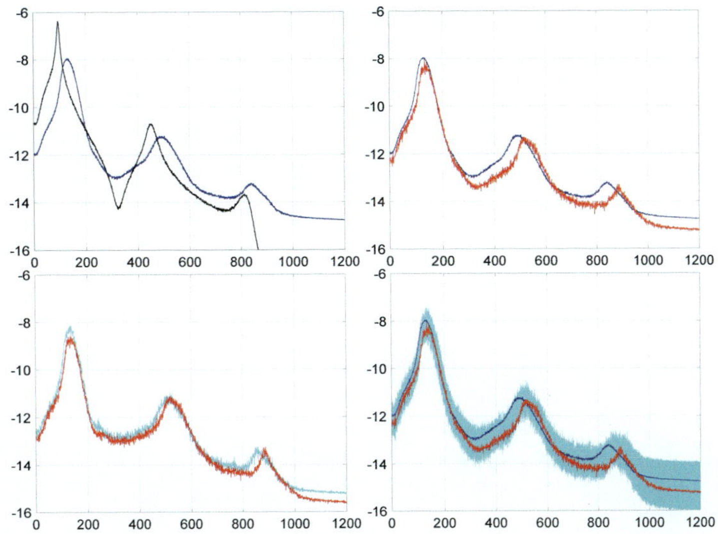

Fig. 3 Analysis of the responses as a function of the frequency at a given observation point. Left upper figure (a) : mean responses of the linearized and non linear system issued from the training set. Right upper figure (b) : mean nonlinear responses issued from both training and target data sets. Left lower figure (c) : mean nonlinear responses issued from both target and learning data sets. Right lower figure (d) : confidence region of the nonlinear response issued from the learning dataset and the corresponding mean nonlinear responses issued from both target and training data sets.

Conclusion

We have proposed a methodology for updating a high-dimensional nonlinear computational model of a nozzle structure under the constraints of limited and partially observable data. The model's complexity and computational cost are driven by uncertain parameters, geometric nonlinearities, and stochastic external forces. Using the PLoM approach, the updated surrogate model is used for estimating conditional statistics of the observations and shows its capability of capturing the characteristics of the target dataset, even for unobserved variables.

References

1. Capiez-Lernout, E. and Soize, C. "Nonlinear stochastic dynamics of detuned bladed disks with uncertain mistuning and detuning optimization using a probabilistic machine learning tool". *International Journal of Non-Linear Mechanics*, 143:104023 (2022)
2. Soize, C. and Ghanem, R. "Data-driven probability concentration and sampling on manifold". *Journal of Computational Physics*, 321:242–258 (2016)
3. Soize, C. and Ghanem, R. "Probabilistic-learning-based stochastic surrogate model from small incomplete datasets for nonlinear dynamical systems,". *Computer Methods in Applied Mechanics and Engineering*, 418:116498 (2023)
4. Capiez-Lernout, E., Ezvan, O., and Soize, C. "Updating nonlinear stochastic dynamics of an uncertain nozzle model using probabilistic learning with partial observability and incomplete dataset". *ASME Journal of Computing and Information Science in Engineering*, 24:061006 (2024)

Chapter 8
Industrial Applications of Nonlinear Modal Modeling: Two Case Studies

Benjamin R Pacini

Abstract Nonlinear structural dynamics testing, identification and modeling is becoming increasingly popular. While many methods exist, the focus of this work is nonlinear modal modeling, a framework for capturing the nonlinear behavior of a structure on a mode-by-mode basis by augmenting a linear modal model with amplitude-dependent stiffness and damping elements placed in parallel with their linear counterparts. Once identified, this highly reduced order model can be used to simulate and predict the response of a structure to arbitrary inputs in a computationally efficient manner. This work describes the application of the nonlinear modal modeling test and identification process on two industrial structures. For each structure, the linear modal analysis and the nonlinear identification are presented. A truth test is conducted on each structure, the results of which are compared to corresponding simulation results using the linear and nonlinear modal models. The nonlinear modal models are shown to better match the measured responses of both structures.

Keywords nonlinear dynamics · system identification · modal modeling · structural dynamics · industrial applications

Introduction

Essential to the robust design of structures is a model that accurately captures the relevant dynamics in its service environments. Historically, linear models have been used in this regard to great success. They are computationally efficient and can be updated via dynamic properties identified from modal tests [1]. However, if there are significant nonlinear physics active during service environments, then linear modelling may not be able to sufficiently replicate the true response of the structure, detracting from their utility in qualification decisions. Thus, there are circumstance where a nonlinear structural dynamics model is required. There are many frameworks for nonlinear modeling, testing, and identification, such as nonlinear normal modes [2] and spectral manifolds [3], but this work focuses on nonlinear modal models (NLMMs) [4, 5]. NLMMs are comprised of single degree of freedom (SDOF) oscillators which each represent a mode of vibration of the structure and have nonlinear springs and dampers that attach the modal mass to ground. As has been demonstrated in [6, 7] and will be shown in this work, NLMMs are able to capture the amplitude-dependent behavior of nonlinear, industrial structures and can be used to replicate the response to arbitrary inputs in a computationally efficient manner.

This work presents two case studies where NLMMs were applied to industrial structures. The first was a structure with mounted components and bolted joints, and an NLMM was desired that could capture the amplitude-dependent response within a specified bandwidth. The second case study utilized a structure with multiple bolted joints, and the dynamic behavior of a specific mode was desired up to approximately 300gs. Each case study is introduced in greater detail where the testing and NLMM identification is described. Each NLMM is then evaluated on its ability to match the responses obtained via time simulation to those measured from corresponding truth tests on each structure.

The remainder of this work is organized as follows. Section 2 describes NLMMs in greater detail including assumptions and limitations. The methods for testing and identification of NLMMs are presented in Section 3. The two industrial case

Sandia National Laboratories is a multimission laboratory managed and operated by National Technology & Engineering Solutions of Sandia, LLC, a wholly owned subsidiary of Honeywell International Inc., for the U.S. Department of Energy's National Nuclear Security Administration under contract DE-NA0003525.

Benjamin R Pacini
Sandia National Laboratories, P.O. Box 5800 – MS0557, Albuquerque, NM, 87185
e-mail: brpacin@sandia.gov

Ludovic Renson et al. (eds.), *Nonlinear Structures & Systems, Vol. 1*, Conference Proceedings of the Society for Experimental Mechanics Series, https://doi.org/10.13052/9788743801467-8

studies where NLMMs were used to capture their nonlinear dynamics are discussed in Sections 4.1 and 4.2. A summary of these works is provided in Section 5.

Nonlinear Modal Models

Nonlinear modal models are the same as their linear counterparts except for additional amplitude-dependent stiffness and damping elements for modes deemed to be nonlinear, see Fig. 1. The number of modes retained in the NLMM is determined by the bandwidth of interest. The physical and modal coordinates of the structure are related through:

$$\{x(t)\} = \mathbf{\Phi}\{q(t)\} \quad (1)$$

$$\{q(t)\} = \mathbf{\Phi}^{+}\{x(t)\} \quad (2)$$

with $\{x(t)\}$ and $\{q(t)\}$ the displacement of the structure in physical and modal coordinates, respectively, Φ is the mass-normalized mode shape matrix, and the "+" superscript is the Moore-Penrose Pseudo-inverse. Similarly, the excitation force in modal $\{f_q(t)\}$ and physical $\{f(t)\}$ coordinates are related through the following transformation:

$$\{f_q(t)\} = \mathbf{\Phi}^{\mathrm{T}}\{f(t)\} \quad (3)$$

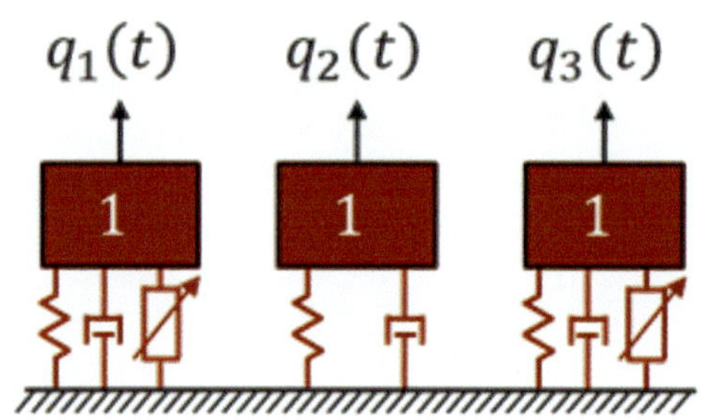

Fig. 1 Nonlinear modal model

Each mode j can be described by the following equation of motion:

$$\ddot{q}_j(t) + c_{0,j}\dot{q}_j(t) + k_{0,j}q_j(t) + f_{r,nl,j}(q_j(t), \dot{q}_j(t)) = f_{q,j}(t) \quad (4)$$

where $k_{0,j}$ and $c_{0,j}$ are the linear spring and damper coefficients, respectively, and are defined by the corresponding natural frequency, $\omega_{n,j}$, and damping ratio, ζ_j:

$$k_{0,j} = \omega_{n,j}^2 \quad (5)$$

$$c_{0,j} = 2\zeta_j\omega_{n,j} \quad (6)$$

The nonlinear modal restoring force is $f_{r,nl,j}(q_j(t), \dot{q}_j(t))$ and is the subject of the nonlinear identification (see Section 3.3). This force combines all of the nonlinear effects of mode j into a single force, so this approach to nonlinear structural dynamics obtains the overall amplitude-dependent stiffness and damping forces as opposed to identifying specific physics associated with a particular nonlinear source.

The work presented herein makes two assumptions. The first is that the mode shapes do not change with amplitude. This allows the use of the linear mode shape matrix to be used to transform between physical and modal coordinates. Second, the modes are assumed to be uncoupled. In other words, the response of one mode does not influence or excite the response of another.

Test Methods and Identification

This section discusses the test and identification methods for case studies presented in this work. The identification of the NLMM for both case studies utilized data from sine beat excitation [6] (Section 3.1), but nonlinear force appropriation [2, 8] (Section 3.2) was conducted on the structure for Case Study 2 to explore a potential modal interaction.

Sine Beats

Sine beat excitation [6] utilizes a modal shaker to concentrate a transient force in the bandwidth around a mode of interest, see Fig. 2. It is created by modulating a sinewave by an amplitude windowing function $a(t)$:

$$v(t) = a(t)\sin(\omega_c t) \tag{7}$$

where $v(t)$ is the sine beat and ω_c is the center frequency. Note that $v(t)$ is the voltage signal that is sent to the shaker and is not the force applied to the structure; the resulting force is a function of both the shaker and structure dynamics.

The peak of $V(\omega)$ (the Fourier Transform of $v(t)$) occurs at ω_c and the width of its main lobe is directly related to $a(t)$ [6]. Since $a(t)$ is a user-defined function, the input energy of a sine beat can be customized to provide forcing of a single, targeted mode while minimally exciting the others.

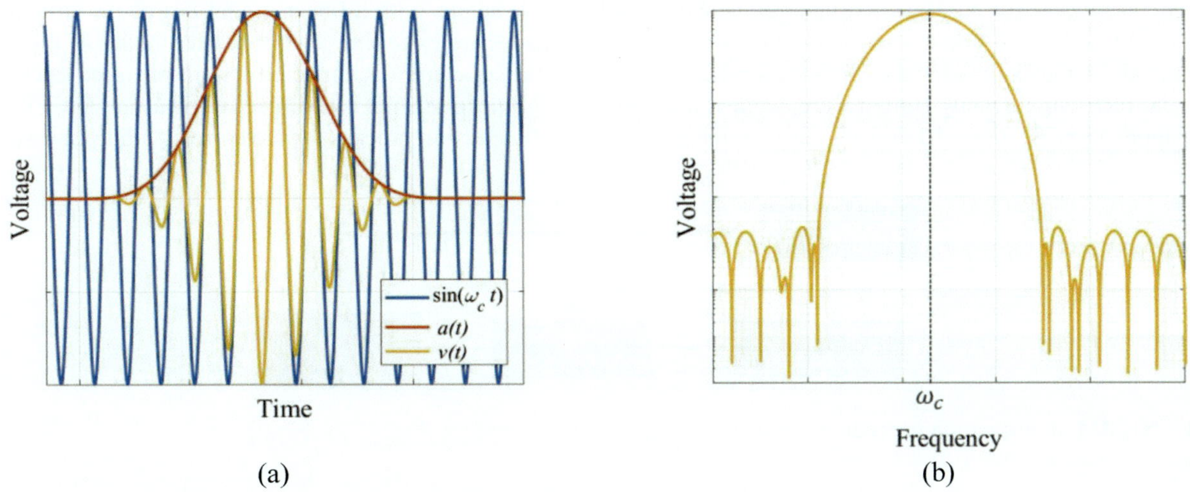

Fig. 2 Sine beat in (a) time domain and (b) frequency domain

In this work, each nonlinear mode was individually excited using sine beats. The frequency ω_c was selected as the linear natural frequency of the target mode. The windowing function $a(t)$ was a Blackman-Harris window whose parameters were selected such that either (1) the first local minima on either side of the peak in $V(\omega_c)$ occurred at adjacent modes, or (2) the width of the main lobe in $V(\omega)$ encompassed the same bandwidth as the imaginary part of the target mode in the drive point frequency response function (FRF).

Nonlinear Force Appropriation

Nonlinear force appropriation (NFA) testing (also known as phase resonance testing) is a technique that utilizes closed loop control to maintain a structure at phase resonance while the amplitude of excitation is slowly increased [2, 8]. For this work, phase resonance was defined as 90 degrees relative phase between the drive point acceleration response and excitation force at the fundamental frequency; in other words, the primary phase resonance [9, 10] was achieved at each excitation level. Fig. 3 shows a block diagram of the NFA test technique. A voltage sinewave of constant amplitude V and frequency ω is sent to a modal shaker which induces a force $f(t)$ on the structure. The corresponding drive point acceleration $\ddot{x}_{dp}(t)$ and $f(t)$ are measured by the Phase-Locked-Loop (PLL) controller. The PLL controller adjusts ω based on the computed error between 90 degrees and the relative phase between $\ddot{x}_{dp}(t)$ and $f(t)$. The updated voltage signal is generated and sent to the shaker. This cycle continues until phase resonance is achieved, and data is collected from all of the accelerometers on the structure. The energy input to the structure is then increased by increasing V, and the above process repeats capturing the entire backbone of the structure over a chosen amplitude range.

This technique characterizes a single mode from low (ideally linear) levels up to a desired high energy level. While not performed in this work, the data used from NFA tests can be used for nonlinear modal model identification. Additionally, modal interactions are more obviously excited and observed with NFA testing than sine beats. This testing technique was conducted in Case Study 2 to show a potential modal interaction (see Section 4.2.2).

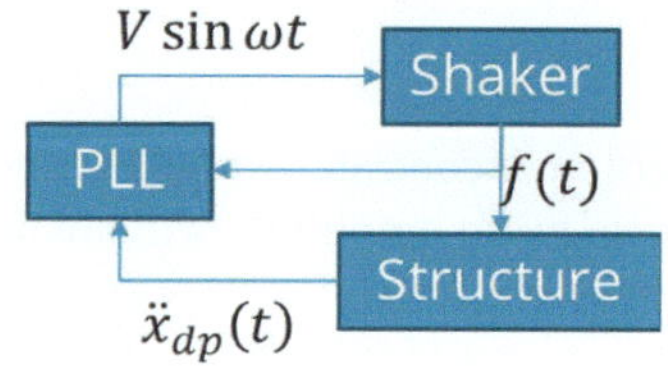

Fig. 3 Force appropriation block diagram

Nonlinear Modal Model Identification

The general process for NLMM identification is shown in Fig. 4. It begins with a linear modal test of the structure from which the linear natural frequencies, damping ratios, and mode shapes are extracted. The nonlinear modes of the structure are detected through comparing the FRFs from experiments where the structure was impacted at multiple force levels. Modes where the FRFs significantly shifted were deemed nonlinear. A signification shift is a subjective measure and will depend on the application. Sections 4.1 and 4.2 describe the nonlinear mode selection process for the corresponding case studies.

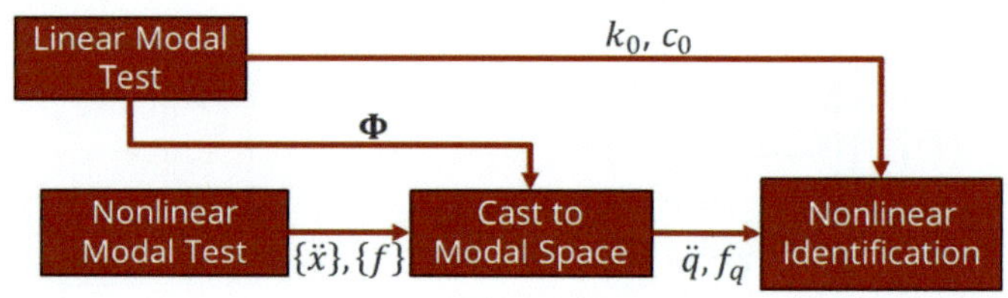

Fig. 4 NLMM identification procedure

The nonlinear modal test is then conducted for each nonlinear mode. For the case studies in this work, this comprised the sine beats. The result of this experiment are the measured accelerations $\{\ddot{x}\}$ measured in physical coordinates, and the excitation force $\{f\}$. Only single input excitation was used for this work, so only the element of $\{f\}$ corresponding to the drive point DOF is non-zero. These measurements are then cast to modal space using Eqs. (2) and (3). The identification is performed on the mode of interest.

The nonlinear identification process described below is the same as that described in [6] and is similar to the process presented in [11]. The mode of interest is now described by Eq. (4). The linear stiffness and damping coefficients (k_0 and c_0) are computed from Eqs. (5) and (6) using the results from the linear modal test. The modal velocity and velocity and displacement are determined via numerical integration of $\ddot{q}\,(t)$. Therefore, the only unknown is the nonlinear restoring force $f_{r,nl}\,(q\,(t)\,,\,\dot{q}\,(t))$ which is assumed to be the sum the nonlinear stiffness and damping restoring forces ($f_{r,nl,s}\,(q)$ and $f_{r,nl,d}\,(\dot{q})$, respectively):

$$f_{r,nl}\,(q\,(t)\,,\,\dot{q}\,(t)) = f_{r,nl,s}\,(q) + f_{r,nl,d}\,(\dot{q}) \tag{8}$$

While $f_{r,nl,s}\,(q)$ and $f_{r,nl,d}\,(\dot{q})$ can be parameterized in many ways, symmetric polynomials were used for both case studies described in this work:

$$f_{r,nl,s}\,(q) = \sum_{n=1}^{N_s} k_n q |q|^{n-1} \tag{9}$$

$$f_{r,nl,d}\,(\dot{q}) = \sum_{n=1}^{N_d} c_n \dot{q} |\dot{q}|^{n-1} \tag{10}$$

N_s and N_d are the maximum selected polynomial order for $f_{r,nl,s}\,(q)$ and $f_{r,nl,d}\,(\dot{q})$, respectively, and k_n and c_n are the associated polynomial coefficients. The absolute values of q and $\dot{q}$ are used to ensure that the identified restoring forces are symmetric, an assumption made for all nonlinear modes in both case studies.

Eqs (8) through (10) are substituted into Eq. (4). The result is rearranged and cast to the frequency domain to yield

$$\begin{bmatrix} \mathcal{F}(q\,|q|) & \ldots & \mathcal{F}\left(q|q|^{N_s-1}\right) & \mathcal{F}(\dot{q}\,|\dot{q}|) & \ldots & \mathcal{F}\left(\dot{q}|\dot{q}|^{N_d-1}\right) \end{bmatrix} \begin{bmatrix} k_1 \\ \vdots \\ k_{N_s} \\ c_1 \\ \vdots \\ c_{N_d} \end{bmatrix} = \left[\mathcal{F}\left(f_q - \ddot{q} - c_0\dot{q} - k_0 q\right)\right] \tag{11}$$

where $\mathcal{F}()$ is the Fourier Transform and represent columns of data in the above matrices. The polynomial coefficients for the nonlinear restoring forces can be solved for in a least-squares since using:

$$\begin{bmatrix} k_1 \\ \vdots \\ k_{N_s} \\ c_1 \\ \vdots \\ c_{N_d} \end{bmatrix} = \begin{bmatrix} \mathcal{F}(q\,|q|) & \ldots & \mathcal{F}\left(q|q|^{N_s-1}\right) & \mathcal{F}(\dot{q}\,|\dot{q}|) & \ldots & \mathcal{F}\left(\dot{q}|\dot{q}|^{N_d-1}\right) \end{bmatrix}^{+} \left[\mathcal{F}\left(f_q - \ddot{q} - c_0\dot{q} - k_0 q\right)\right] \tag{12}$$

where the "+" is the Moore-Penrose Pseudoinverse. The parameterized $f_{r,nl}\left(q\left(t\right),\ \dot{q}\left(t\right)\right)$ is now identified for the mode of interest. Note that data from multiple tests can be stacked within each matrix on the right-hand-side of Eq. (12). For both of the case studies in this work, sine beats at multiple levels were conducted and the corresponding data were all used to fit the parameters.

This entire process is repeated for each mode of interest that is deemed nonlinear. All modes (linear and nonlinear) are then concatenated into a nonlinear modal model of the entire structure:

$$\mathbf{I}\left\{\ddot{q}(t)\right\} + \mathbf{C_0}\left\{\dot{q}(t)\right\} + \mathbf{K_0}\left\{q(t)\right\} + \left\{f_{r,nl}\left(\left\{q(t)\right\},\left\{\dot{q(t)}\right\}\right)\right\} = \left\{f_q(t)\right\} \tag{13}$$

where $\mathbf{I}$ is the identity matrix, $\mathbf{C_0}$ is a diagonal linear damping matrix with entries of individual $c_{0,j}$ for each mode j, $\mathbf{K_0}$ is a diagonal linear stiffness matrix with entries of individual $k_{0,j}$ for each mode j, and $\left\{f_{r,nl}\left(\left\{q(t)\right\},\left\{\dot{q(t)}\right\}\right)\right\}$ is the vector of nonlinear restoring forces defined by Eq. (8) for each nonlinear mode. In this formulation, row j of Eq. (13) corresponds to Eq (4) for each mode j.

The response of the entire structure, $\{x\left(t\right)\}$, to an arbitrary force $\{f\left(t\right)\}$ can be simulated using Eq. (13) by first casting the force in physical coordinates to modal coordinates using Eq. (3). With assumed modal displacement and velocity initial conditions and $\{f_q(t)\}$, Eq. (13) is used to obtain $\{q(t)\}$ via time simulation. These modal responses are then cast back to physical coordinates, $\{x\left(t\right)\}$, using Eq. (1). These $\{x\left(t\right)\}$ are the simulated response of the structure using the identified NLMM and incorporate the amplitude-dependent stiffness and damping due to the nonlinear modes.

Generating an explicit nonlinear model of an industrial structure where the physics are exactly captured is difficult to update via system identification experiments. Moreover, the computational cost to simulate the response of a structure to arbitrary inputs with this model is prohibitively expensive. Conversely, the NLMM is highly computationally efficient and, while not exactly modelling the underlying physics, is able to capture the amplitude dependence of the stiffness and damping of each nonlinear mode. Therefore, the NLMM approach is attractive for industrial applications.

Case Studies: Industrial Applications of the NLMM

In Sections 4.1 and 4.2, the NLMM identification process described in Section 3.3 was implemented on two industrial structures. After conducting linear modal analysis, sine beats were used to excite each structure to high levels. The NLMMs were then identified, and truth tests conducted. The accuracy of the NLMM was evaluated by comparing its predicted response to the truth test forcing to that measured from the experiment. Additionally, for Case Study 2, NFA testing was conducted to evaluate a potential modal interaction. This latter result was not used explicitly in the NLMM identification for this case but is shown to demonstrate that modal interactions occur in industry. It also motivates future research to incorporate this nonlinear behavior into the NLMM process.

Case Study 1

The objective of Cast Study 1 was to conduct nonlinear system identification of the structure and obtain a NLMM. The test article is shown in Fig. 5 and is comprised of many components mounted to a case. The source of the nonlinearities arises from the interfaces between the components and the case.

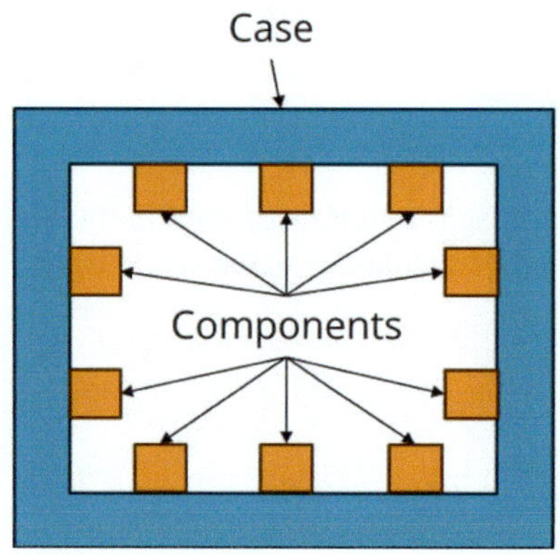

Fig. 5 Case Study 1 test article

NLMM Identification

There were 18 modes within the bandwidth of interest, six rigid and twelve elastic. The rigid body modes were computed using the companion finite element model and the elastic modes were extracted from linear modal testing, see Fig. 6. Fig. 6a compares the measured principle complex mode indicator function (CMIF) from the linear modal tests with that resynthesized from the identified linear modal properties. Fig. 6b shows the auto Modal Assurance Criterion for the extracted linear modes. These results indicate that the linear modal properties accurately represent the low-level low-level dynamics of the test article and that the identified modes are independent. This latter aspect is critical to achieving advantageous conditioning on $\mathbf{\Phi}^+$ when casting the nonlinear modal testing results from physical to modal coordinates.

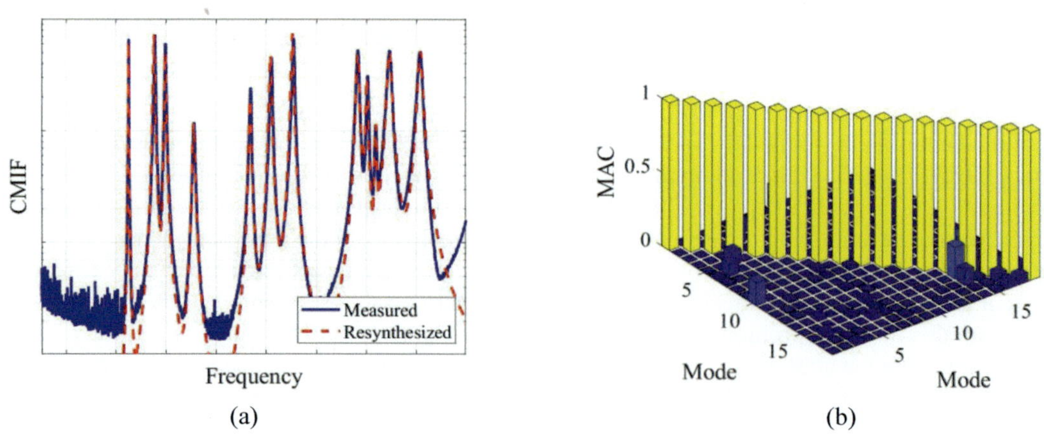

Fig. 6 Case Study 1 linear modal test results; (a) measured vs resynthesized CMIF and (b) auto-MAC

Through multi-level hammer impact testing, modes 7, 11, 12, 13, and 15 showed the largest shifts in the measured FRFs; no objective measure was used in the determination (e.g., a damping change threshold) but was left to the judgement of the author. Fig. 7 shows the CMIF overlays in the bandwidth around the selected nonlinear modes from three different hammer impact levels. Note that all modes show an increase in damping and decrease in frequency.

Sine beats were used to excite each of the nonlinear modes up to as high of level as possible. The limiting factor in this case was the electrical output of the amplifier that powered the shaker. The identification was performed per Section 3.3 with $N_s = N_d = 3$ for all five nonlinear modes. Cubic polynomials were selected as a balance between obtaining accurate representation of the nonlinear restoring force and minimizing the model order to preclude over-fitting. To evaluate

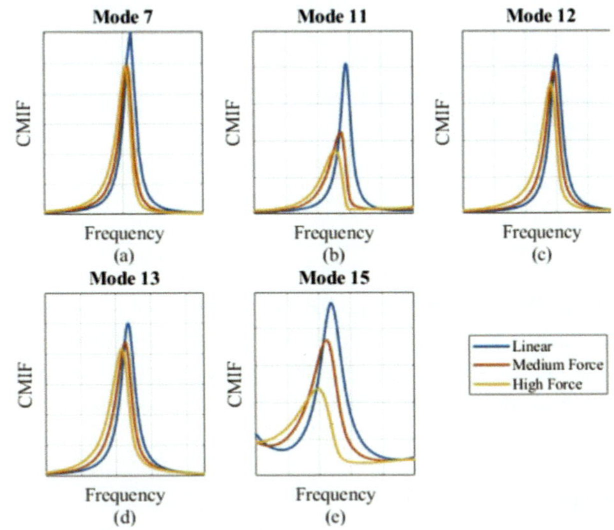

Fig. 7 Case Study 1 CMIFs from multiple input force levels showing nonlinearity of these modes

the accuracy of the identified NLMM of the test article, a truth test was conducted. This was a sine-sweep whose amplitude vs frequency was shaped in order to maximize the response of the structure at each mode. The measured force from this truth test was cast to modal space and used in the NLMM simulation of the test article per Eq. (13). The resulting modal responses were expanded back to physical coordinates in order to compare against the accelerations measured during the truth test.

Fig. 8 shows select time history and frequency domain results from measurements on two of the components of the test article. Along with the measured and NLMM results, the accelerations predicted by a purely linear model are also shown to

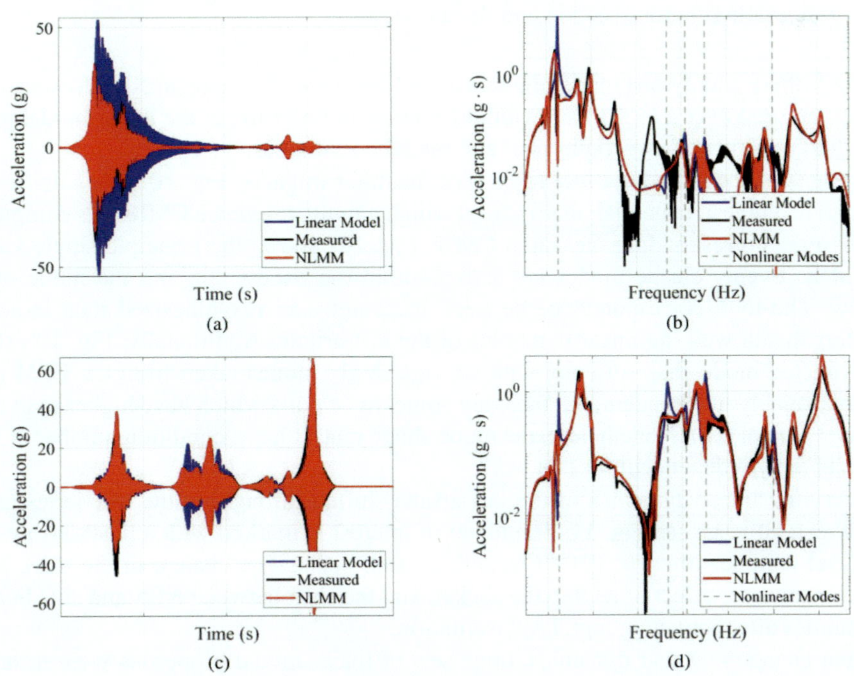

Fig. 8 NLMM vs measurements from truth test, Case Study 1; (a) time history and (b) frequency domain response of component 1 and (c) time history and (d) frequency domain response of component 2. Black data is the measurement from the truth test, the red data is the simulation result from the NLMM, the blue data is the linear simulation result, and the dashed vertical gray lines ((b) and (d) only) indicate the frequencies of the nonlinear modes

demonstrate the improvement achieved with the NLMM. In the frequency domain results, dashed vertical lines indicate the frequencies of the nonlinear modes. In the bandwidth around linear modes, all three data sets agree, confirming the decision to maintain these modes as linear. Conversely, the NLMM is shown to better match the measured results in the vicinity of the nonlinear modes. In some instances, the linear model over-predicts the response by a factor of 2, but the NLMM is able to accurately match the measured response.

Case Study 2

The objective of Case Study 2 was to characterize the nonlinear damping of a single mode of interest up to a high level (300 g at the drive point). The structure was tested in free-free boundary conditions and is shown in Fig. 9. The mode of interest exercised shear motion in the indicated joint between the fixture and the plate. Sine beats were conducted to perform the NLMM identification of the mode of interest (Section 4.2.1). NFA testing was also conducted to investigate a potential modal interaction (Section 4.2.2).

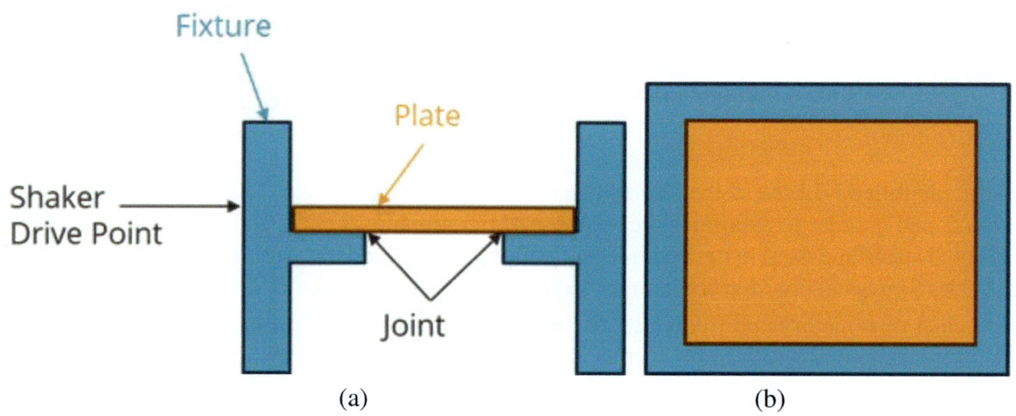

Fig. 9 Case Study 2 test article (a) cross-section and (b) top view

Unlike Case Study 1, there is only one mode of interest for Case Study 2 where an NLMM was desired. However, linear modal testing was still required over a larger bandwidth of interest in order to get the linear modal parameters for the mode of interest as well as Φ to transform between physical and modal coordinates.

Linear modal testing was conducted via low force level hammer impacts. Fig. 10 shows the results. The linear modal parameters were used to resynthesize the FRF matrix from which a multi-reference CMIF was computed. This resynthesized CMIF was compared to the corresponding measured CMIF. Fig. 10a shows the principle single value from the measured and resynthesized CMIF. Twenty elastic modes are active within the bandwidth, and the mode of interest (mode 10) is marked with an arrow. The high correspondence between measured and resynthesized data indicates that the extracted linear modal parameters match well the linear dynamics of the test article. Additionally, Fig. 10b shows the auto-MAC of the experimentally extracted mode shapes (along with six rigid body modes taken from the FEM of the test article). The identified modes are reasonably independent, but there are some mode pairs which have higher than ideal values (e.g., mode 21 and 22 have a MAC value of 0.6). Nonetheless, the mode shape matrix has a condition number of approximately 9 which deemed the modes acceptably independent for this work.

Note that the underlying linear dynamics of the test article shifted throughout the test series as indicted by low-level testing conducted between all tests, see Fig. 11. The mode of interest is marked with a black arrow. The test sequence was as follows: (1) low-level hammer impacts, (2) NFA, and (3) sine beats. The change in the drive point FRF between the hammer and NFA tests shows the effect of adding the shaker, and the shift between NFA and sine beats shows the change in underlying linear dynamics after achieving high level excitation.

Due to the apparent change in linear dynamics, three sets of linear modal properties were fit and used throughout the CS2 work. The first set is the baseline extracted from the low-level hammer taps. A set of linear properties were used to process data from each of the high-level tests. These were constructed from the baseline set, but the natural frequencies, damping ratios, and shapes were updated for the modes well excited from the shaker drive point (i.e., the modes well excited from Fig. 11). The largest differences in linear modal properties for the target mode were between the baseline and the sine beat sets. The differences in the natural frequency and damping ratio of the target mode were approximately 7% and 20%,

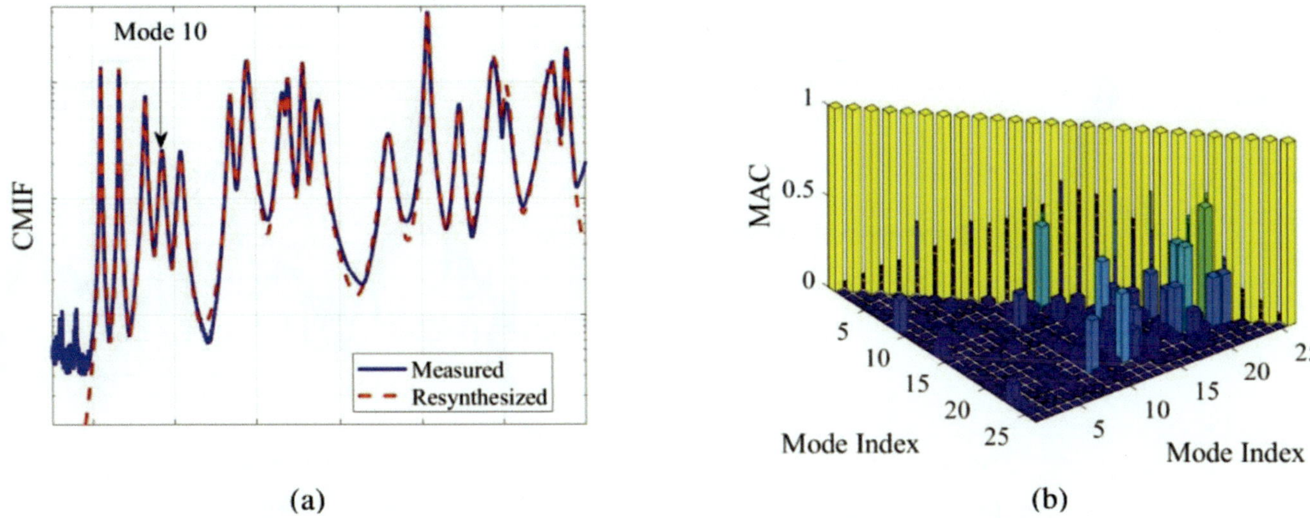

Fig. 10 Case Study 2 linear modal test results; (a) measured (solid lines) vs resynthesized (dashed lines) CMIF and (b) auto-MAC

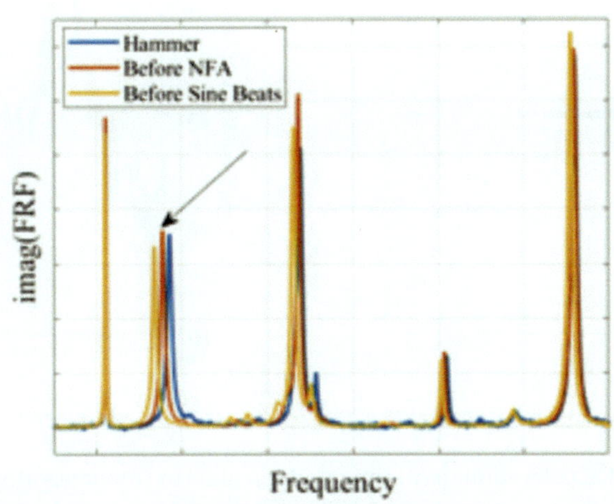

Fig. 11 Drive point FRF comparison at various stages throughout Case Study 2 testing

respectively. The corresponding MAC value for the target mode shape was 0.78. This highlights a practical consideration for nonlinear system characterization in that the underlying linear properties can change throughout the test series. This was accounted for in this work by assigning linear modal properties for each test type which best describes the corresponding underlying linear dynamics.

Nonlinear Modal Model Identification

The NLMM was constructed with 26 modes (six rigid and 20 elastic) where only mode 10 was assumed nonlinear since the objective was to obtain an NLMM of only the target mode. Given the high response levels achieved from the sine beat testing, the order of the nonlinear stiffness and damping restoring force polynomials were $N_s = N_d = 5$ for mode 10. Other modes of the structure were known to be nonlinear, so the evaluation of the NLMM was conducted using narrow-band excitation to minimize the nonlinear effects of the other modes. Thus, the accuracy of the NLMM was assessed using the measurements from the highest-level sine beat. Fig. 12 shows the results for two acceleration DOFs. The first is the drive point (Fig. 12a and b) and the second is the DOF with the largest value in the target mode shape (referred to as the "max DOF" shown in Fig. 12c and d). These figures compare the measured acceleration response with those computed via linear and NLMM simulation.

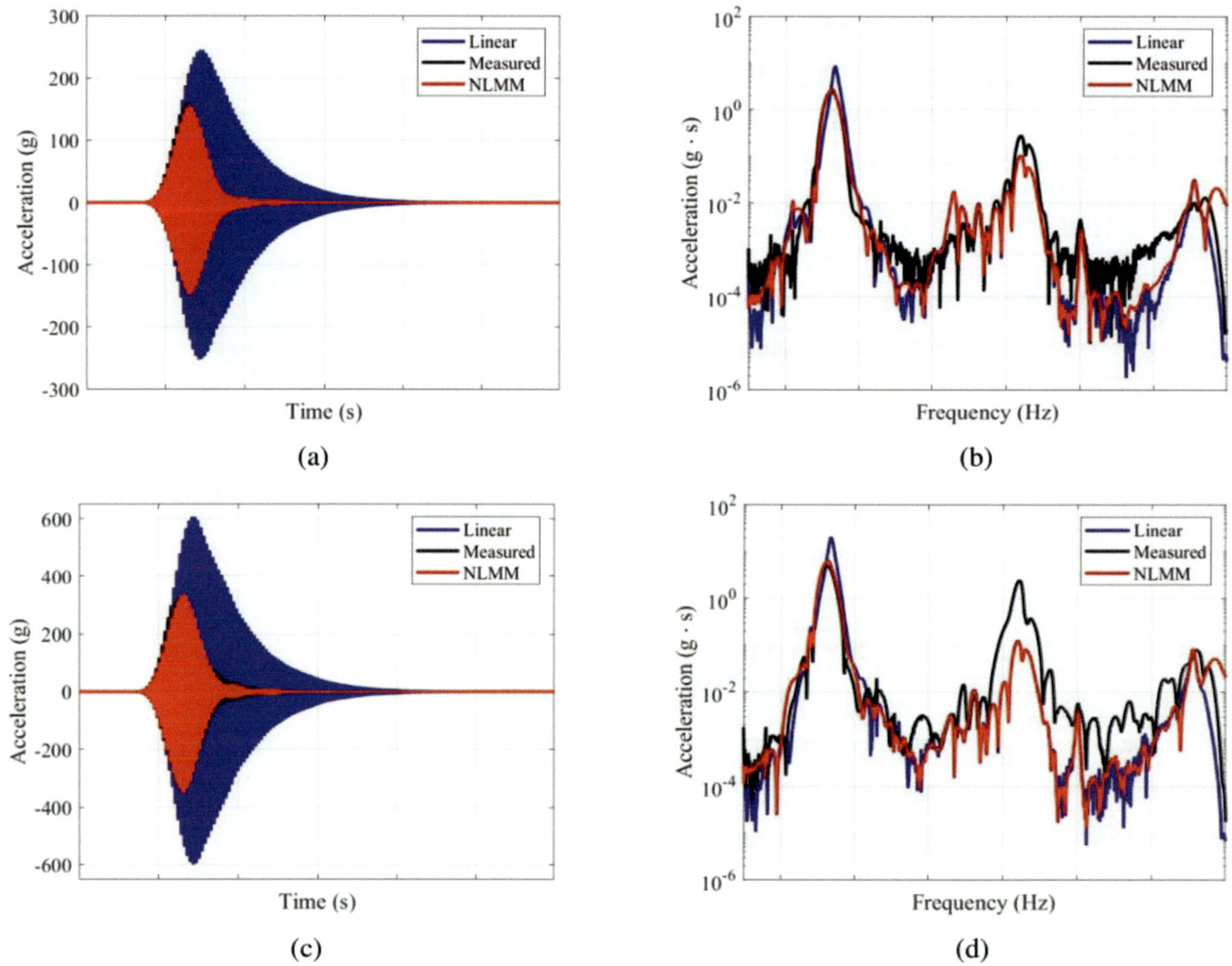

Fig. 12 NLMM results; drive point acceleration in (a) time domain and (b) frequency domain, and the acceleration response of the DOF with the largest shape in the target mode (i.e., "max DOF") in (c) time domain and (d) frequency domain

In both responses, the NLMM matches the measured result well and is a significant improvement in accuracy over the linear model; the linear model over-predicts the peak response by factors of 1.5 and 1.8, respectively, for the measured drive point and max DOF accelerations. Fig. 12d shows that the max DOF has a large second harmonic compared to that shown in Fig. 12b for the drive point. The NLMM does not correctly capture the harmonic at the former DOF, indicating that one of the fundamental assumptions is not fully satisfied by the dynamics of the test article and/or that the chosen model form is not able to adequately capture the response. Based on the results discussed in Section 4.2.2, part of the reason for this higher-frequency discrepancy between measured and NLMM response is that the mode of interest (mode 10) is interacting with a higher-frequency mode.

NFA Results

An NFA test was conducted initially as an additional measure to characterize the mode of interest. However, during the test, a potential modal interaction between modes 10 and 18 was observed. Seventy excitation levels (i.e., values of V from Fig. 3) were selected, and phase resonance was achieved at each. In the plots that follow, the abscissa is labeled with "Excitation Index," corresponding to the shaker voltage indices when phase resonance was achieved.

Fig. 13 shows a subset of the NFA results. The magnitudes of the first 5 harmonics of max DOF are shown in Fig. 13a. The dashed lines correspond to the even harmonics which are shown to be significant with respect to the fundamental frequency.

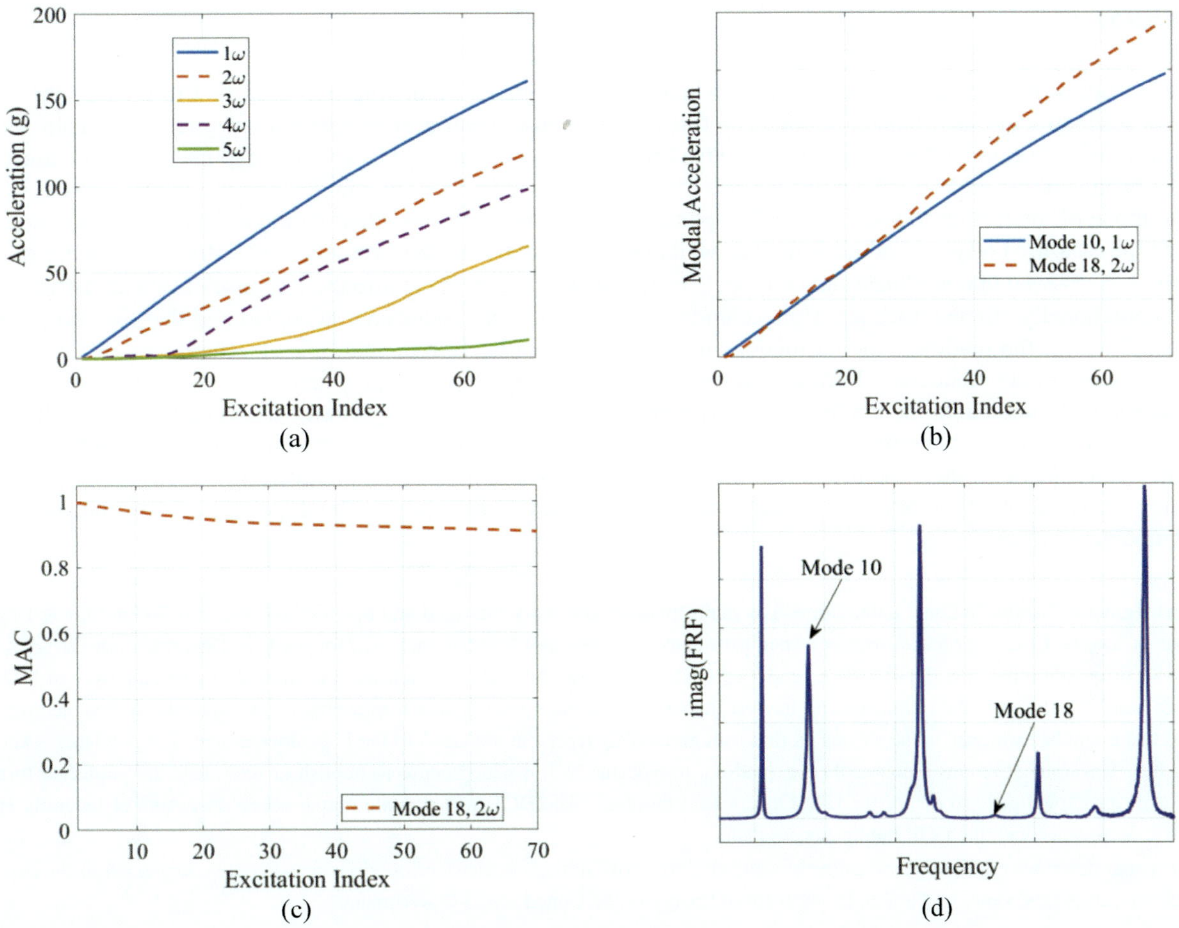

Fig. 13 NFA results showing a modal interaction between modes 10 and 18; (a) max DOF frequency content; (b) significant frequency content for all modes; (c) MAC between mode 18 and the deflection shape of the second harmonic; and (d) imaginary part of the drive point FRF showing that mode 18 is not well excited by the shaker

The linear natural frequency of mode 18 is approximately twice that of mode 10, so the large second harmonic indicates a potential for a modal interaction. The measured accelerations from the NFA test were cast to modal space. The magnitudes of all harmonics of all modes which surpass 50% of that of the fundamental component of the mode of interest (mode 10) are plotted in Fig. 13b. Only the second harmonic of mode 18 exceeds this threshold and even exceeds the amplitude of the fundamental component of the mode 10. Thus mode 18 is well excited and is comprised on mainly of the second harmonic. Furthermore, the MAC between the linear shape of mode 18 and the operational deflection shape of the second harmonic is shown in are plotted in Fig. 13c. The minimum MAC value is 90, indicating a high correlation between these two shapes and that the second harmonic is deforming in the shape of mode 18. No other linear mode shape had a MAC value larger than 30. Lastly, shakers can produce a second harmonic in force through their own nonlinear electrodynamics [12], so there is potential that, given the high amplitude forcing, mode 18 is being driven by the nonlinearity of the shaker. Fig. 13d shows the imaginary part of the drive point FRF with modes 10 and 18 indicated. This FRF shows that mode 18 is poorly excited by the drive location, and therefore it is unlikely to be driven to such high amplitudes via any shaker nonlinearity.

These plots indicate that during the NFA test of mode 10 a large second harmonic that deformed in the shape of mode 18 was active in the measured response. Additionally, the amplitude of mode 18 is equal to or greater than that of the mode targeted by the NFA test. It was also shown that shaker nonlinearity was not the likely source of mode 18 excitation. Therefore, these results indicate that a modal interaction occurred between modes 10 and 18, driven by the nonlinear physics activated by the joint of interest. These results corroborate the observation from Fig. 12d that the NLMM is unable to replicate the second harmonic response of max DOF in the sine beat because modal interactions violate one of the underlying assumptions of the NLMM. This motivates future work to include additional nonlinear forcing terms in the NLMM formulation to include modal interactions as this work has shown that this nonlinear phenomenon can occur in industrial structures.

Summary

Two case studies from industry were presented where nonlinear modal modelling was used first to identify the nonlinear modes of each structure and then to predict the full-field response to an arbitrary excitation force. The main sources of nonlinearity in both case studies were bolted joints. The objective for Case Study 1 was to capture all the nonlinear modes within a bandwidth of interest. The aim for Case Study 2 was only to identify and develop a nonlinear modal model for a single mode of interest up to high levels of response. In both cases, the nonlinear modal model was shown to generally replicate the measured response of the structures well and proved to be an improvement over a strictly linear approach.

This work showed that nonlinear modal models were able to accurately capture the response of two industrial structures in a computationally efficient manner. This technique assumes that the modes are uncoupled and that they do not interact. For Case Study 2, the nonlinear modal model simulation less accurately replicated the structural response at the second harmonic of the mode of interest. Further investigation via nonlinear force appropriation testing of the mode of interest revealed a modal interaction: the nonlinearity activated in the mode of interest generated a second harmonic which excited a higher frequency mode. Therefore, the decreased accuracy of the nonlinear modal model simulation results at this higher frequency is attributed to the structure violating an underlying assumption of the nonlinear modal modeling process. This motivates the future work of developing identification techniques to incorporate this type of behavior into the nonlinear modal modeling process.

Acknowledgments Sandia National Laboratories is a multi-mission laboratory managed and operated by National Technology & Engineering Solutions of Sandia, LLC (NTESS), a wholly owned subsidiary of Honeywell International Inc., for the U.S. Department of Energy's National Nuclear Security Administration (DOE/NNSA) under contract DE-NA0003525. This written work is authored by an employee of NTESS. The employee, not NTESS, owns the right, title and interest in and to the written work and is responsible for its contents. Any subjective views or opinions that might be expressed in the written work do not necessarily represent the views of the U.S. Government. The publisher acknowledges that the U.S. Government retains a non-exclusive, paid-up, irrevocable, world-wide license to publish or reproduce the published form of this written work or allow others to do so, for U.S. Government purposes. The DOE will provide public access to results of federally sponsored research in accordance with the DOE Public Access Plan.

This paper describes objective technical results and analysis. Any subjective views or opinions that might be expressed in the paper do not necessarily represent the views of the U.S. Department of Energy or the United States Government.

References

1. D. J. Ewins, Modal Testing: Theory, Practice, and Application, 2 ed., Bognor Regis, West Sussex, England: John Wiley & Sons, 2009. W409W9360
2. M. Peeters, G. Kerschen and J. Golinval, "Dynamic testing of nonlinear vibrating structures using nonlinear normal modes," Journal of Sound and Vibration, vol. 330, pp. 486-509, 2011. W409W9360
3. G. Haller and S. Ponsioen, "Nonlinear normal modes and spectral submanifolds: existence, uniqueness and use in model reduction," Nonlinear Dynamics, vol. 86, pp. 1493-1534, 2016. W409W9360
4. R. L. Mayes, B. R. Pacini and D. R. Roettgen, "A modal model to simulate typical structural dynamic nonlinearity," in Dynamics of Coupled Structures, Volume 4: Proceedings of the 34th IMAC, A Conference and Exposition on Structural Dynamics 2016, Orlando, FL, 2016. W409W9360
5. M. F. Platten, J. R. Wright, G. Dimitriadis and J. E. Cooper, "Identification of multi-degree of freedom non-linear systems using an extended modal space model," Mechanical Systems and Signal Processing, vol. 23, no. 1, pp. 8-29, 2009. W409W9360
6. B. R. Pacini, R. L. Mayes, B. C. Owens and R. A. Schultz, "Nonlinear finite element model updating, part I: experimental techniques and nonlinear modal model parameter extraction," in Dynamics of Coupled Structures, Volume 4: Proceedings of the 35th IMAC, A Conference and Exposition on Structural Dynamics 2017, Garden Grove, CA, 2017. W409W9360
7. B. C. Owens, R. A. Schultz, B. R. Pacini and R. L. Mayes, "Nonlinear Finite Element Model Updating, Part II: Implementation and Simulation," in Dynamics of Coupled Structures, Volume 4: Proceedings of the 35th IMAC, A Conference and Exposition on Structural Dynamics 2017, Garden Grove, CA, 2017. W409W9360
8. S. Peter and R. I. Leine, "Excitation power quantities in phase resonance testing of nonlinear systems with phase-locked-loop excitation," Mechanical Systems and Signal Processing, vol. 96, pp. 139-158, 2017. W409W9360
9. M. Volvert and G. Kerschen, "Phase resonance nonlinear modes of mechanical systems," Journal of Sound and Vibration, vol. 511, 2021. W409W9360
10. M. Volvert and G. Kerschen, "Resonant phase lags of a Duffing oscillator," International Journal of Non-Linear Mechanics, vol. 146, 2022. W409W9360
11. T. Breunung, L. Cilenti, J. You and B. Balachandran, "Robust Identification of Nonlinear Oscillators from Frequency Response Data," in Nonlinear Structures & Systems, Volume 1. SEM 2023. Conference Proceedings of the Society for Experimental Mechanics Series, Austin, TX, 2023. W409W9360

12. G. R. Tomlinson, "Force distortion in resonance testing of structures with electro-dynamic vibration exciters," Journal of Sound and Vibration, vol. 63, no. 3, pp. 337-350, 1979. W409W9360
13. G. Kerschen, K. Worden, A. F. Vakakis and J.-C. Golinval, "Past, present and future of nonlinear system identification in structural dynamics," Mechanical systems and signal processing, vol. 20, no. 3, pp. 505-592, 2006. W409W9360
14. J.-P. Noël and G. Kerschen, "Nonlinear system identification in structural dynamics: 10 more years of progress," Mechanical Systems and Signal Processing , vol. 83, pp. 2-35, 2017. W409W9360

Chapter 9
Decoupling of Nonlinear Substructures using Quasi-linear Frequency Response Functions

Furkan K. Çelik, Taylan Karaağaçlı, M. Bülent Özer, and H. Nevzat Özgüven

Abstract Although decoupling methods based on measured frequency response functions (FRFs) have been extensively studied for linear systems, the development of decoupling methods for nonlinear systems is still in its infancy. A recently developed nonlinear system identification technique, namely the Response-Controlled Stepped-Sine Test (RCT) method, showed that nonlinear systems, in general, exhibit quasi-linear behavior independent of the location, type, and severity of nonlinearity provided that the displacement amplitude of a specific coordinate (generally excitation point) is kept constant in a single input sine testing with closed-loop control. This finding opened the way for adapting linear modal analysis and substructuring techniques to nonlinear systems. This study extends the previously developed FRF Decoupling Method for Nonlinear Systems (FDM-NS), which is limited to a localized nonlinearity, to the case of multiple nonlinearities distributed all over the coupled structure. After obtaining the quasi-linear FRFs of the coupled nonlinear structure and the known nonlinear substructure using RCT, linear decoupling formulations are applied. These formulations are used to decouple the structure at single or multiple connection points to predict the quasi-linear FRFs of the unknown nonlinear substructure. Then, response-level dependent nonlinear modal model parameters are identified by applying linear modal analysis methods to these quasi-linear FRFs. Finally, the nonlinear modal model can be used to synthesize the constant force FRFs of the unknown nonlinear substructure. The extended approach is verified using case studies, and its accuracy and limitations are discussed.

Keywords Nonlinear Structural Dynamics · Nonlinear Decoupling · Nonlinear Substructuring · Substructuring · Response-Controlled Testing

Introduction

Decoupling techniques aim to determine the dynamic characteristics of unknown substructures or joints within larger assemblies without additional experimental or computational effort. Extensive research on substructuring [1–3] has demonstrated that decoupling can be applied across frequency and modal domains. Studies have elaborated on the theoretical foundations of substructuring and have demonstrated its application through various linear case studies, e.g. [3–6].

Recent studies also applied coupling/decoupling techniques to nonlinear systems. For example, [7–9] implemented linear structural modification techniques to nonlinear systems using response-controlled testing. In [10], FDM-NS was proposed to decouple nonlinear systems with a localized nonlinearity in the frequency domain. Another method conducted coupling of nonlinear numerical systems using optimization [11]. One of the early works showed the applicability of modal coupling on numerical systems with nonlinear connection elements [12] .

With the increasing interest and need for nonlinear mechanical system identification, this area has improved rapidly in the last decade. New generation modal identification methods such as Control-Based Continuation (CBC) [13] , Phase-Locked Loop (PLL) [14] , and Response-Controlled stepped-sine Testing (RCT) [15] have been proposed and validated in

Furkan K. Çelik · M. Bülent Özer · H. Nevzat Özgüven
Department of Mechanical Engineering Middle East Technical University, 06800, Ankara, Türkiye
e-mail: furkan.celik@metu.edu.tr; ozerb@metu.edu.tr; ozguven@metu.edu.tr

Furkan K. Çelik · Taylan Karaağaçlı
The Scientific & Technological Research Council of Turkey, Defence Industries Research and Development Institute, TÜBİTAK SAGE, P. K. 16, 06484 Mamak, Ankara, Türkiye
e-mail: furkan.celik@metu.edu.tr; taylan.karaagacli@tubitak.gov.tr

Ludovic Renson et al. (eds.), *Nonlinear Structures & Systems, Vol. 1*, Conference Proceedings of the Society for Experimental Mechanics Series, https://doi.org/10.13052/9788743801467-9

various theoretical and experimental cases. Among these techniques, the application of RCT on nonlinear structures results in quasi-linear FRFs, which opens the way for using linear substructuring methods with nonlinear systems.

The theory of the method discussed in this paper is quite straightforward [16] and built on quasi-linear FRFs, which can be obtained by employing RCT. The method makes it possible to apply the decoupling techniques for linear structures to nonlinear structures exhibiting nonlinearities in any or all substructures. In that respect, it extends FDM-NS [10] to nonlinear systems regardless of the location, type, or severity of the nonlinear elements. The application of the method on a real structure with weak and local nonlinearity [17] gave promising results. However, the applications and limitations of the method are not studied extensively. This paper aims to provide an investigation on the applicability, accuracy and limitations of the method. Firstly, the underlying theory of RCT and linear decoupling will be presented, and then examples of decoupling from a single and multiple connection degree of freedoms (DoFs) will be presented to reach conclusions about the accuracy and limitations of the method.

Theory

The method suggested in [7,10] was restricted to cases where the nonlinearity is localized at the excitation DoF, or the nonlinearity is between two DoFs where the relative displacement of these DoFs is kept constant with closed loop control. This restriction can be bypassed by the application of RCT, which is used to identify various other real system nonlinearities, such as a piezo-actuator shown in Fig. 1 [18] , a control fin actuation mechanism with high friction [19] and a bolted lap-jointed structure [20] .

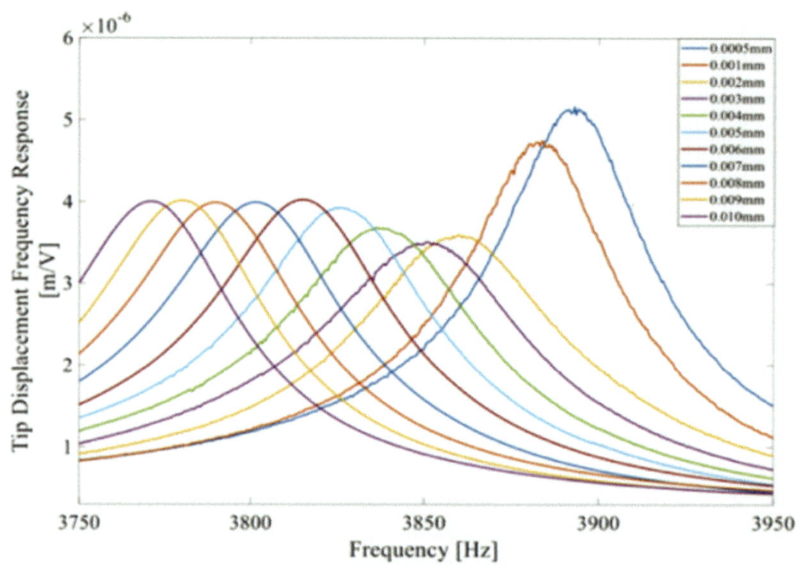

Fig. 1 Quasi-linear constant-response FRFs of Nonlinear Piezoelectric Actuators measured by RCT

The equation of motion of a nonlinear system with structural damping can be written in the frequency domain as follows,

$$-\omega^2 \boldsymbol{MX} + i\boldsymbol{HX} + \boldsymbol{KX} + \boldsymbol{\Delta}(\boldsymbol{X})\boldsymbol{X} = \boldsymbol{F} \tag{1}$$

Herein, $\boldsymbol{M}$, $\boldsymbol{K}$, and $\boldsymbol{H}$ are the underlying linear system's mass, stiffness, and structural damping matrices, respectively. $\boldsymbol{\Delta}$ denotes the response level dependent nonlinearity matrix [21] . $\boldsymbol{X}$ is the response amplitude vector, $\boldsymbol{F}$ is the harmonic excitation amplitude vector, and ω is the excitation frequency.

Eq. (1) assumes that the system is under harmonic excitation and that the effect of sub- and super-harmonics is negligible. By keeping any DoF of the system under a constant harmonic displacement amplitude, the displacement amplitudes of all DoFs remain nearly constant around the resonance region due to nonlinear normal mode theory [22] . This approach ensures that the nonlinearity matrix [21] behaves as an equivalent stiffness or damping matrix, making the system quasi-linear. Subsequently, classical modal identification techniques like peak-picking are applied to the resulting response-level-dependent quasi-linear FRFs, yielding a response-level-dependent modal model of the system. The constant-force frequency responses of the nonlinear system can then be synthesized by solving the following equation with the Newton-Raphson method and the arc-length continuation algorithm.

$$X_j(q_r) = \frac{\overline{A}_{jkr}(q_r)\,F_k}{\overline{\omega_r^2}(q_r) - \omega^2 + i2\overline{\xi}(q_r)\,\omega\overline{\omega_r}(q_r)} \tag{2}$$

Herein, X_j denotes the response amplitude of the jth DoF, F_k denotes the harmonic forcing amplitude at k^{th} DoF, $\overline{A}_{jkr}$, $\overline{\omega_r}$, $\overline{\xi}$ and q_r denote the modal constant, natural frequency, modal damping ratio, and modal amplitude of the rth mode, respectively. A more detailed explanation of the theoretical background and derivations can be found in [15] .

The quasi-linear FRFs can be used to apply structural modifications to a nonlinear system or couple it with another structure [7–9]. They can also be used to decouple a nonlinear structure from a single or multiple connection points [16] . Linear dual decoupling method can be used to decouple the structure. Dual decoupling assumes force equilibrium from the outset. Linear FRFs can be directly used in these equations, but nonlinear systems must also satisfy compatibility conditions because they contain response-level dependent elements. In a previous work [11] , it is shown that compatibility conditions can be satisfied iteratively. However, since RCT enforces a constant displacement level at the control DoF, the compatibility condition is not required to be satisfied iteratively. Any available linear dual decoupling formulation can be used alongside RCT to decouple the quasi-linear FRFs. In this study, the formulation given in Equation (3) is used. Its derivation can be found in [3] .

$$\boldsymbol{H}^B = \begin{bmatrix} \boldsymbol{H}^C & 0 \\ 0 & -\boldsymbol{H}^A \end{bmatrix} - \begin{bmatrix} \boldsymbol{H}^C & 0 \\ 0 & -\boldsymbol{H}^A \end{bmatrix} \begin{bmatrix} \left(\boldsymbol{B}^C\right)^T \\ \left(\boldsymbol{B}^A\right)^T \end{bmatrix} \left(\begin{bmatrix} \boldsymbol{B}^C & \boldsymbol{B}^A \end{bmatrix} \begin{bmatrix} \boldsymbol{H}^C & 0 \\ 0 & -\boldsymbol{H}^A \end{bmatrix} \begin{bmatrix} \left(\boldsymbol{B}^C\right)^T \\ \left(\boldsymbol{B}^A\right)^T \end{bmatrix} \right)^{-1} \begin{bmatrix} \boldsymbol{B}^C & \boldsymbol{B}^A \end{bmatrix} \begin{bmatrix} \boldsymbol{H}^C & 0 \\ 0 & -\boldsymbol{H}^A \end{bmatrix} \tag{3}$$

where $\mathbf{H^A}$, $\mathbf{H^B}$, and $\mathbf{H^C}$ indicate the FRFs of subsystems A, B, and the coupled system C, respectively. $\mathbf{B^A}$ and $\mathbf{B^C}$ correspond to the Boolean matrices of subsystems A and the coupled system C, respectively.

Systems Coupled From A Single Point

The method is initially demonstrated by using the discrete system shown in Fig. 2. The mass and stiffness values used in this case study are $0.1\ kg$ and $1x10^4\ N/m$, respectively. The structure has 1% structural damping, and the cubic stiffness value is $k_{cubic} = 1x10^7 N/m^3$.The mass at the connection point is divided into two equal parts.

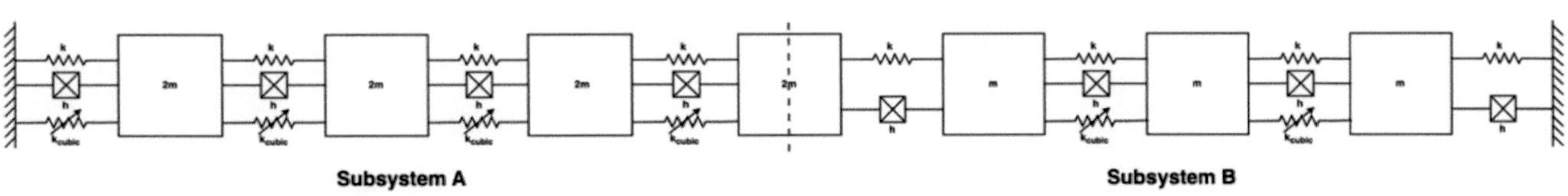

Fig. 2 Lumped coupled system with distributed cubic stiffness nonlinearities

Simulated RCT experiments are performed on both the coupled system C and subsystem A while maintaining the connection DoF (shown as a red dashed line in Fig. 2) at constant displacement amplitude levels and applying force to the same DoF, to decouple subsystem B from subsystem A. These simulations result in quasi-linear FRFs for the coupled system C and subsystem A. The quasi-linear FRFs at different displacement amplitude levels are then used with the dual decoupling formulation given in Eq. (3) to obtain the quasi-linear FRFs of subsystem B. The modal identification using these quasi-linear FRFs provides the modal parameters for each response level. A spline-fitted, response-level dependent modal model is then used in Eq. (2) to synthesize constant-force FRFs of the system, as shown in Fig. 3.

The synthesized constant force FRFs of Subsystem B are validated by comparing them with the ones directly obtained from constant-force simulations of Subsystem B, as shown in Fig. 3. The subscripts in the FRF correspond to the decoupling DoF in Fig. 2, which is named as the first DoF of Subsystem B. As can be seen from Fig. 3, the match between the synthesized and directly simulated FRFs is excellent. In this scenario, the linear modes of systems C and A are at 95.71 rad/s and 87.25 rad/s, respectively, and are relatively near the identified mode of subsystem B. In a real engineering system,

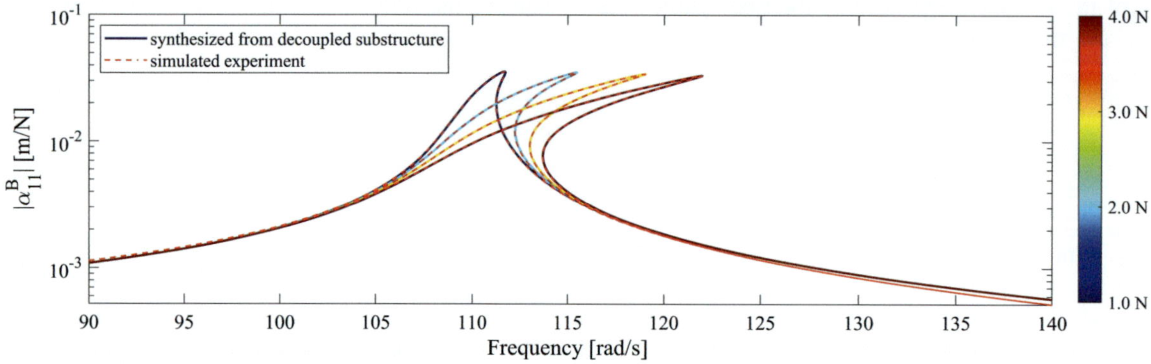

Fig. 3 Constant force direct point FRFs of Subsystem B

as the resonance frequencies of the subsystems become further away from each other, issues such as increased error as well as noisy appearance of the decoupled quasi-linear FRFs may occur due to noise becoming more dominant at off-resonance frequencies. Another practical challenge will be achieving the required force levels to reach higher values of constant displacement amplitudes at frequencies where the receptance values are significantly lower than at resonance.

Systems Connected From Multiple Points

Lumped parameter system

The method is also tested when two substructures are connected from multiple connection points. To see the effect of the strength of nonlinearity on the inaccuracies, the nonlinearities are attached in two configurations to the system, as shown in Fig. 4. In the first configuration, there is only the green nonlinearity (between mass 9-12). The second configuration has both the green (between mass 9-12) and blue (between mass 13-14) nonlinearities. The parameters of the system are given in Table 1. The structure has 1% structural damping, and the cubic stiffness is taken as $k_{cubic} = 1 \times 10^7 N/m^3$. DoFs 6, 7, and 8 are the connection DoFs that connect subsystem B to subsystem A.

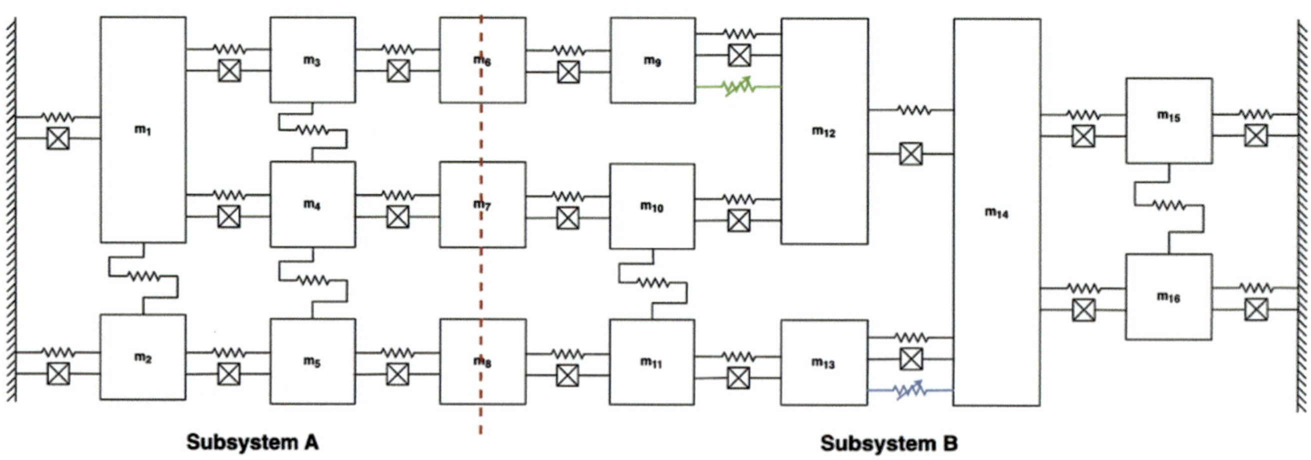

Fig. 4 Coupled nonlinear system with three connections (DoFs)

The displacement of the 6th DoF is kept constant while external excitation forces are applied in three separate sets of simulations at the 6th, 7th, and 8th DoFs. These simulations are designed to capture the quasi-linear FRFs of both subsystem A and the coupled system including H^A_{77}, H^C_{77}, H^A_{88} and H^C_{88}. These FRF matrices are then used in Eq. 3 to determine the FRFs of subsystem B.

Table 1 Properties of the lumped coupled system with multiple connections.

Index	1-5	6-8	9-16
Mass [kg]	1	1.5	1
Stiffness [N/m]	10000	10000	10000

In the case of multiple connection points, errors in the FRF estimates of Subsystem B arise because the system is connected through an nxn matrix where n is the number of connection DoFs (in this example, n = 3). The constant force FRF for the 2nd configuration is shown in Fig. 5. Zoom-box around the resonance peaks in Fig. 5 aims to show the amplitude errors more clearly. The comparison of two versions of the system for all force levels is given in Table 2.

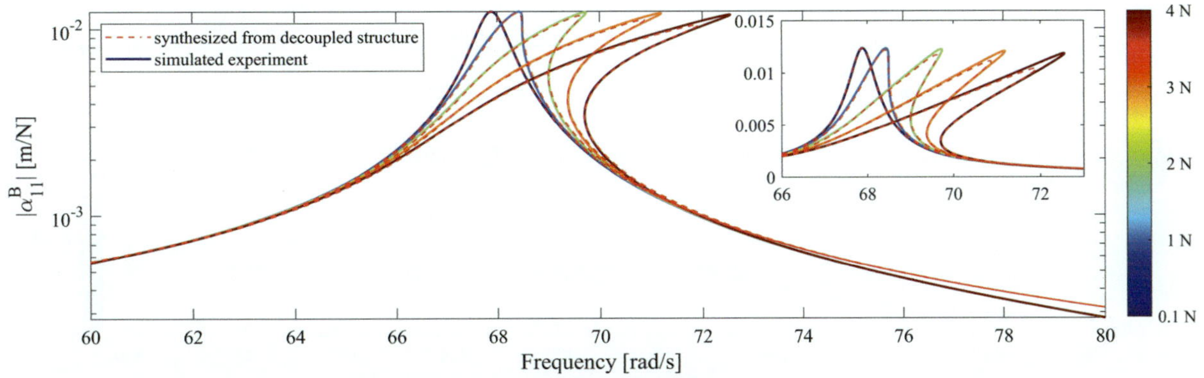

Fig. 5 Constant force FRF comparison for configuration 2 with two nonlinearities

Table 2 Natural frequency and peak level errors of the two configurations of synthesized nonlinear constant force FRFs.

-	Force [N]	0.1	1	2	3	4
Configuration 1	Error in Natural Frequency [%]	0.001	0.033	0.127	0.282	0.415
	Error in Peak [%]	0.019	1.170	3.441	5.296	6.624
Configuration 2	Error in Natural Frequency [%]	0.001	0.027	0.150	0.443	0.880
	Error in Peak [%]	0.020	1.342	4.318	7.642	11.728

The linear modes of system C and A are at 71.31 rad/s and 73.66 rad/s, respectively, and are near the identified mode of subsystem B. Table 2 shows that the errors increase with higher force levels. Also, when the number of nonlinearities in the system increase, the errors also increase.

Continuous system

The effectiveness of the method is also tested on the continuous beam shown in Fig. 6, coupled from a single connection point, but two DoFs (translational and rotational). Since rotational forcing cannot be applied experimentally, and measuring rotational displacement is also very difficult, D'Ambrogio and Fregolent [6] suggested modifying the dual decoupling procedure for continuous systems. One of these methods is the mixed interface decoupling. This method uses some of the DoFs from the connection interface and some of the DoFs from the internal nodes of the known substructure.

To decouple substructure B from the coupled system C, RCT is applied to the translational DoFs of the 5th and 9th nodes for displacement levels 0.01 mm to 0.4 mm. However, in both simulated experiments, the displacement of the translational DoF of the 9th node is kept constant while changing the force location. The decoupled quasi-linear FRFs are used to obtain the modal model of the nonlinear cantilever beam. The obtained modal model is used to synthesize the constant force FRFs in Fig. 7.

As the internal DoF is also used to decouple the nonlinear beam, the errors given in Table 3 are relatively low compared to the system in Fig. 4, despite the higher degree of nonlinearity. The low error shows that using a mixed interface decoupling with RCT can effectively decouple nonlinear systems, whether continuous or lumped.

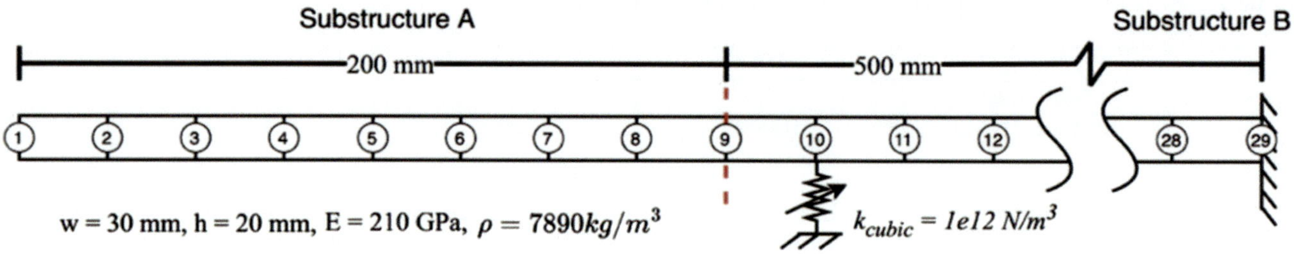

Fig. 6 Nonlinear beam subjected to decoupling

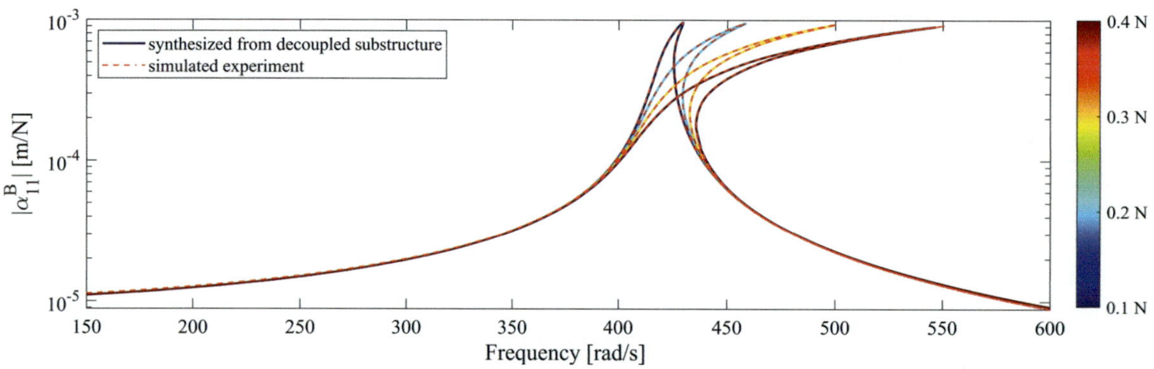

Fig. 7 Constant force FRF comparison of decoupled nonlinear beam

Table 3 Natural frequency and peak level errors of the synthesized nonlinear constant force FRF of the decoupled nonlinear beam.

Force [N]	0.1	0.2	0.3	0.4
Error in Natural Frequency [%]	0.003	0.020	0.279	0.789
Error in Peak [%]	0.132	0.529	1.130	1.906

Conclusion

This study examines the applicability, accuracy, and limitations of the method proposed in [16] to extend FDM-NS [10] , originally developed for systems with localized nonlinearities, to address multiple nonlinearities in one or both substructures of a coupled system. The method's accuracy in estimating a decoupled subsystem's constant force and constant amplitude frequency response functions is demonstrated through numerical simulation case studies. The results are generally highly successful, even for higher levels of nonlinearity. While the error in frequency response estimation increases with the degree of nonlinearity in multiple DoF decoupling, incorporating internal DoFs in the process significantly reduces these errors. A key limitation of the method arises from the relationship between the resonance frequency regions of the substructures and the coupled system. When the resonance frequency of the unknown substructure falls in off-resonance regions of the coupled system, noise may complicate the estimation of the unknown substructure. Additionally, achieving the required force levels in frequency regions where the receptance is lower than near-resonance values may present practical challenges. Nevertheless, the case studies, including those not reported here, show promising results, suggesting that the method holds significant potential for practical applications.

References

1. D'Ambrogio, W., and Fregolent, A., Direct decoupling of substructures using Primal and dual formulation. Conference Proceedings of the Society for Experimental Mechanics Series, 47–76., 2011.

2. de Klerk, D., Rixen, D. J., Voormeeren,S., General Framework for Dynamic Substructuring: History, Review and Classification of Techniques., AIAA Journal 46.5., 2008.
3. W. D'Ambrogio, A. Fregolent, The Role of Interface DOFs in Decoupling of Substructures Based on the Dual Domain Decomposition, Mechanical Systems and Signal Processing, 24 (7) 2035-2048., 2010.
4. van der Seijs, M. V., Pasma, E. A., van den Bosch, D. D., and Wernsen, M. W. M., A benchmark structure of validation of experimental substructuring, transfer path analysis and source characterisation techniques., 35th IMAC, 2017.
5. Mahmoudi, A. E., Rixen, D. J., and Meyer, C. H., Comparison of different approaches to include connection elements into frequency-based substructuring., Experimental Techniques, 44(4), 425–433., 2020.
6. D'Ambrogio, W., and Fregolent, A., Are Rotational DoFs Essential in Substructure Decoupling?., Proceedings of the 32nd IMAC, A Conference and Exposition on Structural Dynamics., 2014.
7. Arslan, Ö. and Özgüven, H. N., Modal Identification of Nonlinear Structures and the Use of Modal Model in Structural Dynamic Analysis, Proceedings of the 26th International Modal Analysis Conference, The Society of Experimental Mechanics, Orlando, Florida, 2008.
8. Ekinci, E. C., Özer, M. B., Özgüven, H. N., A Novel Approach for Local Structural Modification of Nonlinear Structures., Proceedings of the 41st International Modal Analysis Conference, The Society of Experimental Mechanics, Austin, Texas. Springer, 2023.
9. Ekinci, E. C., Karaağaçlı, T., Çelik, F. K., Özer, M. B., and Özgüven, H. N., Extension and Experimental Verification of an Efficient Re-Analysis Method for Modified Nonlinear Structures., Proceedings of the 42nd International Modal Analysis Conference, The Society of Experimental Mechanics, Orlando, Florida. Springer, 2024.
10. Kalaycıoğlu, T., Özgüven, H. N., FRF decoupling of nonlinear systems, Mechanical Systems and Signal Processing, 102, 230–244. ,2018.
11. Tepe, Ç., Ciğeroğlu, E., Structural Coupling of Two-Nonlinear Structures, Proceedings of the 33rd IMAC, A Conference and Exposition on Structural Dynamics, 2015.
12. Cömert, M.D., Özgüven, H.N., A Method for Forced Harmonic Response of Substructures Coupled by Nonlinear Elements, IMAC 13, 1995.
13. Renson, L., Gonzalez-Buelga, A., Barton, D. A. W., and Neild, S. A., Robust identification of backbone curves using control-based continuation., Journal of Sound and Vibration, 367, 145–158., 2016.
14. Peter, S., Scheel, M., Krack, M., and Leine, R. I., Synthesis of nonlinear frequency responses with experimentally extracted nonlinear modes., Mechanical Systems and Signal Processing, 101, 498–515., 2018.
15. Karaağaçlı, T., Özgüven, H. N., Experimental modal analysis of nonlinear systems by using response-controlled stepped-sine testing, Mechanical Systems and Signal Processing, 146, 107023, 2021.
16. Özgüven, H. N., Advanced Topics in Structural Dynamics: Seminar 3: Nonlinear Structural Modification, Coupling and Decoupling, Class and Seminar Notes, Technical University of Munich, Fall 2022.
17. Di Manno, M., Linear and nonlinear mechanical joint identification through substructure decoupling, Doctoral dissertation, Sapienza University of Rome, 2023.
18. Koyuncu, A., Karaağaçlı, T., Şahin, M., Özgüven, H.N., Experimental modal analysis of nonlinear amplified piezoelectric actuators by using response-controlled stepped-sine testing, Exp. Mech., 62, pp. 1579–1594, 2022.
19. Karaağaçlı, T., Özgüven, H.N., Experimental modal analysis of structures with high nonlinear damping by using response-controlled stepped-sine testing, Proceedings of the 41st International Modal Analysis Conference (IMAC), Austin, TX, USA, 2023.
20. Gürbüz, M.F., Karaağaçlı, T., Özer, M.B., Özgüven, H.N., Bypassing the repeatability issue in nonlinear experimental modal analysis of jointed structures by using the RCT-HFS framework, Proceedings of the 41st International Modal Analysis Conference (IMAC), Austin, TX, USA, 2023.
21. Tanrıkulu, Ö., Kuran, B., Özgüven, H.N., Imregün, M., Forced harmonic response analysis of nonlinear structures using describing functions, AIAA J., 31 (7), pp. 1313–1320, 1993.
22. Szemplinska-Stupnicka, W., The modified single mode method in the investigations of the resonant vibrations of nonlinear systems, J. Sound Vib., 63(4), pp. 475-489, 1979.

Chapter 10
An Overview of ANSYS Harmonic Balance Method: Current Capabilities and Future Directions

Michael Kwarta

Abstract Nonlinear structural dynamics have gained increasing interest in the industry due to several key factors, including: (i) the necessity to redesign structures for greater weight efficiency, thereby reducing fuel consumption, (ii) the need for accurate prediction of the dynamics of structures oscillating at high amplitudes, and/or (iii) the importance of modeling the effects of bolted connections on a system's nonlinear response. Although nonlinear structures can be simulated using brute-force transient solvers, the computational time required is often inefficient for practical applications. The Harmonic Balance Method (HBM) has emerged as one of the most popular and effective methods for simulating the forced periodic responses of nonlinear systems. When coupled with a continuation algorithm, HBM can efficiently compute key characteristics of nonlinear systems, such as nonlinear frequency response functions and nonlinear normal modes. This work presents the current capabilities of the Ansys-HBM toolkit, which was recently incorporated into the Ansys portfolio. It discusses the physical features contributing to the nonlinear response of mechanical systems, which are supported by Ansys-HBM at the time of this publication. Additionally, features particular to HBM, such as providing initial guesses and scaling the solution vector, are described. The toolkit is evaluated using benchmark examples. Finally, the publication concludes with a discussion on future enhancements driven by industry requests that will be integrated into upcoming software releases.

Keywords Harmonic Balance Method · Nonlinear Mechanical Vibrations · Nonlinear Response Function · Benchmark Testing · Industry Applications

Introduction

Nonlinear structural dynamics have gained increasing interest in the industry due to several driving factors, including:

(i) the necessity to redesign structures for greater weight efficiency, thereby reducing fuel consumption,
(ii) the need for accurate prediction of the dynamics of structures oscillating at high amplitudes, and/or
(iii) the importance of modeling the effects of bolted connections on a system's nonlinear response.

The nonlinear characteristics of a structure can arise from local or distributed nonlinearities. Examples of the former include a bolted joint connection [1] or a small fracture in the system [2], while the latter may be exhibited by curved beams or plates due to high-amplitude oscillations [3], [4], [5], [6]. Even though the structures with nonlinear characteristics can be simulated using brute-force transient solvers, the computation time required for these simulations is often inefficient.

The Harmonic Balance Method (HBM) [7] is one of the most popular and effective techniques for simulating the forced periodic responses of nonlinear systems. When coupled with a continuation algorithm, HBM can calculate key characteristics that describe a nonlinear system, such as the nonlinear frequency response and the nonlinear normal mode [8].

Michael Kwarta
Ansys, Inc., Southpointe 2600 Ansys Drive, Canonsburg, PA, 15317, USA
e-mail: michael.kwarta@ansys.com

Ludovic Renson et al. (eds.), *Nonlinear Structures & Systems, Vol. 1*, Conference Proceedings of the Society for Experimental Mechanics Series, https://doi.org/10.13052/9788743801467-10

The objective of this work is to present the current capabilities of the Ansys-HBM toolkit, which was recently included in the Ansys portfolio. This paper describes the physical phenomena that cause the nonlinear response of mechanical systems supported by Ansys-HBM at the time of this publication. Furthermore, the numerical capabilities specific to HBM and supported by the software are also described.

Harmonic Balance Method Overview

The Harmonic Balance Method (HBM) is a computational technique used to analyze nonlinear dynamical systems in the frequency domain. It enables the determination of steady-state periodic responses by solving the equations of motion associated with these systems. Recently integrated into Ansys' portfolio, HBM facilitates high-accuracy and high-performance simulations.

In Ansys-HBM analysis, a model is typically divided into linear and nonlinear components. A fundamental harmonic balance equation is established and solved using the Newton-Raphson (NR) routine combined with continuation algorithms. This approach allows for the calculation of nonlinear response functions, making HBM particularly valuable for engineers and researchers working with complex mechanical systems.

At the time of this publication, HBM analysis supports specific nonlinearities, including elements modeling nonlinear stiffness, damping, and node-to-node contact elements. The linear components of the system may be reduced using component mode synthesis (CMS) to enhance efficiency. CMS techniques, such as the Craig-Bampton method [9], effectively split the structure into smaller components while maintaining their interface degrees of freedom. For a more comprehensive description of Ansys-HBM, including the derivation of the fundamental equation, convergence criteria, best practices, and step-by-step examples, please refer to the Ansys-HBM Analysis Guide [10].

Overview of the Convergence Improvement Techniques

Achieving convergence in nonlinear analyses can sometimes be challenging. Ansys-HBM offers several techniques to improve convergence when issues arise.

Initial Guess Adjustment

When HBM analysis encounters convergence problems while calculating the first point on the nonlinear response function (RF), one effective strategy is to optimize the initial guess. By default, the nonlinear multi-harmonic solver uses zero initial condition for this calculation, which can lead to convergence failures. To mitigate this, users can provide a more informed initial guess, derived from a static analysis or a previous HBM run. This approach can significantly improve the likelihood of successful convergence.

Scaling the Solution Vector

Another technique to enhance convergence is scaling the augmented multi-harmonic solution vector. In a Newton-Raphson routine, discrepancies in the orders of magnitude among the solution vector components can lead to numerical roundoff errors, which can extend convergence times or cause failures. This is particularly relevant in scenarios where the amplitude of the forcing frequency differs significantly from the response amplitude.

To address this numerical issue, users can linearly scale each component of the solution vector by a scalar value or by values specified in a scaling vector. In many cases, applying one scaling value for displacements and another for the forcing frequency is sufficient to improve convergence rates.

Assuming the augmented multi-harmonic vector of unknowns $(\overline{u}^{+})$ is expressed as shown in Eq. (1):

$$\overline{\mathbf{u}}^{+} = \begin{bmatrix} \overline{\mathbf{u}} \\ \omega \end{bmatrix}, \tag{1}$$

where $\overline{\mathbf{u}}$ is the multi-harmonic solution vector and ω is the forcing frequency, Ansys-HBM allows for the scaling of selected components of this vector. The scaled vector $\overline{\mathbf{u}}^{+}$ can be represented as shown in Eq. (2):

$$\overline{\mathbf{u}}^{+} = \begin{bmatrix} \overrightarrow{\mathbf{a}} \circ \overline{\mathbf{u}} \\ b \times \omega \end{bmatrix}, \tag{2}$$

where $\overrightarrow{\mathbf{a}}$ is a vector of the same length as $\overline{\mathbf{u}}$, and b is a scalar. This scaling approach helps stabilize the solution process and improve convergence, as illustrated in Figure 1.

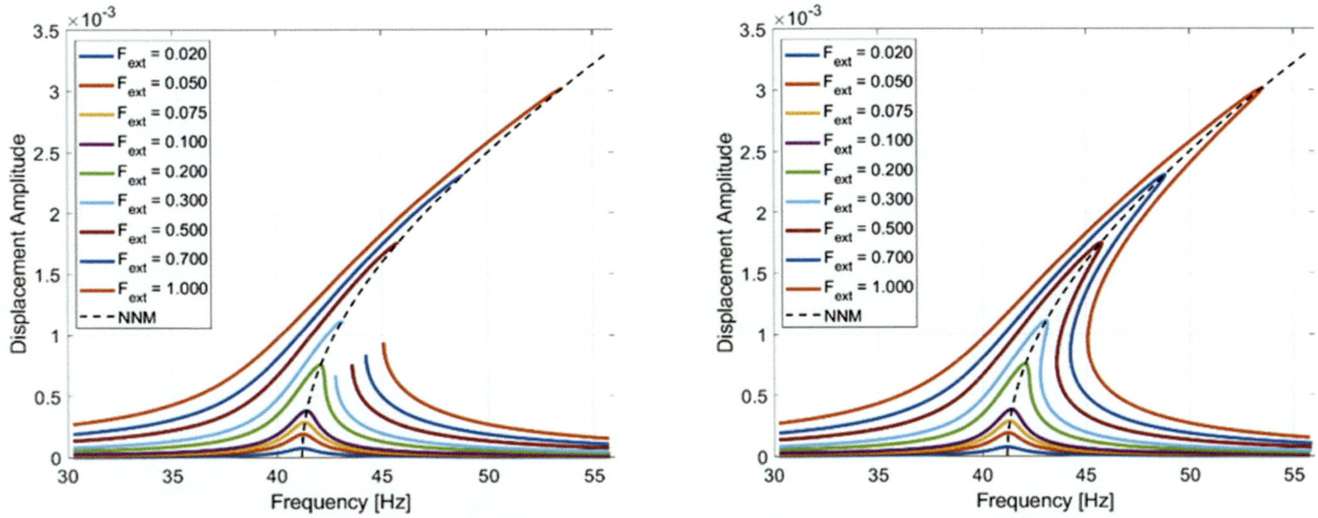

Fig. 1 Solution to a case study from [11] obtained using Ansys-HBM. In this particular study, if solution vector scaling is not applied, the algorithm struggles to achieve convergence near the boundaries between the stable and unstable regions of the nonlinear frequency response functions (chart on the left). The chart on the right shows the solution after applying solution vector scaling, which significantly improves convergence.

For more details on these techniques and their implementation, please refer to the Ansys-HBM Analysis Guide [10].

Application of the Ansys-HBM to a Benchmark Example

In this benchmark example, the Harmonic Balance Method (HBM) is applied to a Brake-Reuss cantilever beam. This structure consists of two beams featuring a frictional contact surface between them, connected by three prestressed bolted joints. A point input force is applied to one beam, while the motion of the corner of the other beam is monitored. Since this example involves a Brake-Reuss cantilever beam, one end of the assembly is fixed. The geometry of the beam in this study is similar to that described in [12]. The specific geometry, boundary conditions, and loads are illustrated in Figure 2.

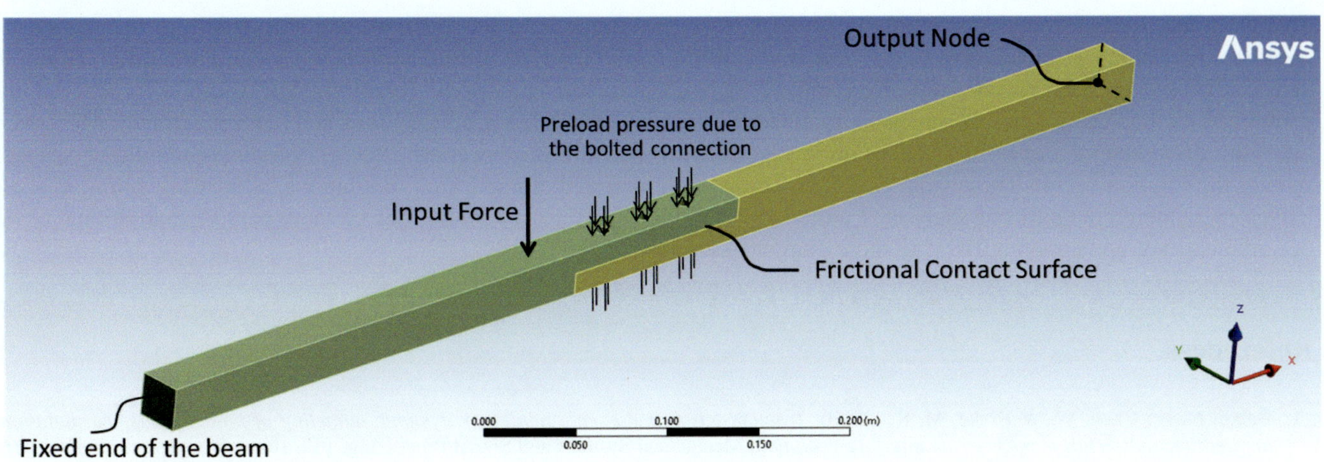

Fig. 2 Geometry, boundary conditions, and load setup for the Brake-Reuss cantilever beam model used in this benchmark example.

Figure 3 presents sample results obtained using the Ansys-HBM. The graph displays frequency responses (FRs) acquired near the second nonlinear bending mode of the cantilever beam, computed for various input force magnitudes. Notably, the

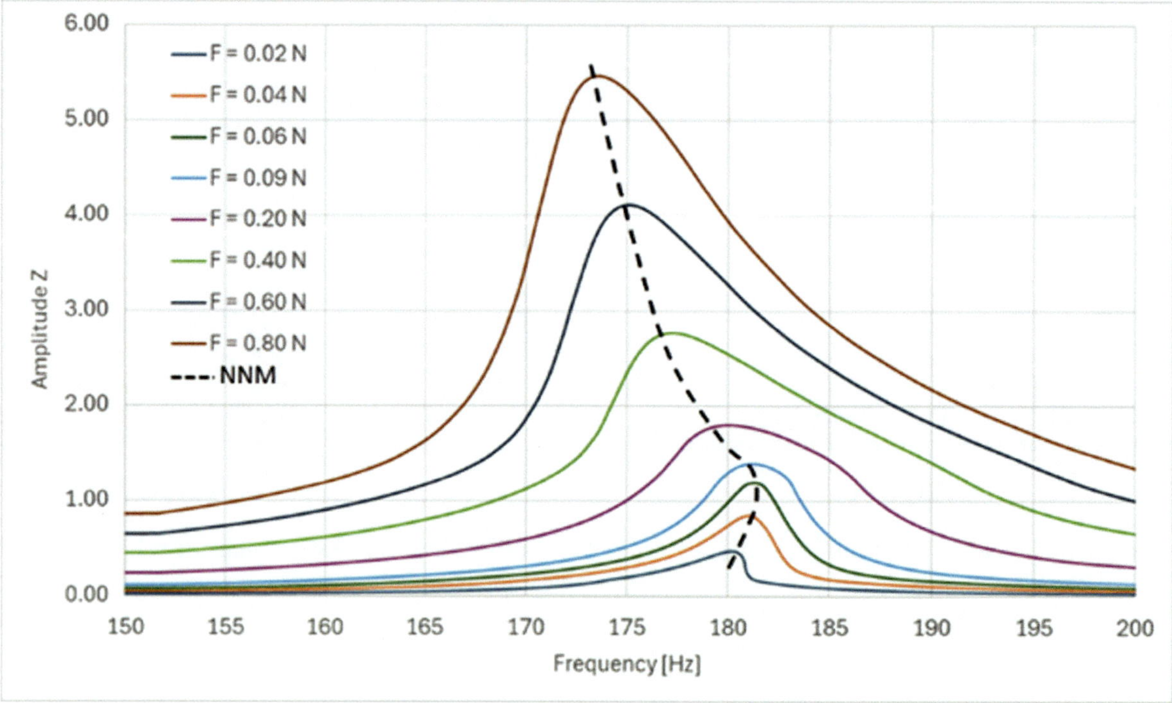

Fig. 3 Frequency response (FR) curves for the Brake-Reuss cantilever beam near the second nonlinear bending mode, obtained for various input force magnitudes. The nonlinear normal mode (NNM) curve illustrates the amplitude-dependent natural frequency and its stiffening-softening characteristic.

plot reveals that the nonlinear natural frequency, represented by the nonlinear normal mode (NNM) curve, is amplitude-dependent and exhibits a stiffening-softening characteristic.

Conclusions

This work has presented an overview of the ANSYS Harmonic Balance Method (HBM) and its capabilities for simulating nonlinear oscillating mechanical systems. The application of ANSYS-HBM to the Brake-Reuss cantilever beam example demonstrated its effectiveness in capturing amplitude-dependent nonlinear response functions. The results show that the implemented method can accurately represent nonlinearities in systems featuring bolted joints and contact surfaces. Moreover, this work discussed various techniques for improving convergence in HBM analyses, such as optimizing initial guesses and scaling the solution vector. These strategies enhance the reliability and efficiency of simulations, addressing common challenges encountered in nonlinear analyses. Future work will focus on further refining this method, exploring additional types of nonlinearities, and applying HBM to more complex systems.

References

1. Balaji, N. N., Chen, W., & Brake, M. R. (2020). *Traction-based multi-scale nonlinear dynamic modeling of bolted joints: Formulation, application, and trends in micro-scale interface evolution*. Mechanical Systems and Signal Processing, 139, 106615, https://doi.org/10.1016/j.ymssp.2020.106615.
2. Di Maio, D., Bruinsma, S., & Tinga, T. (2021). *Diagnostics based on continuous scanning LDV measurements and RASTAR analysis method*. Experimental Techniques, 45, 411-428, https://doi.org/10.1007/s40799-020-00406-4.
3. Kwarta, M., & Allen, M. S. (2022). *Nonlinear Normal Mode backbone estimation with near-resonant steady state inputs*. Mechanical Systems and Signal Processing, 162, 108046, https://doi.org/10.1016/j.ymssp.2021.108046.
4. Kwarta, M., & Allen, M. S. (2023). *Nonlinear Identification through eXtended Outputs (NIXO) with numerical and experimental validation using geometrically nonlinear structures*. Mechanical Systems and Signal Processing, *200*, 110542, https://doi.org/10.1016/j.ymssp.2023.110542.

5. Kwarta, M., & Allen, M. S. (2024). *NIXO-Based identification of the dominant terms in a nonlinear equation of motion of structures with geometric nonlinearity*. Journal of Sound and Vibration, *568*, 117900, https://doi.org/10.1016/j.jsv.2023.117900.
6. Breunung, T., Kwarta, M., Allen, M.S. (2024). *Evaluating New Nonlinear System Identification Methods on Curved Beams.* In: Brake, M.R., Renson, L., Kuether, R.J., Tiso, P. (eds) Nonlinear Structures & Systems, Vol. 1. IMAC 2024. Conference Proceedings of the Society for Experimental Mechanics Series. Springer, Cham. https://doi.org/10.1007/978-3-031-69409-7_20.
7. Krack, M., & Gross, J. (2019). *Harmonic balance for nonlinear vibration problems* (Vol. 1, pp. 26-28). Cham: Springer International Publishing, https://doi.org/10.1007/978-3-030-14023-6.
8. Rosenberg, R. M. (1960). *Normal Modes of Nonlinear Dual-Mode Systems.* ASME. *J. Appl. Mech.* June 1960; 27(2): 263–268. https://doi.org/10.1115/1.3643948.
9. Craig, R. R., & and Bampton, M. D. D. (1968). *Coupling of Substructures for Dynamic Analysis*, AIAA Journal, Vol.12, pp1313-1319, https://doi.org/10.2514/3.4741.
10. Ansys® Harmonic Balance Method Analysis Guide, Release 2024R2, July 2024, Help System, ANSYS, Inc.
11. Nagesh, M. (2021). *Nonlinear modal testing and system modeling techniques* (PhD dissertation, University of Cincinnati).
12. Lacayo, R., Pesaresi, L., Groß, J., Fochler, D., Armand, J., Salles, L., Schwingshackl, C., Allen, M., & Brake, M., *Nonlinear modeling of structures with bolted joints: A comparison of two approaches based on a time-domain and frequency-domain solver*, Mechanical Systems and Signal Processing, Volume 114, 2019, Pages 413-438, ISSN 0888-3270, https://doi.org/10.1016/j.ymssp.2018.05.033.

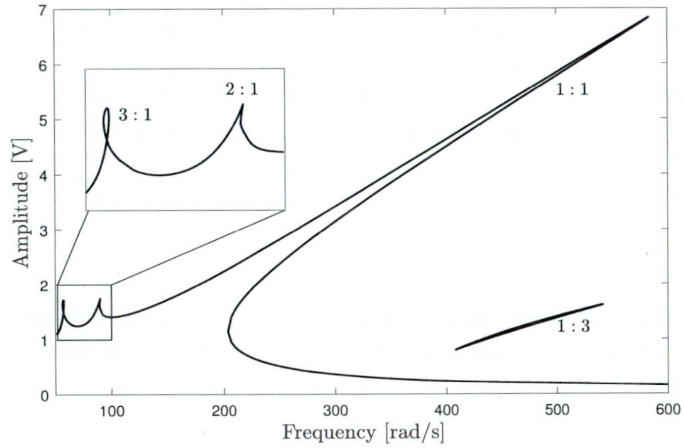

Fig. 5 Bifurcation diagram of the Duffing oscillator with a forcing amplitude of 2 [V]. The FRC was identified using ACBC and x-ACBC. Stable and unstable open-loop solutions are represented in the same way.

the main branch at a frequency near three times the natural frequency. The isolated subharmonic resonance was detected and fully identified experimentally, a remarkable result. A complete isolated subharmonic resonance had already been identified in [9] using an ad hoc method to detect the isola. The x-ACBC algorithm offers a more robust detection approach.

Figure 6 compares the FRC identified with the (x-)ACBC methods with swept sine tests for a 2 [V] forcing amplitude. The (x-)ACBC method matches swept-up and -down sine results in stable solutions and identifies unstable solutions or complex resonance shapes, which open-loop tests cannot. Nonlinear hysteresis is observed by reversing the swept sine test,

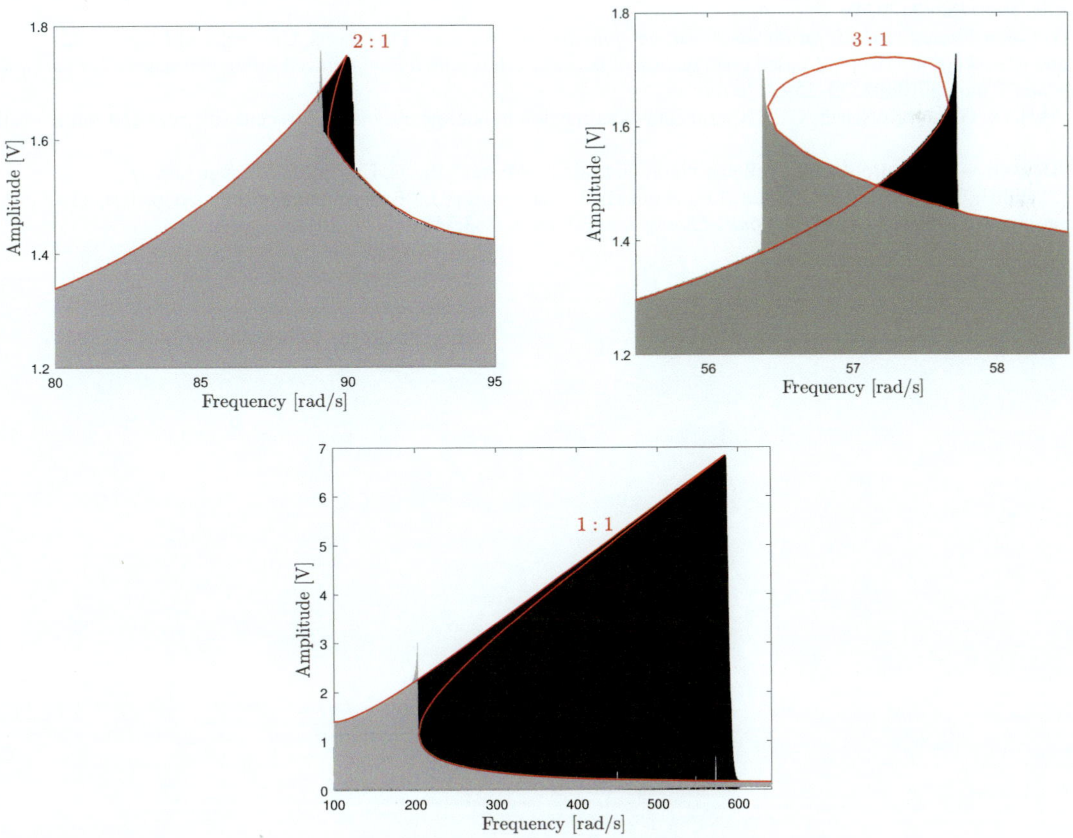

Fig. 6 FRC around the $2 : 1$, $3 : 1$, and $1 : 1$ resonances of the electronic Duffing oscillator at $p = 2$ [V]. The (x-)ACBC results (—) are compared with swept-up (—) and swept-down (—) sine tests.

revealing distinct branches in regions with multiple solutions. The jump phenomenon manifests when the system transitions from multiple possible solutions to a single solution.
Close to fold bifurcations, identification becomes more challenging because the Fourier coefficients of the merging branches are closer in value, suggesting that a reduced integral gain may be needed for the non-fundamental control.

Conclusion

This work introduced an extended version of the ACBC method, a novel online, derivative-free method for experimental continuation. By controlling the non-fundamental resonant harmonic, x-ACBC can detect and identify complex superharmonic and subharmonic resonance branches. The method was demonstrated experimentally using an electronic Duffing system. Beside the fundamental resonance, three secondary resonances, namely 3:1, 2:1 and 1:3 resonances, could be completely characterized.

Acknowledgments Alexandre Spits and Ghislain Raze are a research fellow and a postdoctoral researcher, respectively, of the Fonds de la Recherche Scientifique - FNRS, which is gratefully acknowledged.

References

1. Peter, S. and Leine, R.I. "Excitation power quantities in phase resonance testing of nonlinear systems with phase-locked loop excitation". *Mechanical Systems and Signal Processing*, 96:139–158 (2017)
2. Barton, D. and Sieber, J. "Systematic experimental exploration of bifurcations with noninvasive control". *Physical review E.*, 87:052916 (2012)
3. Volvert, M. and Kerschen, G. "Phase resonance nonlinear modes of mechanical systems". *Journal of Sound and Vibration*, 511:116355 (2021)
4. Zhou, T. and Kerschen, G. "Identification of secondary resonances of nonlinear systems using phase-locked loop testing". *Journal of Sound and Vibration*, 590:118549 (2024)
5. Abeloos, G. *Control-based methods for the identification of nonlinear structures*. PhD thesis, University of Liège (2022)
6. Sieber, J. and Krauskopf, B. "Control-based continuation of periodic orbits with a time-delayed difference scheme". *International Journal of Bifurcation and Chaos*, 17(08):2579–2593 (2007)
7. Raze, G., Abeloos, G., and Kerschen, G. "Experimental continuation in nonlinear dynamics: recent advances and future challenges". *arXiv* (2024)
8. Spits, A. "Derivative-less control-based nonlinear vibration testing". Master's thesis, University of Liège (2024)
9. Bureau, E., Schilder, F., Elmegård, M., Santos, I.F., Thomsen, J.J., and Starke, J. "Experimental bifurcation analysis of an impact oscillator – tuning a non-invasive control scheme". *Journal of Sound and Vibration*, 333:5464–5474 (2014)

Chapter 12
Modelling a Nonlinear Single-Degree-of-freedom System from Experimental Data using Lagrangian Neural Networks

Alan Xavier and Ludovic Renson

Abstract Accurately modelling the dynamic behaviour of nonlinear structures is challenging due to the wide range of potential nonlinearities and dynamic phenomena that they can exhibit. Physics-guided machine learning (PGML) has emerged as an attractive way to combine prior knowledge with data to solve a wide range of complex, nonlinear problems in science and engineering. Lagrangian Neural Networks (LNNs) are a particular PGML approach that models nonlinear systems' Lagrangian functions using artificial neural networks (NNs). The Euler-Lagrange equation is then reconstructed through automatic differentiation (AD) to derive the equations of motion, enforcing physical consistency during training.

In this work, we use LNNs to model the nonlinear vibrations of mechanical structures. We now explore the applicability of our methodology to a nonlinear SDOF system from experimental data. We analyse the physical consistency of the trained model and interpret the identified stiffness and damping nonlinearities from the partial derivatives of the potential energy and dissipation functions.

Keywords Physics-guided Machine Learning · Lagrangian Neural Networks · Control-Based Continuation

Introduction

Physics-guided machine learning (PGML) combines physics principles and domain knowledge with data-driven approaches. The goal of PGML is to create models that outperform both physics-based and data-driven ML models when used in isolation, and address common limitations. Cranmer et al. [1] proposed Lagrangian Neural Networks (LNNs), where the Lagrangian, $\mathscr{L} = \mathscr{T} - \mathscr{V}$, is parameterized by a neural network to model the action of a mechanical system, i.e. $\mathscr{L} \approx \mathscr{L}_\theta$, where θ are trainable weights. By applying Hamilton's principle, the Euler-Lagrange equation

$$\frac{d}{dt}\left(\frac{\partial \mathscr{L}}{\partial \dot{q}_j}\right) - \frac{\partial \mathscr{L}}{\partial q_j} = 0 \tag{1}$$

provide the system's equations of motion (EoM). A convenient vector form of this equation (1) [1] is.

$$\frac{d}{dt}\frac{\partial \mathscr{L}}{\partial \dot{q}_j} = \frac{\partial \mathscr{L}}{\partial q_j} \quad \rightarrow \quad \frac{d}{dt}\nabla_{\dot{q}_j}\mathscr{L} = \nabla_q \mathscr{L} \quad \rightarrow \quad \left(\nabla_{\dot{q}}\nabla_{\dot{q}}^T \mathscr{L}\right)\ddot{q} + \left(\nabla_q \nabla_{\dot{q}}^T \mathscr{L}\right)\dot{q} = \nabla_q \mathscr{L} \tag{2}$$

The resulting EoM can then be used to solve for $\ddot{q}$ (3):

$$\ddot{q} \;=\; \left(\nabla_{\dot{q}}\nabla_{\dot{q}}^T \mathscr{L}_\theta\right)^{-1}\left[\nabla_q \mathscr{L}_\theta - \left(\nabla_q \nabla_{\dot{q}}^T \mathscr{L}_\theta\right)\dot{q}\right] \tag{3}$$

where $\mathscr{L}_\theta$ is the Lagrangian parameterized by a neural network. The neural network parameters can be obtained through the minimisation of a loss function comparing measured data, $\ddot{q}_{true}$, and model predictions, $\ddot{q}_{\mathscr{L}NN}$.

To model vibrating structures, the idea of LNNs was extended to deal with systems with external excitation and non-conservative forces [2]. Incorporating an external excitation $F(t)$ and a dissipation function $\mathscr{D}$ in (1), using the method described in [3], yields the modified Euler-Lagrange equation:

$$\frac{d}{dt}\left(\frac{\partial \mathscr{L}}{\partial \dot{q}_j}\right) - \frac{\partial \mathscr{L}}{\partial q_j} + \frac{\partial \mathscr{D}}{\partial \dot{q}_j} - \mathscr{F}(t) = 0 \tag{4}$$

Alan Xavier · Ludovic Renson
Department of Mechanical Engineering, Imperial College London, South Kensington Campus, London SW7 2AZ, UK
e-mail: aax23@ic.ac.uk; a.xavier23@imperial.ac.uk; l.renson@imperial.ac.uk

Ludovic Renson et al. (eds.), *Nonlinear Structures & Systems, Vol. 1*, Conference Proceedings of the Society for Experimental Mechanics Series, https://doi.org/10.13052/9788743801467-12

The loss function used to minimize any discrepancy between the model and $\ddot{q}_{true}$ is now given by

$$\hat{\ddot{q}} = \left(\nabla_{\dot{q}}\nabla_{\dot{q}}^{T}\mathscr{L}_{\theta}\right)^{-1}\left[\nabla_{q}\mathscr{L}_{\theta} - \left(\nabla_{q}\nabla_{\dot{q}}^{T}\mathscr{L}_{\theta}\right)\dot{q} - \nabla_{\dot{q}}\mathscr{D}_{\theta} + \mathscr{F}(t)\right], \tag{5}$$

where the damping and conservative dynamics of the system are modelled using separate neural networks, $\mathscr{D}_{\theta}$ and $\mathscr{L}_{\theta}$, respectively. Additional constraints are then introduced for the system at rest, namely that $\mathscr{L}(0,0) = 0$ and $\mathscr{D}(0) = 0$. The loss function can then be written as:

$$loss = \lambda_0 \cdot ||\hat{\ddot{q}} - \ddot{q}|| + \sum_{i=1}^{n} \lambda_i \cdot \mathscr{G}_i, \tag{6}$$

where $\mathscr{G}_i$ is the i-th constraint in the loss function with the associated i-th weight, λ_i, balancing the loss terms. This allows us to enforce any prior knowledge about the system to improve our network's performance. The method was successfully demonstrated numerically on a Duffing oscillator in [2].

Here, we test this LNN methodology on the experimental data collected on an SDOF oscillator with an adjustable softening-hardening restoring force. The experimental setup is shown in figure 1. The training data was then obtained using Control-based Continuation (CBC). Details on the experimental setup and the CBC algorithms used can be found in [4]. The experimental data set is a collection of so-called S-curves measured at fixed excitation frequencies, ω, between 23.7 Hz and 25.4 Hz in steps of 0.1 Hz. In total, 780 periodic responses were measured, each consisting of a time series with 2000 samples. The displacements, x, y, velocities, $\dot{x}, \dot{y}$ and accelerations, $\ddot{x}, \ddot{y}$, for the mass and base respectively, were directly available from the experimental data. We divide the 780 continuation samples into 700 training and 80 test samples, as shown in figure 2a.

Compared to our previous application of LNNs, the experimental system considered here has a base excitation. Assuming that x is the displacement of the mass and y is the displacement of the base, we can write the EoM as

$$m\ddot{x} + c[\dot{x} - \dot{y}] + k[x - y] + f_{\text{NL}}(x - y, \dot{x} - \dot{y}) = 0, \tag{7}$$

where m, k, c are the mass, stiffness and damping coefficients respectively. We further simplify (7) by letting $q = x - y$ and dividing throughout by m as shown in equation (8):

$$m[\ddot{x} - \ddot{y}] + c[\dot{x} - \dot{y}] + k[x - y] + f_{\text{NL}}(x - y, \dot{x} - \dot{y}) = -m\ddot{y} \quad \Rightarrow \quad \ddot{q} + \frac{c}{m}\dot{q} + \frac{k}{m}q + \frac{1}{m}f_{\text{NL}}(q, \dot{q}) = -\ddot{y} \tag{8}$$

The prediction of the accelerations from the Lagrangian, $\mathscr{L}_{\theta}(q, \dot{q})$, and damping, $\mathscr{D}_{\theta}(\dot{q})$, functions is thus given by

$$\hat{\ddot{q}} = \left(\nabla_{\dot{q}}\nabla_{\dot{q}}^{T}\mathscr{L}_{\theta}\right)^{-1}\left[\nabla_{q}\mathscr{L}_{\theta} - \left(\nabla_{q}\nabla_{\dot{q}}^{T}\mathscr{L}_{\theta}\right)\dot{q} - \nabla_{\dot{q}}\mathscr{D}_{\theta} - \ddot{y}\right]. \tag{9}$$

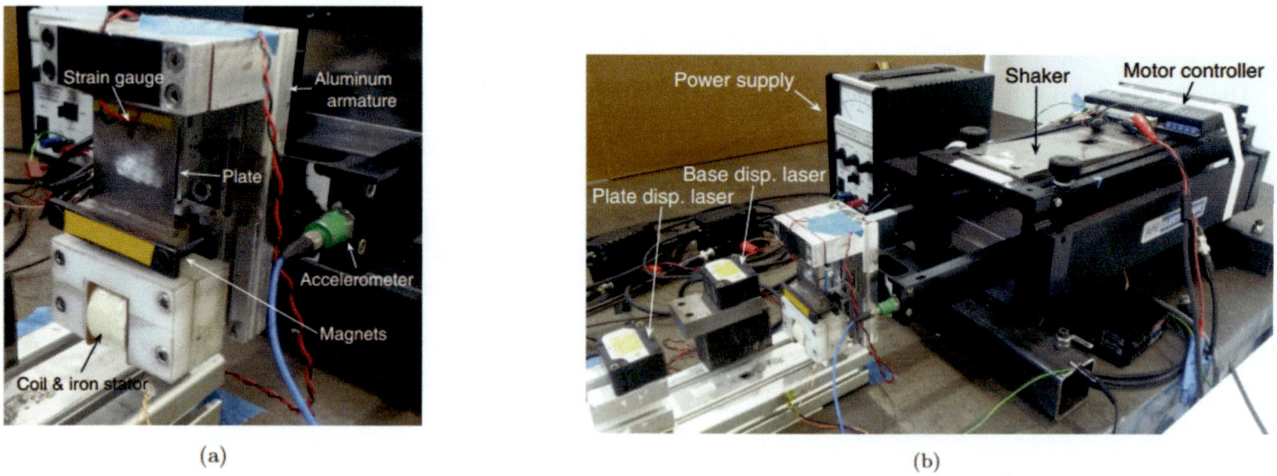

Fig. 1 (a) Picture of the nonlinear oscillator and (b) experimental setup for the SDOF nonlinear system with base excitation [4].

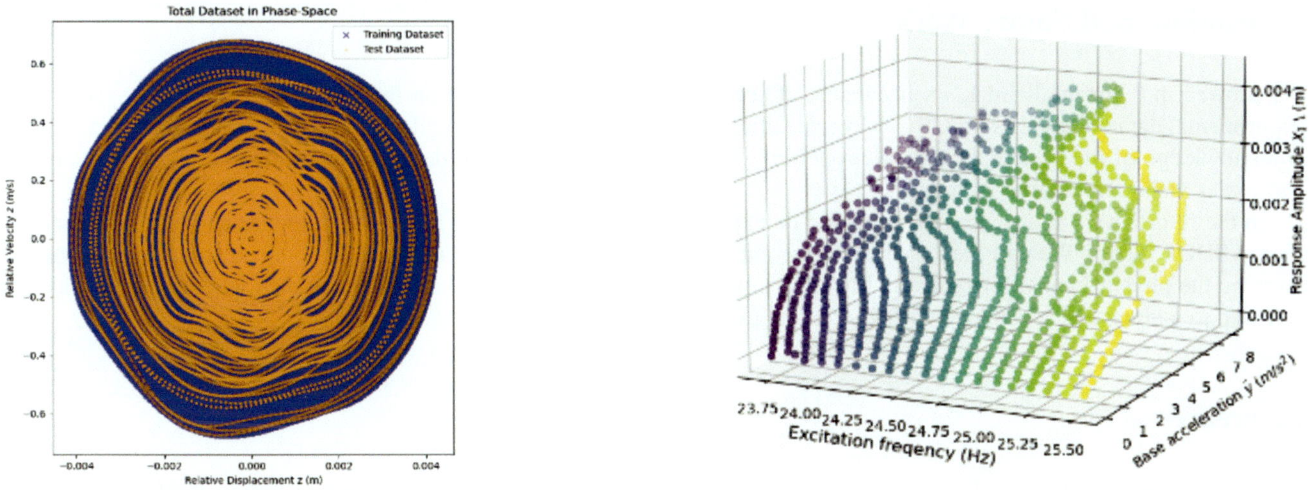

(a) (b)

Fig. 2 (a) Experimental data in phase-space. Blue (x)'s represent the training data, while the orange (•) represents our test data. (b) Forced response of the system, where the (•) are the amplitude of steady-state periodic responses measured during a series of continuations in amplitude (S-curves).

Our LNN is made up of two distinct NNs, one for the Lagrangian, $\mathscr{L}_\theta(q,\dot{q})$, and the other for the damping function, $\mathscr{D}_\theta(\dot{q})$. Each network has its own set of adjustable parameters, θ_i, and they were trained simultaneously to compute equations (5).

Both $\mathscr{L}_\theta$ and $\mathscr{D}_\theta$ have 16 hidden units across 2 hidden layers and use the SiLu activation function. The output layer does not use an activation function. Similarly, no physical bias was used during training. A normalizing layer was included in this network to enhance training stability, and was not learnt during training, but depends only on the data available. The model was trained for 100 epochs, with the Adam optimizer and a learning rate of 10^{-3}.

In figure 3, we present a comparison of the predicted relative acceleration, $\hat{\ddot{q}}$, with the true acceleration, $\ddot{q}$, across different excitation conditions ($\ddot{y}$) within the test dataset. This comparison reveals a qualitatively good agreement between the predicted and true values. However, the LNN fails to capture the higher harmonics in the signal.

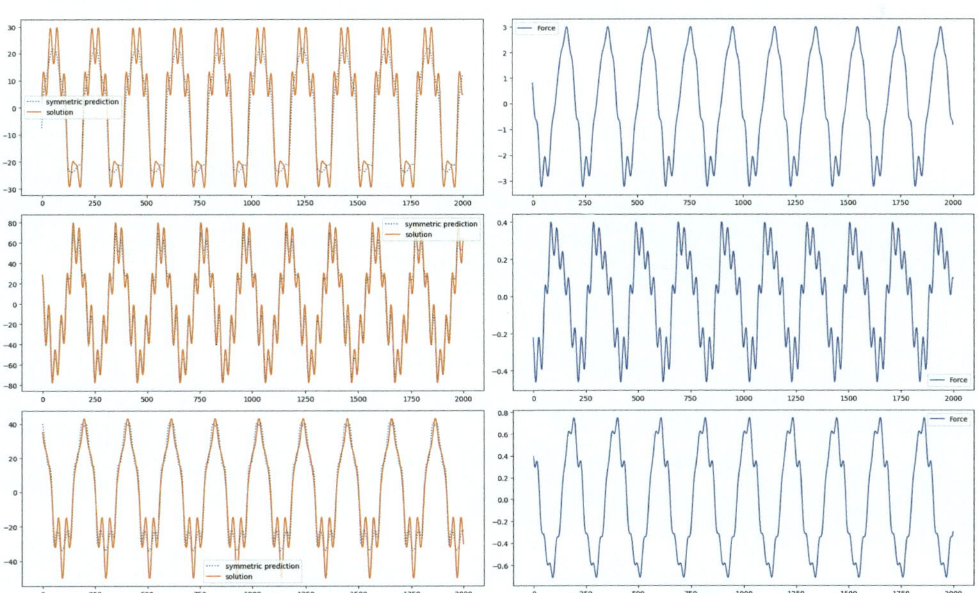

Fig. 3 The predicted relative acceleration, $\hat{\ddot{q}}$, vs the ground truth acceleration, $\ddot{q}$, by the LNN (left) and the applied excitation $\ddot{y}$.

Conclusions and Future Work

We showed that our modified LNN is capable of constructing equations of motion for a base-excited SDOF system from experimental data obtained using CBC. More specifically, the obtained model provides a good qualitative agreement without any knowledge of the system's physical properties, such as mass, stiffness and damping, but fails to reproduce the higher harmonics visible in the true dynamics.

In the future, we will use the LNNs trained on experimental data to trace the frequency response and bifurcation curves for the SDOF model. We will then compare these curves to the ones obtained using the data obtained from experiments. To improve the physical consistency of the LNN, separating NNs for predicting the kinetic and potential energies will also be assessed to address the dependence on different input features.

References

1. Cranmer, M.D., Greydanus, S., Hoyer, S., Battaglia, P.W., Spergel, D.N., and Ho, S. "Lagrangian neural networks". *CoRR*, abs/2003.04630 (2020)
2. M. J. Szydlowski, C.S. and Renson, L. "Modeling nonlinear structures using physics-guided, machine-learnt models". In *Proceedings of the 41st IMAC, A Conference and Exposition on Structural Dynamics 2023* (2023)
3. Meirovitch, L. *Methods of Analytical Dynamics*. Advanced engineering series. Dover Publications (2010)
4. Renson, L., Barton, D.A.W., and Neild, S.A. "Experimental tracking of limit-point bifurcations and backbone curves using control-based continuation". *International Journal of Bifurcation and Chaos*, 27(01):1730002 (2017)

Chapter 13
Computing Periodic Responses of Highly Flexible Beams Modelled under the SE(3) Lie Group Framework with Multiple Shooting

Amir K. Bagheri, Valentin Sonneville, and Ludovic Renson

Abstract Flexible mechanical structures with large displacements and rotations exhibit geometric nonlinearities and require accurate numerical modelling. Geometrically exact beam theory addresses these challenges by maintaining strain measures consistent with equilibrium equations in deformed configurations with finite rotations. However, standard finite element models using the $SO(3)$ Lie group face challenges with rotation parametrisation, non-commutative behaviour, and shear locking, leading to inaccuracies.

Beam formulations based on the Special Euclidean Lie Group $SE(3)$ resolve these issues by coupling rotations and positions with a local frame approach. This ensures strain invariance under rigid body motion and eliminates shear locking through a nonlinear finite element interpolation using the exponential map.

This work computes nonlinear normal modes of 3D beams modelled under the $SE(3)$ framework, applying multiple shooting and pseudo-arclength continuation methods. The multiple shooting method, overcomes limitations of the single shooting method, such as limited domain of attraction and high sensitivity to initial conditions. By discretising the periodic orbit into partitions and integrating over shorter intervals, it minimises error propagation and reduces branch jumps, particularly in systems with Unstable Periodic Orbits (UPOs). Additionally, parallel implementation of this method offers significant computational time savings, making it ideal for large finite element models.

Keywords $SE(3)$ Lie Group · Multiple Shooting · Geometric Nonlinearity · Nonlinear Normal Modes · Continuation

Introduction

The single shooting method is a common approach for solving boundary value problems, valued for its simplicity and ease of implementation alongside existing simulation codes. It also allows for the computation of stability via the monodromy matrix. However, it has notable drawbacks, including a limited domain of attraction for unstable orbits and a tendency for error growth over long integration periods. This can lead to difficulties in locating UPOs and may result in branch jumps when solution branches exist in close proximity, as highlighted by Lust [1] and Lakshmi [2].

To address these challenges, Keller [3] introduced the multiple shooting method, which discretises periodic orbits into smaller partitions, enabling separate time simulations for each partition. This strategy increases the domain of attraction, minimises error propagation, and reduces the likelihood of branch jumps. The method also enhances accuracy in detecting UPOs and can handle rapidly diverging systems more effectively. Parallel implementation further offers significant computational time savings, making it ideal for large multi-degree-of-freedom finite element models.

The multiple shooting method has garnered attention in various fields. Early work by Cebeci and Keller [4] demonstrated its potential to reduce sensitivity to initial conditions, improving solution stability. Bock and Plitt [5] applied it to optimal control problems, while Al-Mdallal et al. [6] showed its efficiency in solving fractional boundary value problems. Gonzales

Amir K. Bagheri · Ludovic Renson
Department of Mechanical Engineering, Imperial College London, London, United Kingdom, SW7 2AZ
e-mail: a.bagheri20@imperial.ac.uk; l.renson@imperial.ac.uk

Valentin Sonneville
Chair of Applied Mechanics, Technical University of Munich, Garching, Germany, D-85748
e-mail: valentin.sonneville@tum.de

Ludovic Renson et al. (eds.), *Nonlinear Structures & Systems, Vol. 1*, Conference Proceedings of the Society for Experimental Mechanics Series, https://doi.org/10.13052/9788743801467-13

and James [7] employed it for computing invariant manifolds in discrete-time systems, demonstrating computational savings, and Artola et al. [8] showcased its utility in nonlinear optimal control of flexible structures. Finally, Schäfer et al. [9] used a variation of the method for nonlinear model predictive control, achieving notable reductions in computational time.

This paper builds on these advancements, applying the multiple shooting method to 3D beam formulations in the Special Euclidean Lie Group $SE(3)$ framework to enhance computational efficiency and accuracy in modelling flexible mechanical structures. The method is used to compute nonlinear normal modes, providing insights into the dynamic behaviour of highly flexible beams under geometric nonlinearities.

Multiple Shooting Formulation

We denote the result of the time integration of the system's equations of motion, over an interval of length T and from an initial condition vector $\boldsymbol{x}_0$ as

$$\phi(\boldsymbol{x}_0, T),\ \phi : \mathbb{R}^M \times \mathbb{R} \to \mathbb{R}^M, \tag{1}$$

We now consider s as a partition of the unit interval such that

$$0 = s_0 < s_1 < \cdots < s_{m-1} < s_m = 1, \tag{2}$$

with

$$\Delta s_i = s_i - s_{i-1}, \tag{3}$$

Note that if the unit interval is split into equal partitions, we have $\Delta s = \frac{1}{m}$ for all values of i. Multiple shooting then solves the following nonlinear system of m equations for finding a periodic orbit

$$\begin{cases} \psi_0 = \phi(\boldsymbol{x}_0, \Delta s_1 T) - \boldsymbol{x}_1 = 0, \\ \psi_1 = \phi(\boldsymbol{x}_1, \Delta s_2 T) - \boldsymbol{x}_2 = 0, \\ \vdots \\ \psi_{m-2} = \phi(\boldsymbol{x}_{m-2}, \Delta s_{m-1} T) - \boldsymbol{x}_{m-1} = 0, \\ \psi_{m-1} = \phi(\boldsymbol{x}_{m-1}, \Delta s_m T) - \boldsymbol{x}_0 = 0 \end{cases} \tag{4}$$

This is shown schematically in Figure 1, which depicts a periodic orbit split into m partitions over the period. For the orbit to be periodic, the time integration from the starting point of each partition and over the length of the partition should bring the system to the starting point of the next partition, within the tolerance specified. For the final partition, the time integration should bring the system back to the first partition in order to close the orbit. To find a periodic solution, the sensitivities of

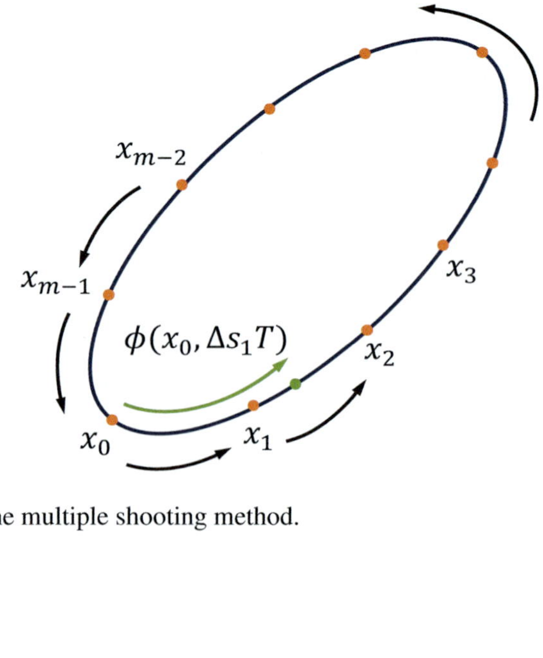

Fig. 1 Schematic illustration of the multiple shooting method.

the periodicity equations with respect to the initial conditions are computed and stored in the following Jacobian matrix:

$$\begin{bmatrix} \mathbf{M}_0 & -\mathbf{I} & & & & \mathbf{b}_0 \\ & \mathbf{M}_1 & -\mathbf{I} & & & \mathbf{b}_1 \\ & & \mathbf{M}_2 & -\mathbf{I} & & \mathbf{b}_2 \\ & & & \ddots & & \vdots \\ -\mathbf{I} & & & & \mathbf{M}_{m-1} & \mathbf{b}_{m-1} \\ \mathbf{B}_0 & \mathbf{B}_1 & \mathbf{B}_2 & \cdots & \mathbf{B}_{m-1} & \mathbf{0} \\ \mathbf{u}_0 & \mathbf{u}_1 & \mathbf{u}_2 & \cdots & \mathbf{u}_{m-1} & \lambda \end{bmatrix} \begin{bmatrix} \Delta \boldsymbol{x}_0 \\ \Delta \boldsymbol{x}_1 \\ \Delta \boldsymbol{x}_2 \\ \vdots \\ \Delta \boldsymbol{x}_{m-1} \\ \Delta T \end{bmatrix} = - \begin{bmatrix} \boldsymbol{\psi}_0 \\ \boldsymbol{\psi}_1 \\ \boldsymbol{\psi}_2 \\ \vdots \\ \boldsymbol{\psi}_{m-1} \\ h(\boldsymbol{X}(0)) \\ 0 \end{bmatrix} \tag{5}$$

where $\mathbf{M}_i$ is the i-th monodromy matrix resulting from the derivative of each $\boldsymbol{\psi}_i$ function with respect to $\boldsymbol{x}_i$, and is given by

$$\mathbf{M}_i = \frac{\partial \phi(\boldsymbol{x}_i, \Delta s_i T)}{\partial \boldsymbol{x}_i} \tag{6}$$

Following the pseudo-arclength continuation method, the phase condition and orthogonality condition are also added to Eq. (5). The linear system of equations is solved iteratively in the Newton scheme to obtain corrections to the initial conditions

In order to validate our multiple shooting formulation and methodology, we apply the algorithm to compute NNMs of the two degree of freedom system with cubic stiffness. The system is depicted in Figure 2, and its equation of motion is

$$\begin{aligned} \ddot{x}_1 + (2x_1 - x_2) + 0.5x_1^3 = 0 \\ \ddot{x}_2 + (2x_2 - x_1) = 0 \end{aligned} \tag{7}$$

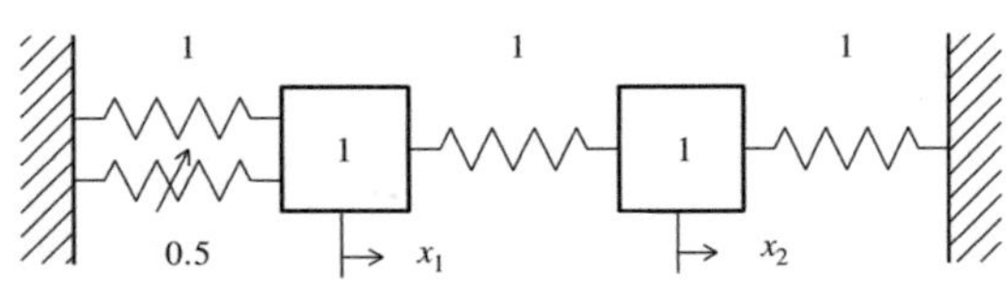

Fig. 2 Schematic illustration of the 2 DoF system [10].

The system above exhibits two NNMs, representing in-phase and out-of-phase motions. The symmetric periodic motion, referred to as $S11+$ in [10], denotes the in-phase motion, and reveals internal resonance branches at higher energy levels. The resonance tongues emanate from the $S11+$ branch and coalesce into the out-of-phase solution branch, demonstrating high-frequency ratio multiples (2:1, 3:1, 4:1, etc.) between the motions.

The FEP for the in-phase branch was computed using the multiple shooting method described, and is shown in Figure 3. Three partitions were chosen, and kept fixed through the continuation process. Zero initial velocity was enforced on both degrees of freedom as the phase condition, and therefore the symmetric NNMs were targeted and captured[1]. Excellent agreement is observed between the multiple shooting and single shooting results through a range of modal interactions, thus validating the formulation of the multiple shooting problem.

Adaptation of the Multiple Shooting Method to Lie Group Formulations

Before the multiple shooting method can be applied to beams modelled on $SE(3)$, it has to be adapted to fit the underlying Lie group framework of the structural model.

We begin by modifying the periodicity condition in Eq. (4). This is due to the nonlinearity of the configuration space of the Lie group manifold, which requires the use of the logarithm map on $SE(3)$ to compute the relative difference between beam configurations. In contrast to the single shooting approach, we need to apply this periodicity condition across all partitions of the multiple shooting method. This results in a Lie group-specific multiple shooting periodicity condition,

[1]These are referred to as "open curve" solutions in the configuration space [10].

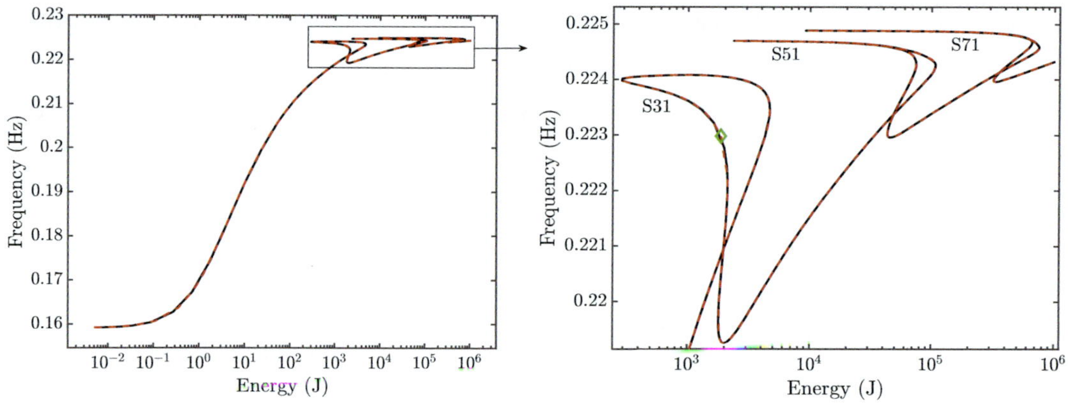

Fig. 3 First NNM of cubic spring. Comparison between single shooting (solid black) and multiple shooting (dashed orange). Symmetric internal resonances with 3:1, 5:1, and 7:1 frequency ratios captured and labelled.

where the beam frames and velocities are compared at the boundaries of each partition to compute the periodicity error. The multiple shooting periodicity function is therefore written as

$$\begin{cases} \boldsymbol{\psi}_0 = \begin{bmatrix} \psi_{\mathbf{H}_0} \\ \psi_{\mathbf{v}_0} \end{bmatrix} = \begin{bmatrix} \log_{SE(3)}(\mathbf{H}_1^{-1}(0)\mathbf{H}_0(\Delta s_1 T))^\vee \\ -\mathbf{T}_{SE(3)}^{-1}(-\psi_{\mathbf{H}_0})\mathbf{v}_1(0) + \mathbf{T}_{SE(3)}^{-1}(\psi_{\mathbf{H}_0})\mathbf{v}_0(\Delta s_1 T) \end{bmatrix}, \\ \boldsymbol{\psi}_1 = \begin{bmatrix} \psi_{\mathbf{H}_1} \\ \psi_{\mathbf{v}_1} \end{bmatrix} = \begin{bmatrix} \log_{SE(3)}(\mathbf{H}_2^{-1}(0)\mathbf{H}_1(\Delta s_2 T))^\vee \\ -\mathbf{T}_{SE(3)}^{-1}(-\psi_{\mathbf{H}_1})\mathbf{v}_2(0) + \mathbf{T}_{SE(3)}^{-1}(\psi_{\mathbf{H}_1})\mathbf{v}_1(\Delta s_2 T) \end{bmatrix}, \\ \qquad \vdots \\ \boldsymbol{\psi}_{m-2} = \begin{bmatrix} \psi_{\mathbf{H}_{m-2}} \\ \psi_{\mathbf{v}_{m-2}} \end{bmatrix} = \begin{bmatrix} \log_{SE(3)}(\mathbf{H}_{m-1}^{-1}(0)\mathbf{H}_{m-2}(\Delta s_{m-1} T))^\vee \\ -\mathbf{T}_{SE(3)}^{-1}(-\psi_{\mathbf{H}_{m-2}})\mathbf{v}_{m-1}(0) + \mathbf{T}_{SE(3)}^{-1}(\psi_{\mathbf{H}_{m-2}})\mathbf{v}_{m-2}(\Delta s_{m-1} T) \end{bmatrix}, \\ \boldsymbol{\psi}_{m-1} = \begin{bmatrix} \psi_{\mathbf{H}_{m-1}} \\ \psi_{\mathbf{v}_{m-1}} \end{bmatrix} = \begin{bmatrix} \log_{SE(3)}(\mathbf{H}_0^{-1}(0)\mathbf{H}_{m-1}(\Delta s_m T))^\vee \\ -\mathbf{T}_{SE(3)}^{-1}(-\psi_{\mathbf{H}_{m-1}})\mathbf{v}_0(0) + \mathbf{T}_{SE(3)}^{-1}(\psi_{\mathbf{H}_{m-1}})\mathbf{v}_{m-1}(\Delta s_m T) \end{bmatrix} \end{cases} \tag{8}$$

where $\mathbf{H}_i(0) \in SE(3)$ and $\mathbf{v}_i(0) \in \mathfrak{se}(3)$ are, respectively, the initial frames and velocities for each partition. The final frames and velocities at the end of the time integration along each partition, are denoted by $\mathbf{H}_i(\Delta s_i T)$ and $\mathbf{v}_i(\Delta s_i T)$ respectively. The $2N \times 1$ vector of initial conditions for each partition will now consist of the N initial increments and N initial velocities, that is $\boldsymbol{x}_i = [\mathbf{h}_i(0) \ \ \mathbf{v}_i(0)]^T$, where the increments give the initial beam configuration through the application of the exponential map and N is the number of degrees of freedom of the model.

Application of Multiple Shooting to 2D Beams

A 2D straight clamped-clamped beam is considered and its undamped periodic solutions (Nonlinear Normal Modes) are computed with multiple shooting and pseudo-arclength continuation. The beam is discretised with 30 elements, and its geometric and material properties are given as: $E = 210\,GPa$, $\rho = 7850\,kg/m^3$, $L = 1\,m$, $A = 10^{-4}\,m^2$. The beam is modelled using the SE Lie group framework in two dimensions [11], namely the $SE(2)$ group, which is the group of rigid body transformations in two dimensions. The frequency energy plot of the first NNM of the beam is shown in Figure 4a which depicts the periodic solution branches computed with single shooting and multiple shooting with 3 partitions. It is seen that the multiple shooting branch captures modal interactions of higher order, evident by the branching of the interaction tongue from the main backbone curve. The time series of a solution on this tongue is depicted in Figure 4b, showing the periodic motion of the beam midpoint across all 3 segments of the multiple shooting partitions. The multiple shooting method is seen to accurately capture the periodic motion of the beam. However, the robustness of the current implementation needs to be improved to ensure convergence along the full range of the backbone curve at higher energy levels, and to avoid high order branching which is a common issue in the continuation of periodic solutions using shooting.

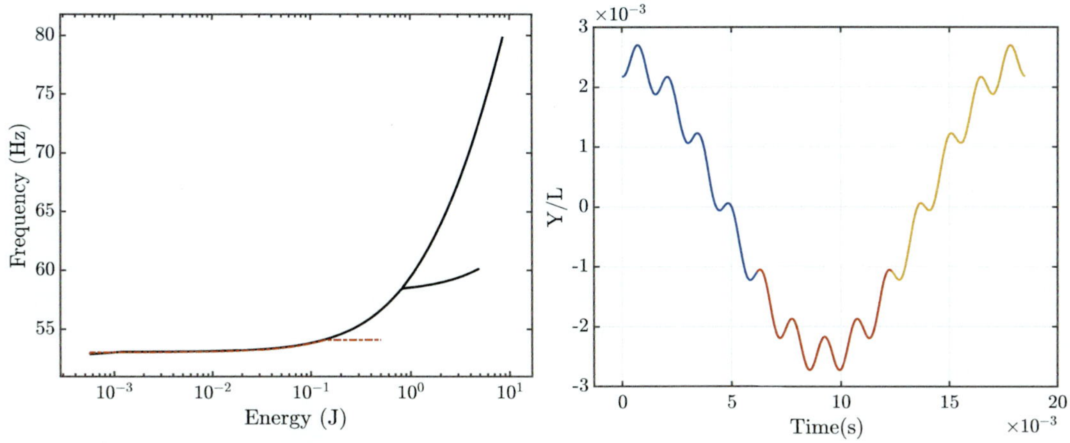

(a) Frequency Energy Plot.

(b) Time Series.

Fig. 4 First NNM of doubly clamped beam. Comparison between single shooting (solid black) and multiple shooting (dashed orange). Time series of a multiple shooting solution on the interaction tongue shown.

References

1. Lust, K. *Numerical bifurcation analysis of periodic solutions of partial differential equations.* PhD thesis, KU Leuven (1997)
2. Lakshmi, M.V., Fantuzzi, G., Chernyshenko, S.I., and Lasagna, D. "Finding unstable periodic orbits: A hybrid approach with polynomial optimization". *Physica D: Nonlinear Phenomena*, 427:133009 (2021)
3. Keller, H.B. *Numerical methods for two-point boundary-value problems.* Dover Publications (2018)
4. Cebeci, T. and Keller, H.B. "Shooting and parallel shooting methods for solving the Falkner-Skan boundary-layer equation". *Journal of Computational Physics*, 7(2):289–300 (1971)
5. Bock, H. and Plitt, K. "A Multiple Shooting Algorithm for Direct Solution of Optimal Control Problems *". *IFAC Proceedings Volumes*, 17(2):1603–1608 (1984)
6. Al-Mdallal, Q.M., Syam, M.I., and Anwar, M. "A collocation-shooting method for solving fractional boundary value problems". *Communications in Nonlinear Science and Numerical Simulation*, 15(12):3814–3822 (2010)
7. Gonzalez, J.L. and Mireles James, J.D. "High-Order Parameterization of Stable/Unstable Manifolds for Long Periodic Orbits of Maps". *SIAM Journal on Applied Dynamical Systems*, 16(3):1748–1795 (2017)
8. Artola, M., Wynn, A., and Palacios, R. "A Nonlinear Modal-Based Framework for Low Computational Cost Optimal Control of 3D Very Flexible Structures". In *2019 18th European Control Conference (ECC)*, pages 3836–3841, Naples, Italy (2019) IEEE.
9. Schäfer, A., Kühl, P., Diehl, M., Schlöder, J., and Bock, H.G. "Fast reduced multiple shooting methods for nonlinear model predictive control". *Chemical Engineering and Processing: Process Intensification*, 46(11):1200–1214 (2007)
10. Peeters, M., Viguié, R., Sérandour, G., Kerschen, G., and Golinval, J.C. "Nonlinear normal modes, part ii: Toward a practical computation using numerical continuation techniques". *Mechanical Systems and Signal Processing*, 23:195–216 (2009)
11. Bagheri, A.K., Sonneville, V., and Renson, L. "Nonlinear normal modes of highly flexible beam structures modelled under the SE(3) Lie group framework". *Nonlinear Dynamics*, 112(3):1641–1659 (2024)

Chapter 14
An Investigation of Nonlinear Solution Techniques for SSM-Reduced Mechanical Systems

Tiago S. Martins, F. Trainotti, L. Kreuzer, and D. J. Rixen

Abstract The parametrization method for invariant manifold reduction, has led to significant advancements in simplifying high-dimensional model reduction via direct computation of reduced dynamics in physical coordinates. Despite these advances, solving reduced systems requires efficient numerical methods. This paper investigates the effectiveness of three solvers—harmonic balance, orthogonal collocation, and shooting—on periodic orbit analysis within reduced-order models. Harmonic balance demonstrated a balance between computational speed and accuracy but was impacted by higher harmonic truncation. Orthogonal collocation, the most efficient for moderate time resolutions, required finer time discretization but scaled well with increasing complexity. Shooting, while initially slower, scaled effectively at higher resolutions but was sensitive to tolerance settings. The comparative analysis provides critical insights into the solvers' performance, offering practical guidance for nonlinear mechanical system analysis.

Keywords spectral submanifolds · reduced-order models · harmonic balance · orthogonal collocation · shooting method

Introduction

The parametrization method, first introduced by Cabré *et al.* [1–3], and later reformulated for engineering applications by Haro *et al.* [4], marked a major advancement in the reduction to invariant manifolds, unifying various approaches under a single framework. Previous methods [5] focused either on invariant manifold computations or relied on normal form theory [6]. The direct parametrization method, shows that both approaches can be derived from the invariance equation, solvable through graph or normal-form styles.

In [7], the parametrization method was adapted to vibratory systems addressing dissipation, while also providing existence and uniqueness theorems for invariant manifolds defined as spectral submanifolds (SSMs), leading to automated reduction methods [8] for two-dimensional manifolds with damping and arbitrary polynomial nonlinearities.

Subsequent advances [9] overcame the limitation of needing to express equations of motion in the modal basis, impractical for large finite element models, by enabling direct computations in physical coordinates. For mechanical systems, normal form-based methods and SSM approaches were developed [10–12] to express reduced dynamics in physical coordinates, incorporating damping and providing high-order automated approximations. These developments were made accessible through `SSMtool 2.0` [13], which automates the computation of high-order SSM approximations.

Futher on this topic, Opreni *et al.* [14] included additional non-autonomous terms for model order reduction in forced-damped mechanical structures with geometric nonlinearities, enabling the prediction of phenomena such as parametric excitation and isolated solutions. Li *et al.* [15] introduced a rigorous method for creating reduced-order models (ROMs) considering configuration constraints. Martin *et al.* [16] presented a direct parametrization method for reducing geometrically nonlinear rotating structures, with a focus on capturing the hardening/softening transition and interpolating ROMs. Axås *et al.* [17] established that the eigenvectors of delay-embedded systems are determined by their eigenvalues, allowing for preemptive identification of multimodal SSMs and their nonlinear interactions. Haller *et al.* [18] expanded the class of SSMs to include mixed stability types and fractional powers, enhancing low-dimensional polynomial models for nonlinear dynamics. Thurnher *et al.* [19] developed multi-index expressions for nonautonomous terms, facilitating resonance tongue extraction and steady-state response formulation in parametrically excited systems. Li *et al.* [20] presented an efficient

Tiago S. Martins · F. Trainotti · L. Kreuzer · D. J. Rixen
Technical University of Munich, Chair of Applied Mechanics, TUM School of Engineering and Design, Department of Mechanical Engineering. Boltzmannstr. 15 85748 Garching, Germany
e-mail: tiago.martins@tum.de; francesco.trainotti@tum.de; leo.kreuzer@tum.de; rixen@tum.de

Ludovic Renson et al. (eds.), *Nonlinear Structures & Systems, Vol. 1*, Conference Proceedings of the Society for Experimental Mechanics Series, https://doi.org/10.13052/9788743801467-14

method for computing forced response surfaces in nonlinear systems, utilizing SSMs for ROMs and multidimensional manifold continuation to extract key features. Haller and Kaundinya [21] derived asymptotic expansions for weakly forced systems, providing techniques for persistent, normally hyperbolic SSMs under varying forces. Cenedese *et al.* [22] presented a data-driven approach to construct SSM-reduced models from finite element simulations, achieving accurate predictions for forced responses. Vizzaccaro *et al.* [23] introduced a novel method to handle superharmonic resonances through expanded parametrizing coordinates. Bettini *et al.* [24] proposed a technique for sparse, nonlinear models on slow attracting spectral submanifolds applicable to various system types. And most recently, Grolet *et al.* [25] formulated a method to transform differential-algebraic equations with non-polynomial nonlinearities into polynomial systems, addressing geometric exactness in finite element structures.

Despite advancements in the computational frameworks presented in these papers, robust solvers are still required to effectively navigate the resultant reduced systems. This paper builds upon previous reduction methods by focusing on the identification of periodic orbits within these reduced systems. The discussion is supported by an example of a ROM of an arch beam generated by `MORFE2.0` [23] and compares three prominent numerical techniques—harmonic balance, orthogonal collocation, and shooting—on their effectiveness in solving the corresponding reduced-order dynamics.

The harmonic balance [26] method analyzes periodic solutions by transforming nonlinear problems into the frequency domain. Orthogonal collocation [27], is based on polynomial approximations and collocation points. Lastly, the shooting method [28] solves the periodic boundary value problems through iterative adjustments of initial conditions. MatCont [29] was used for orthogonal collocation and the other methods were purposely coded for this work.

This study evaluates the three solver methods in the context of reduced-order models to provide valuable insights into the dynamics of nonlinear mechanical systems and offer practical guidance for engineers and researchers. However, it does not exhaustively explore variations and alternatives in nonlinear solver techniques; it only employs the tangent predictor with pseudo-arclength continuation, the alternating frequency-time scheme for harmonic balance, orthogonal collocation at gaussian points, and single shooting.

The analysis begins with the full order dynamical system described by the following first-order autonomous ODE which can encompass a general class of physical systems, including second-order mechanical systems casted to first-order.

$$\dot{\mathbf{x}} = \mathbf{f}(\mathbf{x})\,. \tag{1}$$

Direct Parameterization of Invariant Manifolds

The following is a summary of the method described in [13, 30]. The goal is to generate a reduced dynamical system $\dot{\mathbf{q}} = \mathbf{g}(\mathbf{q})$, i.e. of lower dimension, whose orbits approximate solutions of the full system, close to a stable equilibrium, when mapped through another constructed function, called the manifold parameterization, namely $\mathbf{x} = \mathbf{W}(\mathbf{q})$.

The assumption is that the full dynamics $\mathbf{f}$, the reduced dynamics $\mathbf{g}$ and the manifold parameterization $\mathbf{W}$ are power series. Given $n, m \in \mathbb{N}$ with $m < n$, an open set $U \in \mathbb{R}^n$ containing the origin and an open set $V \in \mathbb{C}^m$ containing the origin, define the functions[1]

$$\begin{aligned} \mathbf{f} : U &\to \mathbb{R}^n & \mathbf{W} : V &\to U & \mathbf{g} : V &\to \mathbb{C}^m \\ \mathbf{x} &\mapsto \mathbf{f}(\mathbf{x}) = \sum_{\kappa\in\mathbb{N}} \mathbf{f}_\kappa \mathbf{x}^{\otimes\kappa} & \mathbf{q} &\mapsto \mathbf{W}(\mathbf{q}) = \sum_{\kappa\in\mathbb{N}} \mathbf{W}_\kappa \mathbf{q}^{\otimes\kappa} & \mathbf{q} &\mapsto \mathbf{g}(\mathbf{q}) = \sum_{\kappa\in\mathbb{N}} \mathbf{g}_\kappa \mathbf{q}^{\otimes\kappa}\,. \end{aligned}$$

Without loss of generality, it is clear that an equilibrium is placed at the origin, as $\mathbf{f}(\mathbf{0}) = \mathbf{0}$. Hence, via direct substitution, one has

$$\frac{\mathrm{d}\mathbf{W}}{\mathrm{d}\mathbf{q}}(\mathbf{q})\, \mathbf{g}(\mathbf{q}) \approx \mathbf{f}(\mathbf{W}(\mathbf{q})) \iff \left(\sum_{\kappa\in\mathbb{N}} \mathbf{W}_\kappa \frac{\mathrm{d}}{\mathrm{d}\mathbf{q}} \mathbf{q}^{\otimes\kappa}\right)\left(\sum_{\sigma\in\mathbb{N}} \mathbf{g}_\sigma \mathbf{q}^{\otimes\sigma}\right) \approx \sum_{\sigma\in\mathbb{N}} \mathbf{f}_\sigma \left(\sum_{\kappa\in\mathbb{N}} \mathbf{W}_\kappa \mathbf{q}^{\otimes\kappa}\right)^{\otimes\sigma}.$$

The method matches all the polynomial terms up to some chosen degree, such as to accumulate all the error in higher order terms, as with the idea that underlies Taylor expansions. Therefore, equating the coefficients of order γ yields

$$\sum_{\kappa=1}^{\gamma} \mathbf{W}_\kappa \, \mathbf{\Gamma}_{\gamma,\kappa} = \sum_{\kappa=1}^{\gamma} \mathbf{f}_\kappa \, \mathbf{\Xi}_{\gamma,\kappa}\,, \tag{2}$$

[1]The Kronecker power denotes the repeated Kronecker product such that: $\mathbf{x}^{\otimes 0} = [1]$, the uni-dimensional identity matrix; and $\mathbf{x}^{\otimes\kappa} = \underbrace{\mathbf{x} \otimes \mathbf{x} \otimes \cdots \otimes \mathbf{x}}_{\kappa \text{ times}}$ for $\kappa \in \mathbb{N}$.

where $\mathbf{\Gamma}_{\gamma,\kappa}$ and $\mathbf{\Xi}_{\gamma,\kappa}$ can be defined by the following recurrence relations

$$\mathbf{\Gamma}_{\gamma,1} = \mathbf{g}_\gamma \qquad \mathbf{\Gamma}_{\gamma+1,\kappa+1} = \mathbf{I}_m \otimes \mathbf{\Gamma}_{\gamma,\kappa} + \mathbf{g}_{\gamma-\kappa+1} \otimes \mathbf{I}_{m^\kappa},$$

$$\mathbf{\Xi}_{\gamma,1} = \mathbf{W}_\gamma \qquad \mathbf{\Xi}_{\gamma,\kappa+1} = \sum_{\sigma=\kappa}^{\gamma-1} \mathbf{W}_{\gamma-\sigma} \otimes \mathbf{\Xi}_{\sigma,\kappa}.$$

Here, $\mathbf{\Gamma}_{\gamma,\kappa}$ is a $m^\kappa \times m^\gamma$ complex matrix whose entries are sums of κ elements of $\mathbf{g}_{\gamma-\kappa+1}$. Additionally, $\mathbf{\Xi}_{\gamma,\kappa}$ is a $n^\kappa \times m^\gamma$ complex matrix whose entries are sums of products involving elements of $\mathbf{W}_1$, $\mathbf{W}_2$, ... and $\mathbf{W}_{\gamma-\kappa+1}$. In particular, it can be seen that $\mathbf{\Xi}_{\gamma,\gamma} = \mathbf{W}_1^{\otimes\gamma}$.

In the literature, $\mathbf{\Gamma}_{\gamma,\kappa}$ and $\mathbf{\Xi}_{\gamma,\kappa}$ are defined with closed form expressions. However, here they are stated via recurrence to suggest an efficient computation algorithm, as is the topic of this work.

Now, regarding the goal of constructing the reduced dynamics $\mathbf{g}$ and the manifold parameterization $\mathbf{W}$, one must solve equation (2) for $\mathbf{W}_\gamma$ and $\mathbf{g}_\gamma$ with γ ranging from 1 up to the desired expansion order, $\gamma_{\max}$. At its simplest case, for order $\gamma = 1$, equation (2) becomes

$$\mathbf{W}_1 \mathbf{g}_1 = \mathbf{f}_1 \mathbf{W}_1, \tag{3}$$

which can be made an eigenvalue problem by imposing $\mathbf{g}_1$ to be diagonal. In that case, $\mathbf{\Gamma}_{\gamma,\gamma}$ is a diagonal matrix comprised of all the possible sums of γ eigenvalues of $\mathbf{g}_1$.

Given $\gamma \geq 2$, by collecting the terms of equation (2) that depend on $\mathbf{W}_\gamma$ and $\mathbf{g}_\gamma$, one defines

$$\mathcal{L}_\gamma(\mathbf{W}_\gamma, \mathbf{g}_\gamma) = \mathbf{W}_\gamma \mathbf{\Gamma}_{\gamma,\gamma} - \mathbf{f}_1 \mathbf{W}_\gamma + \mathbf{W}_1 \mathbf{g}_\gamma,$$

where $\mathcal{L}_\gamma$ is a linear map between complex vector spaces of dimension $nm^\gamma + m^{\gamma+1}$ and nm^γ. Subsequently, accumulating all the remaining terms of (2), one constructs

$$\mathbf{B}_\gamma = \sum_{\kappa=2}^{\gamma} \mathbf{f}_\kappa \mathbf{\Xi}_{\gamma,\kappa} - \sum_{\kappa=2}^{\gamma-1} \mathbf{W}_\kappa \mathbf{\Gamma}_{\gamma,\kappa},$$

which only depends on the coefficients of $\mathbf{W}$ and $\mathbf{g}$ up to order $\gamma - 1$, and on the coefficients of $\mathbf{f}$ up to order γ. Finally, the equivalent homological equation of order $\gamma \geq 2$ becomes

$$\mathcal{L}_\gamma(\mathbf{W}_\gamma, \mathbf{g}_\gamma) = \mathbf{B}_\gamma. \tag{4}$$

Equations (3) and (4) can be used to define a procedure for building the manifold parameterization $\mathbf{W}$ and reduced dynamics $\mathbf{g}$. Namely, non-trivial solutions of (3) produce $\mathbf{W}_1$ and $\mathbf{g}_1$, which are used to construct the linear equation (4) at order 2, that is then solved for $\mathbf{W}_2$ and $\mathbf{g}_2$. All the previous solutions are processed to build the order 3 equation, which can be linearly solved for $\mathbf{W}_3$ and $\mathbf{g}_3$, that can again be recycled to compute order 4, and so forth.

Whenever there are nontrivial intersections between the eigenspace of $\mathbf{\Gamma}_{\gamma,\gamma}$ and the eigenspace of $\mathbf{f}_1$, there are resonance phenomena; nonlinear if $\gamma > 1$. If that intersection is in the eigenspace of $\mathbf{\Gamma}_{1,1} = \mathbf{g}_1$, the span of the m linear modes, there is inner resonance, since the movement of the nonlinear resonance is spanned by the linear modes $\mathbf{W}_1$. Conversely, there is outer resonance which in practice means that equation (4) might be unsolvable [13].

Because the dimension of the codomain of $\mathcal{L}_\gamma$ is smaller than that of its domain, there may be multiple solutions to equation (4), and it is common to parameterize the solution space [14]. The normal-form style of parameterization starts by determining the image of $\mathcal{L}_\gamma(\cdot, \mathbf{0})$, by detecting nonlinear resonances, then solves for $\mathbf{W}_\gamma$ with fixed $\mathbf{g}_\gamma = \mathbf{0}$ by projecting $\mathbf{B}_\gamma$ on said image, and solves for $\mathbf{g}_\gamma$ with fixed $\mathbf{W}_\gamma = \mathbf{0}$ on the complementary subspace. The graph style of parameterization is similar but assumes that all possible nonlinear resonances occur.

Additionally, the fact that the initial problem is real can be leveraged to improve computational efficiency. Specifically, this allows for the construction of $\mathbf{g}$ and $\mathbf{W}$ as mappings between real spaces of dimension m, halving the *real dimension* of the original definition. Henceforth, this works assumes that realification was performed.

Practical issues

There are many large matrices involved in this procedure, therefore it is advised to explore sparsity and symmetry, use smart memory allocation and properly scalable algorithms to combat computational limitations. It can be seen that the size of the matrices increases exponentially with the polynomial degree and polynomially with the number of degrees of freedom as $O(n^\gamma m^\gamma)$. The number of matrices increases factorially since there are $\binom{\gamma+m-1}{m-1}$ monomials of degree γ.

Furthermore, it is important to understand that this is a local method that relies on polynomial expansions. Therefore one must be careful when applying it to nonlinear problems and generating results far from the equilibrium.

External forcing terms are simulated as stable solutions of well chosen ODEs: $\dot{\mathbf{x}}_\mathrm{F} = \mathbf{f}_\mathrm{F}(\mathbf{x}_\mathrm{F})$. The internal dynamics then depend on the internal state and the state of the external forcing: $\dot{\mathbf{x}}_\mathrm{S} = \mathbf{f}_\mathrm{S}(\mathbf{x}_\mathrm{S}, \mathbf{x}_\mathrm{F})$. For example, regarding sinusoidal forcing, for any $\overline{\beta}, \omega, \phi \in \mathbb{R}$

$$\begin{cases} \beta_\mathrm{F}(t) = \overline{\beta} \\ x_\mathrm{F}(t) = \beta_\mathrm{F}(t)\cos(\omega t + \phi) \\ y_\mathrm{F}(t) = \beta_\mathrm{F}(t)\sin(\omega t + \phi) \end{cases} \quad \text{is a stable solution of} \quad \begin{cases} \dot{\beta}_\mathrm{F} = 0 \\ \dot{x}_1 = -\omega y_\mathrm{F} - x_\mathrm{F}\,(x_\mathrm{F}^2 + y_\mathrm{F}^2 - \beta_\mathrm{F}^2) \\ \dot{y}_\mathrm{F} = +\omega x_\mathrm{F} - y_\mathrm{F}\,(x_\mathrm{F}^2 + y_\mathrm{F}^2 - \beta_\mathrm{F}^2) \end{cases} . \tag{5}$$

In practice, this involves appending extra equations to describe the forcing, ensuring $\mathbf{f}_\mathrm{F}(\mathbf{0}) = \mathbf{0}$ to retain the equilibrium at the origin and guaranteeing that the forcing stays close to the equilibrium state: in the example above, by having a small absolute value for $\overline{\beta}$. On this matter, having forcing of multiple frequencies has been discussed [19], but it requires more complex theory and implementation. In the case of `MORFE2.0` [23], that feature was yet unavailable.

Moreover, the choice of style significantly impacts results and efficiency: the graph style is slower than the normal-form style, in general, but more useful in the context of center manifold computation, while the normal style provides simpler reduced dynamics and uniquely describes manifold folding [14]. Also, to avoid outer resonance, all the modes of the full order model, that (nonlinearly) resonate with the master modes up to order $\gamma_{\max}$, must be included from the start.

Nonlinear Solvers for Computing Periodic Solutions

Harmonic Balance via Alternating Frequency-Time

One first assumes that $\mathbf{q}(t)$ is a Fourier series, which already guarantees periodicity, restricted to the harmonics in a previously chosen finite set of non-negative integers $\mathcal{H}$:

$$\mathbf{q}(t) = \mathrm{Re} \sum_{h \in \mathcal{H}} \mathbf{Q}_h \exp(\mathrm{i} h \omega t) .$$

This allows expressing $\dot{\mathbf{q}}(t)$ and $\mathbf{g}(\mathbf{q}(t))$ as Fourier series with coefficients $\mathrm{i}h\omega\mathbf{Q}_h$ and $\mathbf{G}_h$ for harmonic $h \in \mathcal{H}$, respectively. Via the alternating frequency-time scheme, one has

$$\mathbf{G}_h = \mathrm{rDFT}(\{\mathbf{g}(\mathbf{q}_k)\}_{k=1}^N)_h \quad \text{with} \quad \mathbf{q}_k = \mathrm{irDFT}(\{\mathbf{Q}_h\}_{h\in\mathcal{H}})_k ,$$

where rDFT and irDFT are the real discrete Fourier transform and its inverse. The computation of $\mathbf{G}_h$ can be made exact for polynomial nonlinearities (the case of this work) by having enough time samples: $N > (\gamma_{\max} + 1)\, h_{\max}$, as in [31]. In practice, the computation of these functions is done with divide and conquer algorithms like the fast Fourier transform. This way, the ODE of the reduced dynamics, $\dot{\mathbf{q}} = \mathbf{g}(\mathbf{q})$, is transformed into the family of algebraic equations

$$\mathrm{i} r \omega \mathbf{Q}_r = \mathbf{G}_r(\{\mathbf{Q}_h\}_{h\in\mathcal{H}}) \quad \text{with} \quad r \in \mathcal{H} .$$

It is possible to write an equivalent set of twice as many equations[2] using only real numbers, by separating the real and imaginary parts of all the quantities involved, resulting in $2m|\mathcal{H}|$ real polynomial equations of degree $\leq \max(\gamma_{\max}, 2)$ with $2m|\mathcal{H}| + 1$ real unknowns, including ω.

The irDFT naturally computes $N = 2h_{\max}$ time domain samples, given the set of non-negative harmonics $\mathcal{H}$ and $h_{\max} = \max \mathcal{H}$, which defines the resolution of the harmonic balance method.

Orthogonal Collocation

The T-normalized time interval $[0,1]$ is divided into z test intervals with variable[3] duration, forming the coarse mesh: $0 = \tau_0 < \tau_1 < \cdots < \tau_z = 1$. Each interval is further subdivided into $c+1 \in \mathbb{N}$ subintervals, via c Gaussian collocation points[4],

[2] The equation of harmonic zero is already a real equation. When $0 \in \mathcal{H}$, one gets $2m|\mathcal{H}| - 1$ real polynomial equations with $2m|\mathcal{H}|$ real unknowns, including ω.

[3] In MatCont [29], the coarse mesh is adapted after a certain amount of solution points.

[4] In MatCont [29], the collocation points are the roots of the c-th Legendre polynomial, shifted and scaled.

forming the fine mesh $\{\tau_{\sigma,\kappa}\}$ such that

$$\tau_\sigma < \tau_{\sigma,1} < \cdots < \tau_{\sigma,c} < \tau_{\sigma+1} \quad \text{with} \quad \sigma \in \{0,\ldots,z-1\}.$$

The method assumes that the limit cycle is approximated by a continuous piecewise polynomial in each of the coarse mesh intervals:

$$\mathbf{q}(t=\tau/\omega) = \mathbf{p}(\tau) = \mathbf{p}_\sigma(\tau), \quad \tau \in [\tau_\sigma, \tau_{\sigma+1}] \quad \text{with} \quad \mathbf{p}_\sigma(\tau_{\sigma,c}) = \mathbf{p}_{\sigma+1}(\tau_{\sigma,c}),$$

where $\mathbf{p}_\sigma$ is the degree-c orthogonal polynomial relevant to the σ-th coarse mesh interval. The second condition ensures continuity at the coarse mesh, and imposing periodicity yields $\mathbf{p}_\sigma(\tau_0) = \mathbf{p}_{z-1}(\tau_z)$. Finally, the method aims to ensure that the ansatz complies with the dynamics at each of the $N = cz$ fine mesh points, expressly

$$\omega\mathbf{p}'(\tau_{\sigma,\kappa}) = 2\pi\,\mathbf{g}(\mathbf{p}(\tau_{\sigma,\kappa})) \quad \text{with} \quad \sigma \in \{0,\ldots,z-1\} \quad \text{and} \quad \kappa \in \{1,\ldots,c\}.$$

For a system of dimension m, at each of the z coarse mesh intervals, the local polynomial is defined by $(c+1)m$ coefficients, which yields a total of $(c+1)zm+1$ unknowns accounting for ω. Per coarse mesh interval, there are cm collocation constraints for the dynamics and m continuity (and periodicity) constraints, totaling $(c+1)zm$ polynomial equations of degree $\leq \max(\gamma_{\max}, 2)$.

On this note, harmonic balance can also be viewed as a frequency domain orthogonal collocation method, where complex exponentials form an orthonormal basis, and the residue must vanish at specific harmonics, effectively "collocating" or enforcing the system dynamics at key frequencies, rather than in the time domain.

Single Shooting

Shooting converts the periodic boundary-value problem into an initial-value problem. In this approach, an initial guess for the system's state is made, and the system's equations of motion are integrated over time to compute the state at the end of the period. For a given initial condition $\mathbf{q}_0 = \mathbf{q}(0)$, the flow $\boldsymbol{\varphi}(\mathbf{q}_0, 1)$ maps the initial state to the final state after a time period T by solving the time-normalized differential equations. Namely, $\boldsymbol{\varphi}(\mathbf{q}_0, \cdot) : \mathbb{R} \to \mathbb{R}^m$ solves

$$\boldsymbol{\varphi}'(\mathbf{q}_0, \tau) = T\,\mathbf{g}(\boldsymbol{\varphi}(\mathbf{q}_0, \tau)) \quad \text{and} \quad \boldsymbol{\varphi}(\mathbf{q}_0, 0) = \mathbf{q}_0,$$

where the derivative is with respect to $\tau = t/T$. Moreover, periodicity is stated through the constraint $\boldsymbol{\varphi}(\mathbf{q}_0, 1) = \mathbf{q}_0$, with m equations and $m+1$ unknowns, including T.

The flow is approximated via a direct (transient) time integration method. This work uses the fourth order Runge-Kutta method, chosen for its accuracy and simplicity, with N evenly spaced time steps. At each time increment, the next state becomes a weighted average of four state estimations which are computed by recursively evaluating the time-normalized dynamics $T\mathbf{g}(\cdot)$, which has degree $\gamma_{\max}+1$ on $(\mathbf{q}_0, T)$. Hence, accounting for the four levels of recursion at each of the N time increments, $\boldsymbol{\varphi}(\mathbf{q}_0, T)$ can be interpreted as a polynomial of degree $4N[\gamma_{\max}+1]$.

Numerical Continuation

In all the aforementioned methods, the problem of finding periodic solutions to the reduced dynamics can be reformulated as solving an algebraic equation, $\mathbf{R}(\mathbf{X}) = \mathbf{0}$, where $\mathbf{X}$ is the vector of unknowns (including the frequency) and $\mathbf{R}$ defines the constraints: Fourier coefficients with frequency collocation constraints for harmonic balance, polynomial coefficients with collocation and continuity constraints for orthogonal collocation, and the initial state with the periodicity constraint for shooting. Predictor corrector numerical continuation aims to find a sequence of solution points by using the last solution $\mathbf{X}_{\text{last}}$ and executing a predictor procedure to estimate a new solution $\mathbf{X}_0$ which is then iteratively refined by a corrector procedure. In this work, the tangent prediction method, defined as follows, was used:

$$\mathbf{X}_0 = \mathbf{X}_{\text{last}} + s\mathbf{V} \qquad \text{with } \mathbf{V} \text{ such that} \qquad \left.\frac{d\mathbf{R}}{d\mathbf{X}}\right|_{\text{last}} \mathbf{V} = \mathbf{0}, \qquad \mathbf{V}_{\text{ref}}^{\mathrm{T}}\mathbf{V} > 0, \qquad \|\mathbf{V}\| = 1,$$

where $s > 0$ is the step length and $\mathbf{V}_{\text{ref}}$ is a reference vector for the direction of the continuation, e.g. the secant through the last two solutions. The guess $\mathbf{X}_0$ is assumed to be close to the curve, hence it is used to initialize a Newton based corrector procedure to find a new solution. Since the standard Newton iterations can only be applied to systems with the same number of equations as unknowns, one extra scalar condition is appended: the pseudo-arclength parameterization, $p(\mathbf{X}) = \mathbf{V}^{\mathrm{T}}(\mathbf{X} - \mathbf{X}_0)$. Equating it to zero limits the solution to a hyperplane orthogonal to the predictor vector $\mathbf{V}$ and passing through the guess $\mathbf{X}_0$. Finally, the step length s is adjusted according to the number of iterations in the previous corrector step, in an inverse manner, to place the next guess in the fast convergence region of the Newton corrector.

Application and Discussion of Results

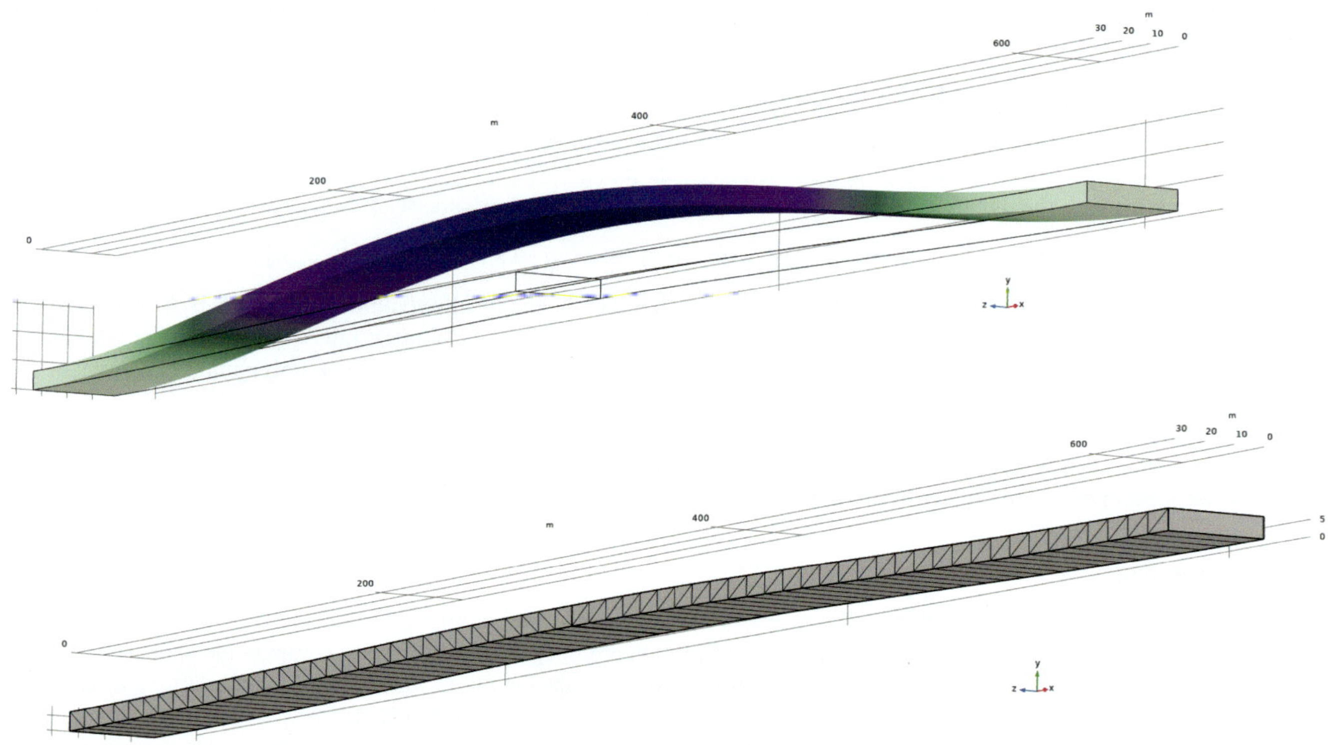

Fig. 1 Full-order finite element model of the clamped-clamped arch-beam. Geometry and mesh on top. Shape of the first linear eigenmode, below. Generated in COMSOL Multiphysics 6.2 [32].

The analysis begins with a full-order finite element model (of second-order elements) of the clamped-clamped arch-beam depicted in figure 1, which does not have preload and is anticipated to exhibit geometric nonlinear behavior. Figure 1 also illustrates the shape of the first linear mode, characterized by an eigenfrequency of $\omega_1 = 1.03308$. In `MORFE2.0` [23], the external force is represented as a small modal excitation of mode 1, with light mass-proportional damping introduced to achieve a damping ratio of 0.1%. The reduction settings are defined as follows: the maximum expansion order is set to $\gamma_{\max} = 7$; one master mode was utilized from the ten computed to identify outer resonances; and the (complex) normal-form parameterization style was chosen.

For this study case, `MORFE2.0` [23] took[5] 110s to reduce the full order model of 2664 degrees of freedom into a ROM with just 2 normal coordinates. There are 2 additional decoupled forcing variables, like those in example (5), which can be directly substituted by sinusoidal functions.

The references in figure 2 were obtained by computing periodic solutions to the reduced dynamics, on the frequency range $\omega \in [0.7\omega_1, 1.3\omega_1]$, via harmonic balance, orthogonal collocation and shooting with a time-resolution of $N = 160$, residual tolerance of $\text{tol} = 10^{-9}$ and maximum allowed step length of $s_{\max} = 0.1$. Figures 3 and 4 were generated by running the solvers with varying parameters—time resolution, continuation step length and tolerance—while only including those that traced smooth frequency-response curves with proper resolution of the resonance. It is relevant to state that the "loop" intersections observed close to the resonance are an artifact of the projection of the frequency-response curves to a bidimensional plot, in fact the intersection corresponds to two different solutions.

Figure 3 cannot be directly used to rank method performance without discussion, as variations may arise from differences in software optimization and user expertise, and most importantly from the particularities of the methods themselves. For instance, figure 4 suggests that harmonic balance yields meaningful results with coarser time resolution, while shooting and especially orthogonal collocation require larger N to compute accurate frequency-response curves.

---

[5] The computer used in this work has an AMD Ryzen 9 7950X 16-Core Processor running at 4.50GHz and 128GB of RAM.

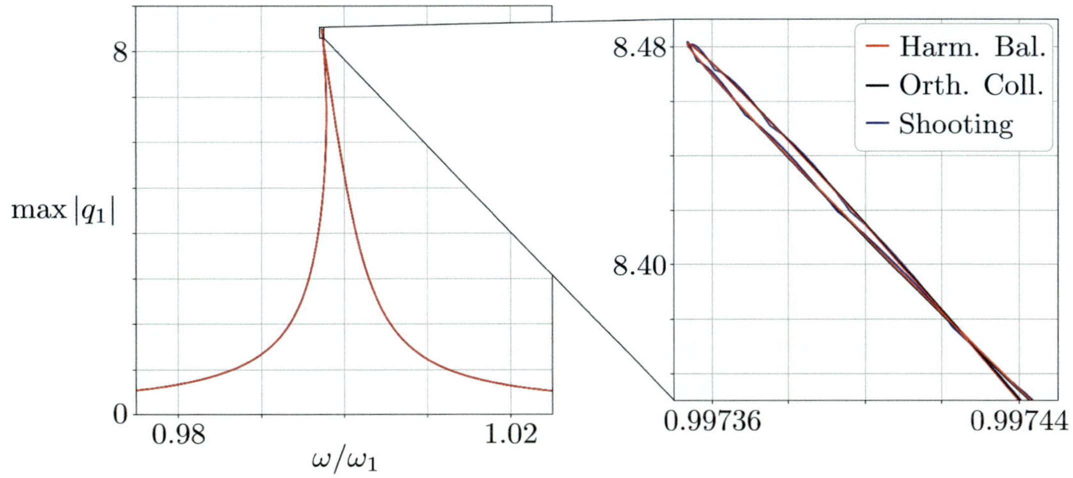

Fig. 2 Reference frequency-response curves of the SSM-reduced arch-beam computed using harmonic balance, orthogonal collocation and shooting with tolerance tol = 10^{-9}, maximum continuation step length $s_{\max} = 0.1$ and time resolution $N = 160$. The left plot is an amplification of the resonance region. The ratio between the frequency and the first linear eigenfrequency is ω/ω_1 and the maximum amplitude of first normal coordinate is $\max|q_1|$.

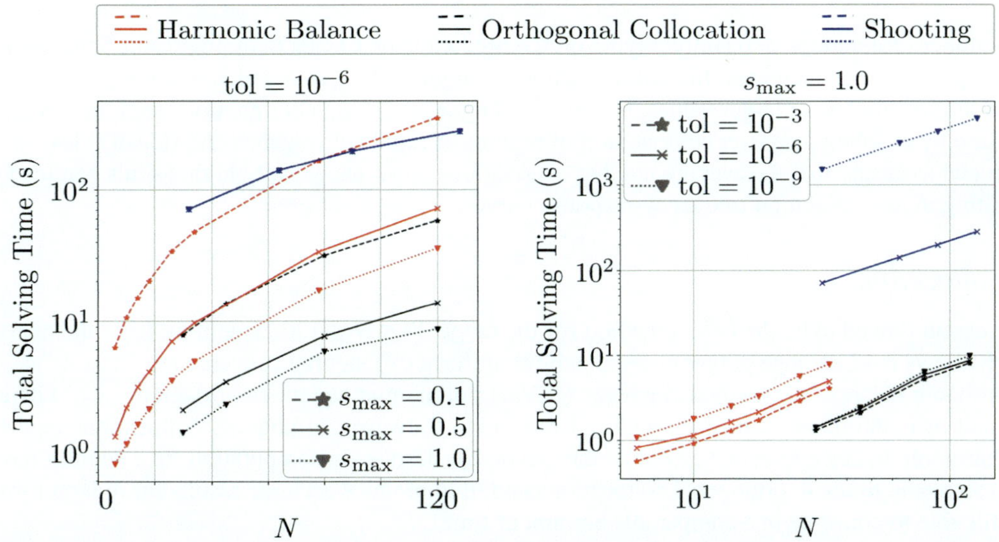

Fig. 3 Comparison of computation times for harmonic balance, orthogonal collocation and shooting regarding the construction of the frequency-response curve of the SSM-reduced arch beam. N indicates the time resolution, $s_{\max}$ is the maximum continuation step length and tol is the tolerance, whose meaning varies by method; the discussion section explains how to make the values comparable. On the left tol = 10^{-6} is fixed, and on the right $s_{\max} = 1.0$.

Harmonic Balance via Alternating Frequency-Time

Harmonic balance[6] achieves qualitatively accurate solutions with few harmonics (small $N = 2h_{\max}$) due to the Fourier series ansatz, which is optimized for finding periodic orbits and exhibits exponential convergence with truncation order $h_{\max}$, for smooth signals. When low truncation orders suffice to describe periodic time responses of low-dimensional smooth systems, the problem size and computation time are significantly reduced, leading to strong performance. On the other hand, including a large amount of harmonics significantly slows down the procedure due to the complexity of performing Newton iterations with larger matrices and the increased number of FFTs required.

[6] As a note, the set of harmonics $\mathcal{H}$ was optimized to include just odd harmonics, because the $\mathbf{g}$ of the system in question is an odd function and the forcing only excites the first harmonic.

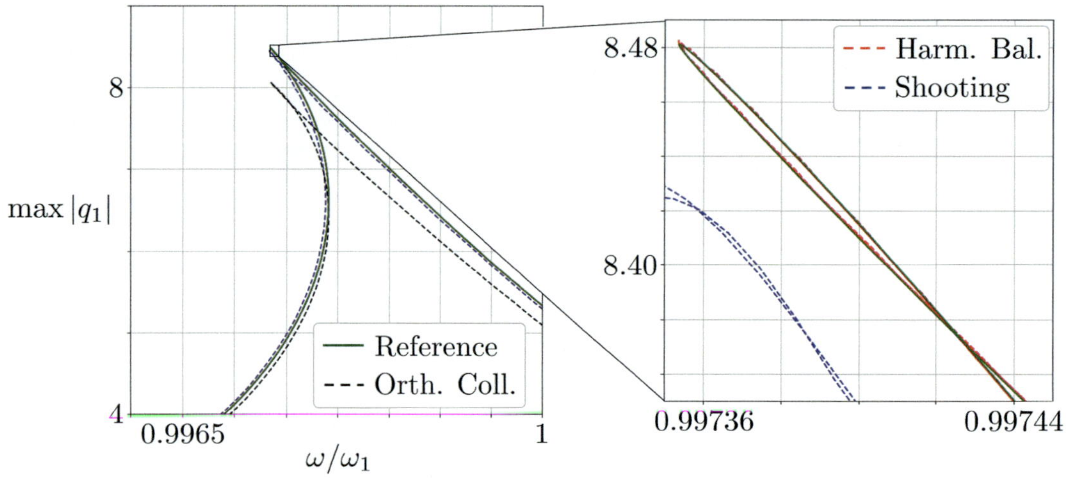

Fig. 4 Comparison to reference of frequency-response curves of the SSM-reduced arch-beam computed using harmonic balance ($N = 10$ and 0.93s), orthogonal collocation ($N = 30$ and 1.3s) and shooting ($N = 25$ and 21.2s) with the minimal parameters required for convergence. The left plot is an amplification of the resonance region. The ratio between the frequency and the first linear eigenfrequency is ω/ω_1 and the maximum amplitude of first normal coordinate is max$|q_1|$.

Targeting a number of iterations of 0 (fully relying on the predictor) or 1 leads to impractically long computation times, particularly as the problem size increases. In contrast, setting a target of 2, 3, or 4 yields relatively consistent computation times, with slight improvements for larger values. Naturally, a larger maximum continuation step length significantly accelerates the procedure by enabling a coarser computation of regions of rapid convergence and (ideally) less intricate features. However, it is crucial to ensure that the average number of iterations is not excessively high, as this can lead to overly large step lengths resulting in low-resolution frequency-response curves.

Orthogonal Collocation

Orthogonal collocation proved to be the fastest method for the range $30 \leq N \leq 120$, benefiting from the analytical nature of the state $\mathbf{q}$, which allows the Lagrange polynomials to exhibit uniform exponential convergence.

Similar to harmonic balance, the methods involves $O(Nm)$ polynomial equations of degree $\gamma_{\max}$, where $N = cz$ is the time resolution and m is the dimension of the reduced dynamics. However, orthogonal collocation requires a finer time resolution than harmonic balance to achieve comparable accuracy, despite its computation time scaling more slowly with increasing N. As shown in figure 4, orthogonal collocation can fail to produce accurate results for N larger than of harmonic balance, while still able to converge in a comparable amount of time.

Remarkably, for fixed maximum step length, changing the tolerance had a very small effect on the speed of collocation, opposite to what happens with shooting.

Single Shooting

The residue in shooting, defined as the difference between final and initial states, $\mathbf{q}(T) - \mathbf{q}(0)$, is not directly comparable to residues in other methods, which relate to the "force" $\mathbf{g}(\mathbf{q})$. To ensure a fair comparison, the shooting solutions were post-processed to yield a corresponding residue magnitude of $\|\mathbf{g}(\mathbf{q}(T)) - \mathbf{g}(\mathbf{q}(0))\|$. Figure 3 illustrates that the shooting method is slower compared to the other two methods for lower time resolutions, and suggests that it might scale better for higher N.

Furthermore, no valid solutions were obtained for tol $= 10^{-3}$ and $s_{\max} \geq 0.1$, as the continuation procedure exhibited erratic behavior, including jumping along the solution curve, backtracking, and becoming trapped in loops of solutions. Instead, for tol $= 10^{-5}$, additional data points revealed times of $27.30, 54.32, 76.36, 108.26$ seconds for $N = 32, 64, 90, 128$, respectively. Shooting was the most sensitive to tol, with the computation time increasing approximately linearly with the number of time steps, for all cases.

The step lengths observed during the shooting runs were, on average, orders of magnitude smaller than the maximum allowed step lengths of 1.0, 0.5, and 0.1, indicating no practical benefit from increasing the step length within this range. Moreover, the optimal goal number of iterations was determined to be 1, as 0 resulted in impractically small step lengths and an excessive number of solution points, while 2 in unstable erratic behavior.

Shooting has to solve $O(m)$ polynomial equations of degree $O(N\gamma_{\max})$. While the number of equations is reduced, they are significantly more complex to manage. In particular, accurately approximating the flow through time integration is a challenging task.

Conclusion

Advancements in the parametrization method, particularly its adaptation to vibratory systems and the introduction of spectral submanifolds (SSMs), have greatly simplified the reduction of high-dimensional models. While the automated computation of reduced dynamics in physical coordinates marks significant progress, robust numerical solvers remain essential.

This study evaluated three solvers—harmonic balance, orthogonal collocation, and shooting—for periodic orbit analysis. Harmonic balance provided an effective trade-off between speed and accuracy but slowed with higher harmonics. Orthogonal collocation, the fastest for moderate resolutions, required finer time steps yet scaled well. Single shooting, while slower initially, scaled efficiently with increasing resolution but was highly sensitive to tolerance.

Acknowledgments The authors acknowledge Alessandra Vizzaccaro for her invaluable help regarding `MORFE2.0` [23].

Disclosure statement No potential conflict of interest is reported by the authors.

References

[1] Xavier Cabré, Ernest Fontich, and Rafael De la Llave. "The parameterization method for invariant manifolds I: manifolds associated to non-resonant subspaces". In: *Indiana University mathematics journal* (2003), pp. 283–328.

[2] Xavier Cabré, Ernest Fontich, and Rafael De la Llave. "The parameterization method for invariant manifolds II: regularity with respect to parameters". In: *Indiana University mathematics journal* (2003), pp. 329–360.

[3] Xavier Cabré, Ernest Fontich, and Rafael De La Llave. "The parameterization method for invariant manifolds III: overview and applications". In: *Journal of Differential Equations* Vol. 218.No. 2 (2005), pp. 444–515.

[4] Alex Haro, Marta Canadell, Jordi-Lluis Figueras, Alejandro Luque, and Josep-Maria Mondelo. "The parameterization method for invariant manifolds". In: *Applied mathematical sciences* Vol. 195 (2016).

[5] John Guckenheimer and Philip Holmes. *Nonlinear oscillations, dynamical systems, and bifurcations of vector fields*. Vol. 42. Springer Science & Business Media, 2013.

[6] Cyril Touzé, Olivier Thomas, and Antoine Chaigne. "Hardening/softening behaviour in non-linear oscillations of structural systems using non-linear normal modes". In: *Journal of Sound and Vibration* Vol. 273.No. 1-2 (2004), pp. 77–101.

[7] George Haller and Sten Ponsioen. "Nonlinear normal modes and spectral submanifolds: existence, uniqueness and use in model reduction". In: *Nonlinear dynamics* Vol. 86 (2016), pp. 1493–1534.

[8] Sten Ponsioen, Tiemo Pedergnana, and George Haller. "Automated computation of autonomous spectral submanifolds for nonlinear modal analysis". In: *Journal of sound and Vibration* Vol. 420 (2018), pp. 269–295.

[9] Jorge Gonzalez, JD Mireles James, and Necibe Tuncer. "Finite element approximation of invariant manifolds by the parameterization method". In: *Partial Differential Equations and Applications* Vol. 3.No. 6 (2022), p. 75.

[10] Alessandra Vizzaccaro, Yichang Shen, Loıc Salles, Jiří Blahoš, and Cyril Touzé. "Direct computation of nonlinear mapping via normal form for reduced-order models of finite element nonlinear structures". In: *Computer Methods in Applied Mechanics and Engineering* Vol. 384 (2021), p. 113957.

[11] Yichang Shen, Alessandra Vizzaccaro, Nassim Kesmia, Ting Yu, Loıc Salles, Olivier Thomas, and Cyril Touzé. "Comparison of reduction methods for finite element geometrically nonlinear beam structures". In: *Vibration* Vol. 4.No. 1 (2021), pp. 175–204.

[12] Andrea Opreni, Alessandra Vizzaccaro, Attilio Frangi, and Cyril Touzé. "Model order reduction based on direct normal form: application to large finite element MEMS structures featuring internal resonance". In: *Nonlinear Dynamics* Vol. 105.No. 2 (2021), pp. 1237–1272.

[13] Shobhit Jain and George Haller. "How to compute invariant manifolds and their reduced dynamics in high-dimensional finite element models". In: *Nonlinear dynamics* Vol. 107.No. 2 (2022), pp. 1417–1450.

[14] Andrea Opreni, Alessandra Vizzaccaro, Cyril Touzé, and Attilio Frangi. "High-order direct parametrisation of invariant manifolds for model order reduction of finite element structures: application to generic forcing terms and parametrically excited systems". In: *Nonlinear Dynamics* Vol. 111.No. 6 (2023), pp. 5401–5447.

[15] Mingwu Li, Shobhit Jain, and George Haller. "Model reduction for constrained mechanical systems via spectral submanifolds". In: *Nonlinear Dynamics* Vol. 111.No. 10 (2023), pp. 8881–8911.

[16] Adrien Martin, Andrea Opreni, Alessandra Vizzaccaro, Marielle Debeurre, Loic Salles, Attilio Frangi, Olivier Thomas, and Cyril Touzé. "Reduced-order modeling of geometrically nonlinear rotating structures using the direct parametrisation of invariant manifolds". In: *Journal of Theoretical, Computational and Applied Mechanics* (2023).

[17] Joar Axås and George Haller. "Model reduction for nonlinearizable dynamics via delay-embedded spectral submanifolds". In: *Nonlinear Dynamics* Vol. 111.No. 24 (2023), pp. 22079–22099.

[18] George Haller, Bálint Kaszás, Aihui Liu, and Joar Axås. "Nonlinear model reduction to fractional and mixed-mode spectral submanifolds". In: *Chaos: An Interdisciplinary Journal of Nonlinear Science* Vol. 33.No. 6 (2023).

[19] Thomas Thurnher, George Haller, and Shobhit Jain. "Nonautonomous spectral submanifolds for model reduction of nonlinear mechanical systems under parametric resonance". In: *Chaos: An Interdisciplinary Journal of Nonlinear Science* Vol. 34.No. 7 (2024).

[20] Mingwu Li, Shobhit Jain, and George Haller. "Fast computation and characterization of forced response surfaces via spectral submanifolds and parameter continuation". In: *Nonlinear Dynamics* (2024), pp. 1–27.

[21] George Haller and Roshan S. Kaundinya. "Nonlinear model reduction to temporally aperiodic spectral submanifolds". In: *Chaos: An Interdisciplinary Journal of Nonlinear Science* Vol. 34.No. 4 (2024).

[22] Mattia Cenedese, Jacopo Marconi, George Haller, and Shobhit Jain. "Data-assisted non-intrusive model reduction for forced nonlinear finite elements models". In: *arXiv preprint arXiv:2311.17865* (2023).

[23] Alessandra Vizzaccaro, Giorgio Gobat, Attilio Frangi, and Cyril Touzé. "Direct parametrisation of invariant manifolds for non-autonomous forced systems including superharmonic resonances". In: *Nonlinear Dynamics* Vol. 112.No. 8 (2024), pp. 6255–6290.

[24] Leonardo Bettini, Mattia Cenedese, and George Haller. "Model reduction to spectral submanifolds in piecewise smooth dynamical systems". In: *International Journal of Non-Linear Mechanics* Vol. 163 (2024), p. 104753.

[25] Aurélien Grolet, Alessandra Vizzaccaro, Marielle Debeurre, and Olivier Thomas. "High order invariant manifold model reduction for systems with non-polynomial non-linearities: geometrically exact finite element structures and validity limit". In: (2024).

[26] Malte Krack and Johann Gross. *Harmonic balance for nonlinear vibration problems*. Vol. 1. Springer, 2019.

[27] Carl De Boor and Blâir Swartz. "Collocation at Gaussian points". In: *SIAM Journal on numerical analysis* Vol. 10.No. 4 (1973), pp. 582–606.

[28] Ehsan Loghman, Ali Kamali, Firooz Bakhtiari-Nejad, Mostafa Abbaszadeh, and Marco Amabili. "On the combined Shooting-Pseudo-Arclength method for finding frequency response of nonlinear fractional-order differential equations". In: *Journal of Sound and Vibration* Vol. 516 (2022), p. 116521.

[29] Annick Dhooge, Willy Govaerts, and Yu A Kuznetsov. "MATCONT: a MATLAB package for numerical bifurcation analysis of ODEs". In: *ACM Transactions on Mathematical Software (TOMS)* Vol. 29.No. 2 (2003), pp. 141–164.

[30] A. Vizzaccaro, A. Opreni, L. Salles, and et al. "High order direct parametrisation of invariant manifolds for model order reduction of finite element structures: application to large amplitude vibrations and uncovering of a folding point". In: *Nonlinear Dynamics* Vol. 110 (2022), pp. 525–571.

[31] Lukas Woiwode, Nidish Narayanaa Balaji, Jonas Kappauf, Fabia Tubita, Louis Guillot, Christophe Vergez, Bruno Cochelin, Aurélien Grolet, and Malte Krack. "Comparison of two algorithms for harmonic balance and path continuation". In: *Mechanical Systems and Signal Processing* Vol. 136 (2020), p. 106503.

[32] COMSOL AB. *AC/DC Module User's Guide*. COMSOL Multiphysics®v. 6.2. Stockholm, Sweden, 2023, pp. 75–84.

Chapter 15
Flutter Analysis of Control Surface Free-play by Using a 3-DOF Airfoil Model

İzzet Uluğ and Ender Cigeroglu

Abstract The phenomenon of flutter in aeroelastic systems remains a critical area of research due to its implications for the safety of aircraft wings. Free-play, modeled by using a piecewise linear approach to examine its effect on the aeroelastic response of a system, is a common nonlinearity in control surfaces, which occurs due to gaps at the joints in the control mechanisms. This paper investigates the flutter characteristics of a 3-DOF airfoil model with free-play between plunge and flap motions. The equation of motion (EOM) is obtained via Lagrange's equations, and Theodorsen approach is employed for aerodynamic loads. The EOM can be solved by various methods, namely the harmonic balance method (HBM) and the describing function method (DFM), which can be used in the frequency domain, as well as time integration methods. In this study, DFM is used due to frequency domain methods being more efficient compared to time domain methods. Flutter instability is determined by varying the airspeed and identifying the equivalent damping of the system where the airspeed that causes the damping value of zero is the flutter speed searched. A parametric study is conducted to evaluate the effects of varying free-play gap sizes on the flutter characteristics. The results indicate that free-play changes the flutter speed and mode, with a reduction in the flutter boundary compared to the linear case. Additionally, the study examines nonlinear flutter boundaries, focusing on the development of limit cycle oscillations (LCOs) due to free-play.

Keywords Flutter · Free-play · Nonlinear Flutter · Limit Cycle Oscillations (LCOs) · Aeroelastic Instability

Introduction

Flutter is a type of aeroelastic instability that arises from the interaction of structural, aerodynamic, and inertial forces, leading to an increase in vibrations. This oscillatory instability occurs when aerodynamic forces couple with the natural vibration modes of the structure, potentially resulting in dangerous, self-sustaining oscillations. Nonlinearities, both structural and aerodynamic, play a critical role in influencing the flutter behavior of a system. One common structural nonlinearity is free-play, which emerges due to gaps or looseness in the hinge mechanisms of control surfaces. Free-play significantly alters the flutter characteristics, potentially introducing limit cycle oscillations (LCOs), as shown in studies by Dowell and Tang [1]. Yu et al. [2] conducted a numerical study analyzing the impact of free-play nonlinearity on flutter and LCOs by using 2-DOF airfoil models, showing how free-play influences flutter behavior and the occurrence of LCOs. This type of structural nonlinearity has been a key focus in many aeroelastic studies, since it changes how the stiffness in control surfaces behaves depending on the displacement. This leads to complex flutter behavior that cannot be captured by linear models alone. For instance, Conner et al. [3] demonstrated how free-play introduces nonlinear dynamics that lead to periodic oscillations and other behaviors beyond the flutter boundary. Similarly, Irani et al. [4] further confirmed that an increase in the gap results in consistent decrease in flutter speed, emphasizing the importance of considering these effects early in the design phases. These structural nonlinearities affect both the flutter speed and the post-flutter behavior of the system, making it critical to account for them in flutter analysis.

Traditionally, time-domain methods have been used to analyze flutter by simulating the response of the system over time. While these methods provide detailed insights into transient and steady-state behaviors, they can be computationally

İzzet Uluğ · Ender Cigeroglu
Department of Mechanical Engineering, Middle East Technical University, Ankara, Türkiye; Center for Gear, Power Transmission Systems and Vibration Research, Middle East Technical University, Ankara, Türkiye
e-mail: ender@metu.edu.tr; izzet.ulug@metu.edu.tr

Ludovic Renson et al. (eds.), *Nonlinear Structures & Systems, Vol. 1*, Conference Proceedings of the Society for Experimental Mechanics Series, https://doi.org/10.13052/9788743801467-15

expensive, especially for systems with nonlinearities like free-play. As shown by Conner et al. [3] and Trickey et al. [5], solving nonlinear equations in the time domain often requires significant computational resources due to the complexity of the simulations. In contrast, frequency-domain methods, such as the DFM, are more efficient for analyzing systems with periodic responses, providing accurate predictions of flutter behavior. This method linearizes the nonlinear system around limit cycles, allowing for the calculation of stability boundaries and LCOs without extensive simulations. Studies by Dowell and Tang [1] and Yurkovich [6] have demonstrated the advantages of frequency-domain approaches in handling nonlinear aeroelastic problems, offering accurate predictions of flutter dynamics. In a recent study, Frey et al. [7] compared time-domain and frequency-domain methods on turbine blades with nonlinear unsteady flow conditions in case, and demonstrated that frequency-domain solvers not only reduce the computational cost but also effectively model aeroelastic instabilities, especially under transonic flow conditions where strong shock interactions impact system dynamics.

The flutter characteristics of an airfoil with free-play nonlinearity at the control surface are explored using a three-degree-of-freedom (3-DOF) model. Free-play nonlinearity is represented by a piecewise linear stiffness element and the behavior of the system is studied in the frequency domain. This approach effectively predicts the flutter boundary and the onset of LCOs, providing a clearer understanding of how free-play affects system stability and post-flutter dynamics.

Mathematical Modelling

The 3-DOF airfoil model used in this study is given in Fig. 1. The airfoil exhibits three degrees of freedom: vertical motion (plunge), rotational motion around a horizontal axis (pitch), and control surface motion (flap), denoted by h, α, and β, respectively. The moments of inertia about the elastic axis for the airfoil and control surface are represented by I_α and I_β, respectively. Nonlinearity is introduced through torsional free-play at the control surface hinge, as previously studied and tested by Conner et al. [2]. The numerical parameters used in this paper were also adopted from Conner et al.'s study. Simple models, like the one shown in Fig. 1, are commonly used in similar analyses for their practicality.

As shown in Fig. 1, the model represents a two-dimensional airfoil placed in a steady horizontal airflow. Key parameters include the freestream velocity U, the semi-chord length b, stiffnesses in pitch, plunge, and control surface hinge directions (k_α, k_h, k_β), as well as the mass of the wing m_w, and the total mass of the system m_T. Additionally, the damping coefficients for the pitch, plunge, and flap motions are denoted as c_α, c_h and c_β, respectively, which describe the energy dissipation in each mode of motion.

The structural behavior is governed by linear bending and torsional springs along the airfoil's elastic axis and control surface hinge line, which simulate the restoring forces from the structure. Additionally, the distances between key points in the system —such as from elastic axis to the mid-chord a the mid-chord to the hinge line c, the elastic axis to the center of mass of the airfoil x_α, and the hinge line to the center of mass of the control surface x_β—are also considered. Finally, the total unsteady aerodynamic forces in the vertical direction, as well as the aerodynamic moments about the elastic axis and flap hinge, are represented by L, M_α and M_β, respectively.

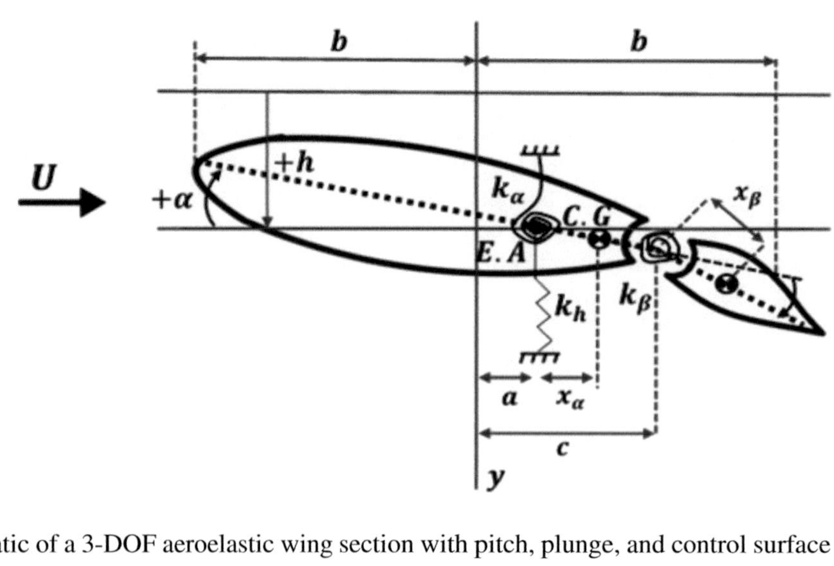

Fig. 1 Schematic of a 3-DOF aeroelastic wing section with pitch, plunge, and control surface spring restraints.

Equation of motion of the 3-DOF airfoil model given in Fig. 1 can be written as follows.

$$\mathbf{M}\left\{\begin{array}{c}\ddot{h}\\ \ddot{\alpha}\\ \ddot{\beta}\end{array}\right\}+\mathbf{C}\left\{\begin{array}{c}\dot{h}\\ \dot{\alpha}\\ \dot{\beta}\end{array}\right\}+\mathbf{K}\left\{\begin{array}{c}h\\ \alpha\\ \beta\end{array}\right\}=\left\{\begin{array}{c}L\left(h,\alpha,\beta\right)\\ M_{\alpha}\left(h,\alpha,\beta\right)\\ M_{\beta}\left(h,\alpha,\beta\right)\end{array}\right\}, \tag{1}$$

where, $\mathbf{M}$, $\mathbf{C}$ and $\mathbf{K}$ are the representatives of structural mass, stiffness, and damping of the system. The same system is considered in Fung [8] and equations of motion of the system given in Eq. (1) are the same. System matrices defined in Eq. (1) can be written explicitly as

$$\mathbf{M}=\begin{bmatrix} m_T & m_{wb}x_{\alpha} & m_{wb}x_{\beta}\\ m_{wb}x_{\alpha} & m_{wb}^2 r_{\alpha}^2 & m_{wb}^2 r_{\beta}^2+m_{wb}^2 x_{\beta}(c-a)\\ m_{wb}x_{\beta} & m_{wb}^2 r_{\beta}^2+m_{wb}^2 x_{\beta}(c-a) & m_{wb}^2 r_{\beta}^2\end{bmatrix} \tag{2a}$$

$$\mathbf{K}=\begin{bmatrix} k_h & 0 & 0\\ 0 & k_{\alpha} & 0\\ 0 & 0 & k_{\beta}(\beta)\end{bmatrix} \tag{2b}$$

$$\mathbf{C}=\begin{bmatrix} c_h & 0 & 0\\ 0 & c_{\alpha} & 0\\ 0 & 0 & c_{\beta}\end{bmatrix} \tag{2c}$$

This model builds on the state-space framework proposed by Edwards et al. [9]. Due to the free-play nonlinearity, which causes a piecewise linear variation in the structural stiffness of the control surface hinge, as illustrated in Fig. 2, the entire system can be described as a nonlinear combination of three linear subsystems. Understanding this nonlinearity is crucial in flutter analysis, as the presence of free-play significantly alters the dynamics of the system. Bai et al. [10] demonstrated that free-play affects flutter speed by causing sudden stiffness changes once the gap is closed, which reduces stability and lowers the flutter onset.

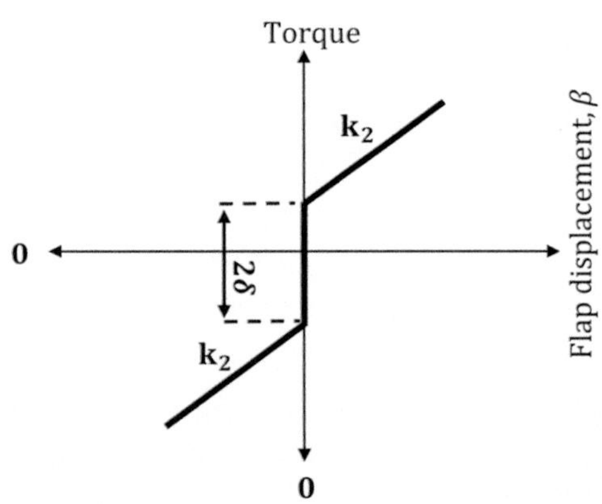

Fig. 2 Torque vs. flap displacement (β) in the presence of free-play nonlinearity.

The aerodynamic lift and moments L, M_α and M_β are calculated using the Theodorsen method, which employs linearized thin airfoil theory to evaluate unsteady aerodynamic forces and moments, as outlined in Theodorsen [11].

$$L\left(h,\alpha,\beta\right)=-\pi\rho b^2\left[\ddot{h}+U\dot{\alpha}-ba\ddot{\alpha}-\frac{U}{\pi}T_4\dot{\beta}-\frac{b}{\pi}T_1\ddot{\beta}\right]-2\pi\rho UbQC\left(k\right), \tag{3}$$

$$M_{\alpha}(h,\alpha,\beta)=\pi\rho b^2\left[ba\ddot{h}-Ub\left(\frac{1}{2}-a\right)\dot{\alpha}-b^2\left(\frac{1}{8}+a^2\right)\ddot{\alpha}-\frac{U^2}{\pi}(T_4+T_{10})\beta\right.$$
$$\left.+\frac{Ub}{\pi}\left\{-T_1+T_8+(c-a)T_4-\frac{1}{2}T_{11}\right\}\dot{\beta}+b^2\pi\left\{T_7+(c-a)T_1\right\}\ddot{\beta}\right]\quad+2\pi\rho b^2\left(a+\frac{1}{2}\right)QC(k), \tag{4}$$

$$M_\beta(h,\alpha,\beta)=\pi\rho b^2\left[\frac{b}{\pi}T_1\ddot{h}+\frac{Ub}{\pi}\left\{2T_9+T_1-\left(a-\frac{1}{2}\right)T_4\right\}\dot{\alpha}-\frac{2b^2}{\pi}T_{13}\ddot{\alpha}-\left(\frac{U}{\pi}\right)^2(T_5-T_4T_{10})W\beta\right]$$
$$+\frac{Ub}{2\pi^2}T_4T_{11}\dot{\beta}+\left(\frac{b}{\pi}\right)^2T_3\ddot{\beta}-\rho Ub^2T_{12}QC(k). \quad (5)$$

where Theodorsen's constants T_j are defined in [11], while $C(k)$ and Q are described as follows

$$C(k)=\frac{H_1^{(2)}(k)}{H_1^{(2)}(k)+iH_0^{(2)}(k)},\ \text{and}\ H_n^{(2)}=J_n(k)-iY_n(k) \quad (6)$$
$$Q=U\alpha+\dot{h}+\dot{\alpha}b\left(\frac{1}{2}-a\right)+\frac{U}{\pi}T_{10}\beta+\frac{b}{2\pi}T_{11}\dot{\beta}. \quad (7)$$

When flutter occurs, i.e. on the onset of flutter, the net damping in the system becomes zero and one of the vibration modes of the system follows a simple harmonic behavior. Due to the small amplitude of these oscillations, linear aerodynamic theory is adequate for modelling the unsteady air loads on a lifting surface in motion. Therefore, h, α and β can be represented as follows

$$\begin{Bmatrix} h\\ \alpha\\ \beta \end{Bmatrix}=\begin{Bmatrix} \hat{h}\\ \widehat{\alpha}\\ \widehat{\beta} \end{Bmatrix}e^{i\omega t}, \quad (8)$$

where ω is frequency of the motion and, $\hat{h}$, $\widehat{\alpha}$ and $\widehat{\beta}$ are complex amplitudes. Moreover, aerodynamic force and moments are functions of the generalized coordinates. By identifying h, α and β within the expressions for L, M_α and M_β, these terms can be arranged into matrices where the stiffness, damping, and mass matrices of the system can rewritten as follows

$$\mathbf{M}_\mathrm{a}=\begin{bmatrix} -\pi\rho b^2 & \pi\rho b^3 a & \rho b^3T_1\\ \pi\rho b^3 a & -\pi\rho b^4\left(\frac{1}{8}+a^2\right) & \rho b^4\left(T_7+(c-a)T_1\right)\\ \rho b^3T_1 & -2\rho b^4T_{13} & \frac{\rho b^4T_3}{\pi} \end{bmatrix} \quad (9a)$$

$$\mathbf{K}_\mathrm{a}=\begin{bmatrix} 0 & -2\pi\rho U^2bC(k) & -2\rho U^2bT_{10}C(k)\\ 0 & 2\pi\rho b^2UC(k)\left(a+\frac{1}{2}\right) & 2\rho b^2UT_{10}C(k)\left(a+\frac{1}{2}\right)-\rho b^2U^2(T_4+T_{10})\\ 0 & -\rho U^2b^2T_{12}C(k) & -\frac{\rho b^2U^2\left(T_5-T_4T_{10}\right)+\rho U^2b^2T_{12}T_{10}C(k)}{\pi} \end{bmatrix} \quad (9b)$$

$$\mathbf{C}_\mathrm{a}=\begin{bmatrix} -2\pi\rho UbC(k) & -\pi\rho b^2U-2\pi\rho Ub^2C(k)\left(\frac{1}{2}-a\right) & \rho b^2UT_4-\rho Ub^2T_{11}C(k)\\ 2\pi\rho b^2\left(a+\frac{1}{2}\right)C(k) & 2\pi\rho b^3C(k)N_2-\pi\rho b^3U\left(\frac{1}{2}-a\right) & \rho b^3UN_1+\rho b^3\left(a+\frac{1}{2}\right)C(k)T_{11}\\ -\rho Ub^2T_{12}C(k) & \rho b^3UN_3-\rho Ub^3T_{12}C(k)\left(\frac{1}{2}-a\right) & \frac{\rho b^3U\left(T_{12}T_{11}C(k)-T_4T_{11}\right)}{2\pi} \end{bmatrix} \quad (9c)$$

where,

$$N_1=\left(-T_1+T_8+(c-a)T_4-\frac{1}{2}T_{11}\right),\quad N_2=\left(a+\frac{1}{2}\right)\left(a-\frac{1}{2}\right),\quad N_3=\left(3T_1-\left(a-\frac{1}{2}\right)T_4\right) \quad (9d)$$

Then, equation of motion can be written as follows:

$$\overline{\mathbf{M}}\ddot{\mathbf{x}}+\overline{\mathbf{C}}\dot{\mathbf{x}}+\overline{\mathbf{K}}\,\mathbf{x}=0, \quad (10)$$

where,

$$\overline{\mathbf{M}}=\mathbf{M}-\mathbf{M}_a,\quad \overline{\mathbf{C}}=\mathbf{C}-\mathbf{C}_a,\quad \overline{\mathbf{K}}=\mathbf{K}-\mathbf{K}_a \quad (11a)$$
$$\mathbf{x}=\{h\ \alpha\ \beta\} \quad (11b)$$

Nonlinear Flutter Solution

The nonlinear flutter solution is obtained by using the state-space approach and solved through incrementally increasing free-stream velocity, as shown by Kösterit et al. [12]. Eigenvalue analysis is conducted within the state-space formulation, where the real parts of the eigenvalues are examined to identify the flutter speed. Depending on the value of eigenvalue λ, behavior of the system may exhibit oscillatory, dissipative, or dispersive behavior. If $\lambda < 0$, the system is dynamically stable; if $\lambda > 0$, it becomes dynamically unstable, leading to divergent oscillations. The point where $\lambda = 0$ defines the 'stability boundary.' In aeroelastic systems, this instability is known as 'flutter', and the corresponding boundary is called the 'flutter boundary'. In flutter cases, the solution typically shows oscillatory or dispersive behavior, both indicating instability, where even a small disturbance can lead to large structural displacements. Hence, the flutter speed is determined as the point where the real part of the eigenvalue transitions from negative to positive. In state space form, equation of motion of the system can be written as follows:

$$\mathbf{A}\dot{\mathbf{q}} + \mathbf{B}\mathbf{q} = 0 \tag{12a}$$

$$\mathbf{A} = \begin{bmatrix} 0 & \overline{\mathbf{M}} \\ \overline{\mathbf{M}} & \overline{\mathbf{C}} \end{bmatrix}, \quad \mathbf{B} = \begin{bmatrix} -\overline{\mathbf{M}} & 0 \\ 0 & \overline{\mathbf{K}} \end{bmatrix}, \quad \mathbf{q} = \begin{Bmatrix} \dot{\mathbf{x}} \\ \mathbf{x} \end{Bmatrix} \tag{12b}$$

Here, q assuming harmonic form for x, q can be represented as follows

$$\mathbf{q} = \mathbf{q}e^{\lambda t}, \tag{13a}$$

Substituting Eq. (13a) into Eq. (12a) the following eigenvalue problem is obtained

$$\lambda \mathbf{A}\mathbf{q} + \mathbf{B}\mathbf{q} = 0. \tag{13b}$$

Eq. (13b) is for the linear problem. For the nonlinear problem, DFM is used to find the nonlinear stiffness matrix. This method, as demonstrated by Iannelli et al. [13], allows for stability assessment without requiring detailed frequency-domain analysis. In their study, Iannelli et al. applied DFM to aeroelastic systems with free-play nonlinearities, addressing uncertainties using structured singular values and integral quadratic constraints. Their work effectively captured the dynamic behavior of aeroelastic systems, providing an efficient way to model and analyze nonlinearities such as free-play in control surfaces. In the presence of free-play nonlinearity, the nonlinear restoring torque acting on the control surface can be written as follows by using the describing function theory

$$T_N = \nu \bullet \widehat{\beta}, \tag{14}$$

where, ν is the describing function which is defined as

$$\nu = \frac{i}{\pi A} \int_0^{2\pi} T_N \left(A sin\left(\psi\right) \right) e^{-i\psi} d\psi. \tag{15}$$

T_N is the nonlinear restoring torque, A is the amplitude of oscillations, $\psi = \omega t$ and $\widehat{\beta}$ is the complex representation of the flap displacement. It should be noted that, in general, ν is complex quantity; however, for free play nonlinearity, it is a real quantity, and it can be obtained as follows

$$\nu\left(A\right) = \begin{cases} \frac{-2k_2}{\pi} \left(\sin^{-1}\left(\frac{\delta}{A}\right) + \frac{\delta}{A}\sqrt{1 - \left(\frac{\delta}{A}\right)^2} \right) + k_2 & A > \delta \\ 0 & A \leq \delta \end{cases}, \tag{16}$$

Here, A is the flap displacement amplitude, k_2 is the torsional stiffness of the hinge mechanism, and δ defines the gap in free-play at one side of the equilibrium resulting total gap of 2δ. Using DFM, the stiffness matrix of the system can be written as follows

$$\overline{\mathbf{K}} = \mathbf{K} - \mathbf{K_a} + \begin{bmatrix} 0 & 0 & 0 \\ 0 & 0 & 0 \\ 0 & 0 & \nu\left(\beta\right) \end{bmatrix}. \tag{17}$$

Results

The flutter characteristics of the system are influenced by several structural and geometric parameters, which affect both flutter speed and frequency. The interaction between these parameters and the aeroelastic response is critical for understanding the stability of the system. The results demonstrate how variations in stiffness, semi-chord length, and hinge line position shape susceptibility to flutter, providing insights into optimizing aeroelastic performance.

Table 1 System parameters used in flutter analysis

Geometric Parameters	
b	$0.127\ [m]$
a	$-0.5\ [m]$
c	$0.5\ [m]$
Mass Parameters	
m_w	$0.62868\ [kg]$
m_T	$1,76435\ [kg]$
Inertial Parameters	
x_a	$0.434\ [m]$
x_b	$0.01996\ [m]]$
I_α	$0.0137\ [kg\ m^2]$
I_β	$0.0003264\ [kg\ m^2]$

First, linear flutter solution is obtained and given in Fig. 3. The real and imaginary parts of the eigenvalue provide the flutter speed and frequency, respectively. From the analysis, it is observed that the plunge mode reaches flutter at a speed of 22.72 m/s where the real part of the eigenvalue reaches zero with a corresponding frequency of 5.74 Hz. The system parameters used in the solution are provided in Table 1.

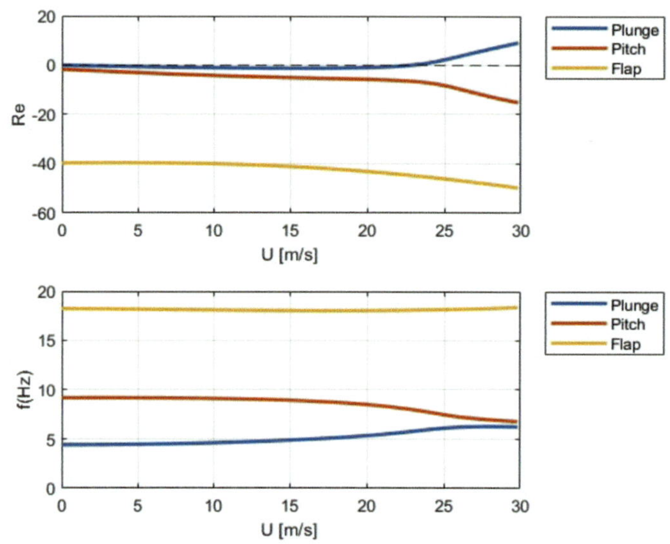

Fig. 3 Variation of real and imaginary parts of the eigenvalues vs. freestream velocity for pitch, plunge, and flap modes.

The nonlinear problem is considered next. As stiffness increases ($k = 10,\ 25,\ 100\ [N/m]$), the flutter speed increases significantly after exceeding the free-play limit. This effect is illustrated in Fig. 4a, which shows the sharp increase in flutter speed, indicating a more stable system with higher stiffness. However, up to around 2 degrees of rotational displacement, the flutter speed remains unchanged. This behavior is a direct result of the free-play that was introduced in the model. The gap allows for initial motion without engaging the stiffness; and as a result, the system does not exhibit any change in flutter speed within this range. Once the rotational displacement exceeds the free-play limit, the stiffness engages, leading to a sharp increase in flutter speed. After this point, the curves become steeper, showing a more abrupt transition to flutter with higher stiffness values. This outcome aligns with the idea that increased stiffness enhances the stability of the system, making it less prone to aeroelastic instabilities. The next analysis focuses on the semi-chord length b as shown in Fig. 4b. As the semi-chord length increases, both the flutter speed and frequency show a slight increase. This can be attributed to changes in the distribution of aerodynamic forces along the airfoil. Although the effect of semi-chord length is more subtle compared to stiffness, it still plays a role in stabilizing the system. Larger semi-chord lengths result in a marginally higher resistance to flutter, particularly when aerodynamic loads are distributed across the control surface.

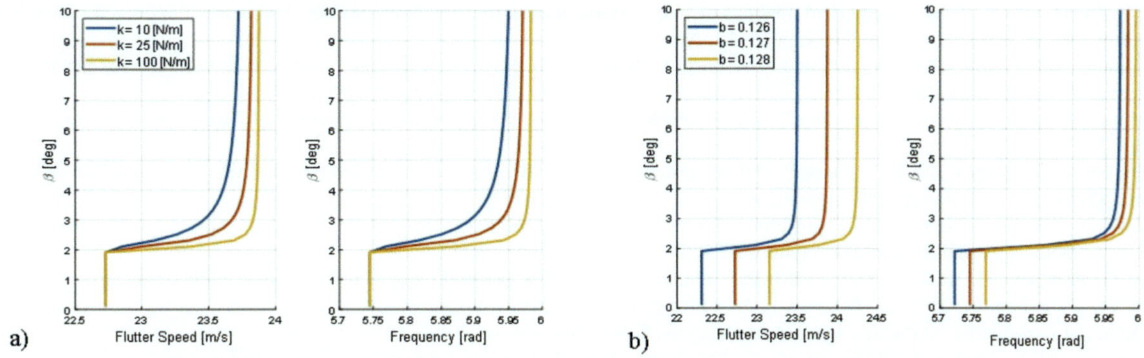

Fig. 4 Variation of flutter speed and frequency with respect to a) stiffness and b) chord length.

When evaluating the hinge line distance, c, a more significant influence on flutter behavior is observed, which is illustrated in Fig. 5. As c increases, both flutter speed and frequency decrease noticeably. This indicates that the system becomes more susceptible to flutter at lower speeds as the hinge line moves further from the mid-chord. This shift alters the torsional dynamics of the airfoil, making the structure more sensitive to aeroelastic forces. The positioning of the hinge line is therefore a critical design consideration for managing flutter risk in practical applications. Lastly, the distance between the elastic axis and the mid-chord, a, shows a similar trend. As a increases, both flutter speed and frequency decrease slightly. This suggests that as the elastic axis moves further from the mid-chord, the system becomes more prone to flutter. While the effect of a is less pronounced than that of stiffness or hinge line position, it still plays an important role in the overall stability of the system. Adjusting this parameter in combination with others can help optimize the aeroelastic performance of the system.

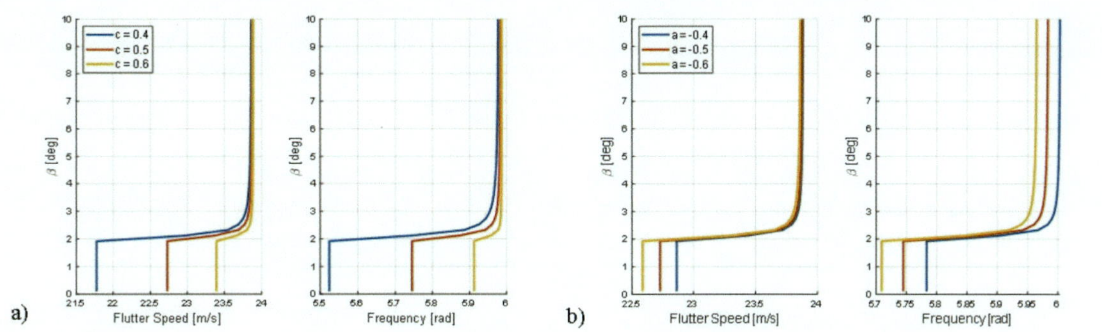

Fig. 5 Variation of flutter speed and frequency with respect to a) hinge line distance and b) elastic axis location.

Conclusion

This study focuses on analyzing the flutter behavior of a 3-DOF airfoil model with control surface free-play. Describing Function Method (DFM) is used to capture the nonlinear dynamics. The results indicate that free-play significantly affects both the flutter speed and the aeroelastic response of the system. Initially, control surface moves within the free-play gap without activating the stiffness, resulting in no change in flutter speed over a specific range of rotational displacement. Once the stiffness engages, there is a sharp increase in flutter speed, indicating a faster transition to instability in stiffer systems. It is observed from free-play affects aeroelastic behavior significantly, highlighting the importance of optimizing structural system parameters to reduce the vulnerability to flutter.

References

1. E. H, Dowell and D. M. Tang, "Aeroelastic response induced by free play, Part 2: Theoretical/experimental correlation analysis," *AIAA Journal*, vol. 49, no. 11, pp. 2543–2554, 2011.

2. Q. Yu, Y. Zhang, and Y. Wang, “Numerical analysis of free-play-induced aeroelastic phenomena: A numerical approach with adaptive step size control,” *Int. J. Aerosp. Eng.*, vol. 2024, pp. 1–13, Apr. 2024.
3. M. D. Conner, D. M. Tang, E. H. Dowell, and L. N. Virgin, “Nonlinear behavior of a typical airfoil section with control surface freeplay: A numerical and experimental study,” *Journal of Fluids and Structures*, vol. 11, no. 1, pp. 89–109, 1997.
4. S. Irani and S. Sazesh, “A new flutter speed analysis method using stochastic approach,” *Journal of Fluids and Structures*, vol. 40, pp. 105–114, 2013.
5. S. T. Trickey, L. N. Virgin, and E. H. Dowell, “Characterizing stability of responses in a nonlinear aeroelastic system,” in 41st Structures, Structural Dynamics, and Materials Conference and Exhibit, Dec. 2000.
6. R. Yurkovich, “Nonlinear Aeroelasticity and Flutter Suppression Using Frequency Domain Techniques,” presented at the 20th International Forum on Aeroelasticity and Structural Dynamics (IFASD), June 2004.
7. C. Frey, G. Ashcroft, H. Kersken, and D. Schlüß, “Flutter analysis of a transonic steam turbine blade with frequency and time-domain solvers,” *Int. J. Turbomach. Propuls. Power*, vol. 4, no. 2, pp. 15, Jun. 2019.
8. Y. C. Fung, An Introduction to the Theory of Aeroelasticity, Dover Publications, 1993.
9. J. W. Edwards, M. P. Conner, and T. W. Strganac, “Aeroelastic dynamic response and control of an airfoil section,” Journal of Aircraft, vol. 16, no. 3, pp. 65–74, 1979.
10. L. Bai, Z. Wu, and C. Yang, “Nonlinear Aeroelastic Characteristics of a Supersonic All-Movable Fin with Single or Multiple Freeplay Nonlinearities,” Aerospace, vol. 10, no. 1, pp. 1–20, 2023.
11. T. Theodorsen, “General theory of aerodynamic instability and the mechanism of flutter,” *NACA Report 496*, 1935.
12. G. Kösterit and E. Ciğeroğlu, “Determination of flutter speed of 2D nonlinear wing by using describing function method and state-space formulation,” *presented at the 40th Conference and Exposition on Structural Dynamics (IMAC)*, Florida, United States, 2023.
13. A. Iannelli, A. Marcos, and M. Lowenberg, “Nonlinear flutter analysis with uncertainties based on Describing Function and Structured Singular Value with an IQC validation,” in *Proc. 7th European Conf. Aeronautics and Aerospace Sciences (EUCASS)*, pp. 1–12, Milan, Italy, 2017.

Chapter 16
The Nonlinear Restoring Force Method: An Overview with Numerical and Experimental Applications

Cristiano Martinelli and Andrea Cammarano

Abstract The stringent CO2 policies and the competitive market are increasing the demand for lightweight high-performance structures in many engineering sectors. Under normal operations, these structures may experience large deformations which may lead the dynamics of the systems into the nonlinear regime. Consequently, methods and approaches for the analysis and the identification of accurate mathematical models, capable of representing the dynamics of such systems, have become of central importance in the current scientific research. This paper proposes an overview of a novel approach for the identification of strongly nonlinear systems, named the Nonlinear Restoring Force (NLRF) method. Based on the classical Restoring Force Surface (RFS) method, this approach separates the nonlinear contribution of the restoring force from the linear one exploiting well-established linear identification methods currently used in industrial practices. This enables the visualisation of the pure nonlinear part of the restoring force surface, facilitating the identification of an appropriate mathematical function capable of capturing the nonlinear behaviour of the system. Firstly the mathematical background of the theory and numerical examples are presented and then, to show the capabilities of the method, nonlinear Reduced Order Models are identified from experimental data. To this end, the dynamics of two experimental test rigs, characterised by geometric (smooth) and localised (non-smooth) nonlinearities, are presented and analysed. Finally, all the identified models are validated against experimental data, using different excitation levels to test their prediction accuracy.

Keywords Nonlinear System Identification· Nonlinear Dynamics· Model Updating· Experimental Analysis· FEA

Introduction

Lightweight high-performance structures are becoming popular in many engineering sectors. The advantages of these structures consist in using less material and obtaining a lightweight structure. Historically the space sector has been interested in reduce the weight of systems, as the cost of space mission is directly proportional to the weight, a good cost predictor [1]. Nonetheless, in recent years, other engineering sectors such as the civil, power, and transport industries [2] have also developed an interest in lightweight structures. This interest is driven by the stringent CO2 policies [3, 4] which force industries to reduce their emissions. Despite these advantages, lightweight structures are characterised by slender sections which can induce the nonlinear dynamic responses when large deformations are achieved.

Nowadays, the identification of linear models via model updating procedures [5] and modal experimental analysis of linear structures [6] are considered well-established practices by the scientific and industrial engineering community. On the contrary, a solid and robust procedure for the identification of nonlinear mechanical systems has not been developed yet, but rather several methods have been proposed. Kerschen, Noël and collaborators [7, 8] provided a systematic review of the different methods for the identification of nonlinear systems: between the different methods, the Restoring Force Surface (RFS) method [9] is considered one of the most well-established identification methods for nonlinear systems. The RFS method has been successfully used for the identification of several nonlinear systems from experimental data: for example, Bonisoli and Vigliani [10] adopted the RFS method to identify the nonlinear parameters of a passive magneto-elastic suspension,

Cristiano Martinelli · Andrea Cammarano
Department of Aeronautical and Astronautical Engineering, University of Southampton, U.K.
e-mail: c.martinelli@soton.ac.uk

Andrea Cammarano
James Watt School of Engineering, University of Glasgow, U.K.
e-mail: A.Cammarano@soton.ac.uk

Ludovic Renson et al. (eds.), *Nonlinear Structures & Systems, Vol. 1*, Conference Proceedings of the Society for Experimental Mechanics Series, https://doi.org/10.13052/9788743801467-16

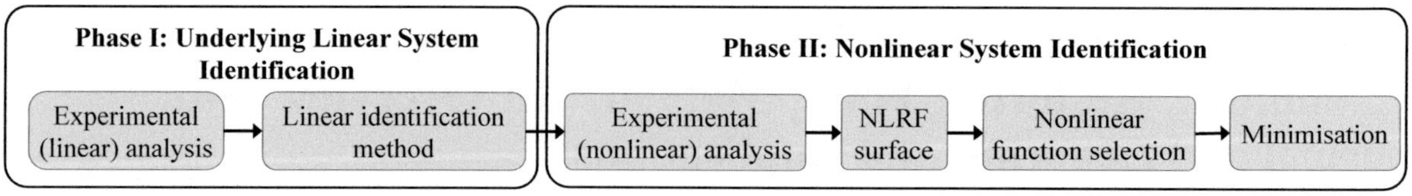

Fig. 1 Nonlinear Restoring Force (NLRF) method: outline of the procedure for the identification of nonlinear models.

Cammarano et al. [11] utilised the RFS method to obtain the nonlinear properties associated to a bistable mechanical system for vibration energy harvesting, and Kerschen et al. [12, 13] adopted the same method to identify a trilinear characteristics of a small real satellite and the nonlinear characteristics between the aircraft wings and fuel tanks.

Although efficient, the classical RFS method allows for the graphical representation of the restoring surface only when single Degree of Freedom (DOFs) systems are representative of the system dynamics and becomes computationally expensive when MDOF structures are considered [14]. This results in a difficult implementation of the method in the current industrial practices which rely on definition of MDOF linear models. To solve these problems, a modified version of the RFS method has been recently proposed by Martinelli et al. [15, 16], named the Nonlinear Restoring Force (NLRF) method. The method is build on existing linear identification methods and aims to identify the linear and nonlinear behaviour of the mechanical structure in a separate way. This allows the method to be easily implemented in the current industrial practices, facilitating the identification of nonlinear mechanical systems and structures. This paper proposes an overview of NLRF method: firstly the methodology is presented, then a numerical example is introduced to show how the method can be applied in practice. The method is then utilised to identify the nonlinear Reduced-Order Models (ROMs) of two experimental test rigs characterised by smooth and non-smooth nonlinearities. Finally, the identified nonlinear models are validated against experimental data that were not used during the identification phase.

Methodology, Background Theory, and Limitations

The methodology of the Nonlinear Restoring Force Surface Method is schematically described by Fig. 1. The procedure is divided into two different phases: in the first phase the underlying linear behaviour the system is identified while in the second phase aims to identify the nonlinear characteristics of the system. The linear behaviour of the system is identified using classical identification methods, such as Half Power method, Circle Fitting Method, or stabilisation diagram. During this phase, experimental linear analysis is performed by applying low amplitude random excitation to the structure. This excitation reduces the effect of smooth nonlinearities, e.g. hardening or softening stiffness, and allows estimating the underlying linear behaviour of the system [6, 17]. At this stage, a linear model representing the underlying linear behaviour of the system is identified using one of the aforementioned techniques and by selecting an appropriate number DOFs. In the second phase, the nonlinear dynamic behaviour of the system is identified. Firstly a nonlinear experimental analysis is performed; this often involves the usage of sinusoidal excitation at specific frequencies or forward/backward frequency sweeps to excite the nonlinear response of the structure. Then, the knowledge of underlying linear system and the experimental nonlinear data are utilised to obtain the nonlinear restoring force using the following expression:

$$\mathbf{NRF} = \mathbf{N}(\dot{\mathbf{x}}, \mathbf{x}) = \mathbf{F} - \mathbf{M}\ddot{\mathbf{x}} - \mathbf{C}\dot{\mathbf{x}} - \mathbf{K}\mathbf{x} \tag{1}$$

where $\mathbf{x} \in \mathbb{R}^{n\times 1}$ represents the displacement vector, $\mathbf{M}$, $\mathbf{C}$, and $\mathbf{K} \in \mathbb{R}^{n\times n}$, denote the linear mass, damping, and stiffness matrices, and $\mathbf{N}(\dot{\mathbf{x}}, \mathbf{x})$ and $\mathbf{F} \in \mathbb{R}^{n\times 1}$ are the nonlinear contribution and the forcing function. Eq. 1 results in a vector of nonlinear restoring forces for each DOF of the selected ROM. These restoring forces can be plotted in a sub-space of coordinate to obtain additional information about the nonlinear behaviour of the system. Specifically the visualisation of the nonlinear restoring force surface allows to:

- Understand if the selected sub-space of coordinates is sufficient to fully describe the nonlinear dynamics of the system [15].
- Identify and characterise localised nonlinearities [16].

These two features provide additional insights about the dynamics of the system and helps to define a suitable nonlinear function, e.g. a polynomial, for the identification of the nonlinear characteristics. An overview of the features is reported in Section Experimental Validation: Identification of Smooth and Non-Smooth Nonlinearities where the nonlinear dynamic behaviour of two experimental test rigs is identified using the proposed method: in the identification of the first test rig (characterised by only smooth nonlinearities), the NLRF method is utilised to understand if the nonlinear characteristics of each DOF of the structure can be fully described using only a sub-set of local coordinates, e.g. to understand if the nonlinear characteristic associated to the first DOF of the structure can be represented using a function of the type $N(x_1, \dot{x}_1)$. The second test rig is instead used to show how the proposed method can be used to identify localised nonlinearities.

One of the limitation of the proposed approach is the need of identifying an underlying linear system: to this end the considered mechanical system must behaves linearly at low amplitude of excitation. Without this feature, the method cannot be applied as the first phase of the procedure would be invalidated. Another limitation of the approach comes from the linear identification methods [6] and consists in the selection of an appropriate linear ROM: by selecting a ROM which does not capture the full dynamics of the system, the identification of the linear behaviour is inaccurate, and this affect the capability of the method in identifying the sought nonlinear characteristics.

Numerical Example: A Three Storey Building with a Localised Nonlinearity

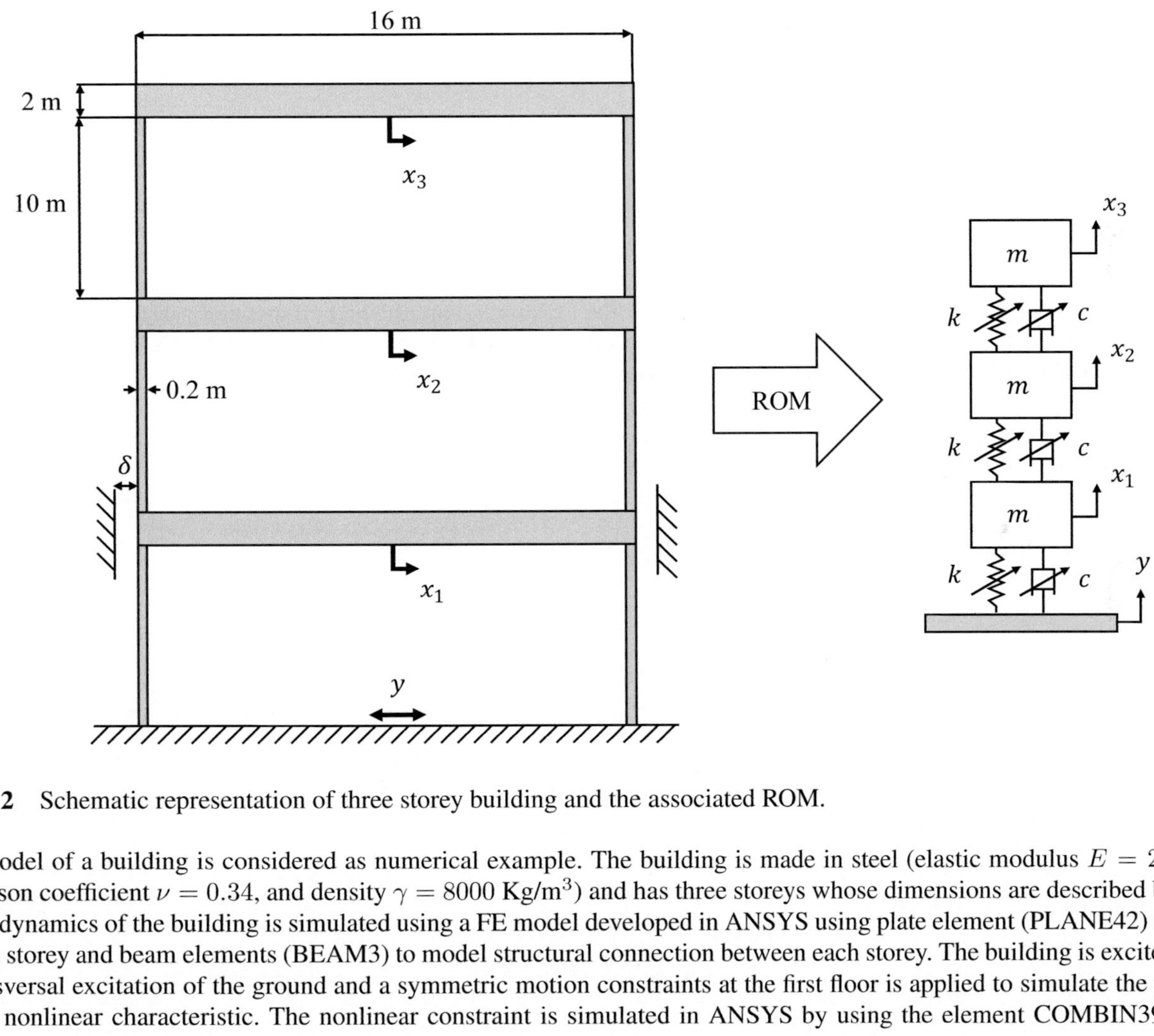

Fig. 2 Schematic representation of three storey building and the associated ROM.

A model of a building is considered as numerical example. The building is made in steel (elastic modulus $E = 210$ GPa, Poisson coefficient $\nu = 0.34$, and density $\gamma = 8000$ Kg/m^3) and has three storeys whose dimensions are described by Fig 2. The dynamics of the building is simulated using a FE model developed in ANSYS using plate element (PLANE42) to model each storey and beam elements (BEAM3) to model structural connection between each storey. The building is excited by the transversal excitation of the ground and a symmetric motion constraints at the first floor is applied to simulate the presence of a nonlinear characteristic. The nonlinear constraint is simulated in ANSYS by using the element COMBIN39 and by imposing a piecewise characteristic of the form:

$$F_p = \begin{cases} k_p(x_1 - \delta) & \text{if } x_1 > \delta \\ 0 & \text{if } -\delta < x_1 < \delta \\ k_p(x_1 + \delta) & \text{if } x_1 < -\delta \end{cases} \tag{2}$$

where δ denotes the semi non-contact gap and k_p the piecewise stiffness. Fig. 2 shows the ROM representing the three storey building: each floor shares the same dimensions and material therefore the stiffness (k), the damping (c), and the mass (m)

of each DOF of the ROM are considered to be the same. The final ROM results in the following equation of motion:

$$m\ddot{x}_1 + c(\dot{x}_1 - \dot{y}) + k(x_1 - y) - c(\dot{x}_2 - \dot{x}_1) - k(x_2 - x_1) + N_1(x, y) = 0 \quad (3a)$$

$$m\ddot{x}_2 + c(\dot{x}_2 - \dot{x}_1) + k(x_2 - x_1) - c(\dot{x}_3 - \dot{x}_2) - k(x_3 - x_2) + N_2(x, y) = 0 \quad (3b)$$

$$m\ddot{x}_3 + c(\dot{x}_3 - \dot{x}_2) + k(x_3 - x_2) + N_3(x, y) = 0 \quad (3c)$$

where $N_k(x, y)$ indicate a general nonlinear terms applied to the k-th DOF.

Linear and nonlinear FE simulations are performed to identify the underlying linear and nonlinear behaviour of the three storey building in time and frequency domain. The modal analysis shows that the first three natural frequencies are found at the following frequencies: 0.234 Hz, 0.657 Hz, 0.951 Hz. The cumulative mass fraction of these three modes displace more than 99% of the mass of the system (considering the first 20 modes). This analysis confirms that a 3 DOFs reduced-model is sufficient to describe the dynamics of the of the building. The Frequency Response Functions (FRFs) of the system are obtain via harmonic analysis: a based displacement Y of 0.01 m is imposed to the base of the model and the response of the system is measured at nodes located in the middle of each floor. The mass m of each floor is estimated by knowing the volume and the density of the material [1]. The stiffness k, instead, is obtained from the value of the natural frequencies which can be estimated via classical experimental modal analysis [6]. The value of the stiffness k is the obtained by solving $\omega_n = f(k)$ where $f(k)$ is obtained by solving the eigenvalues problem associated to the ROM. For the considered problem, the value of the natural frequencies are found:

$$\omega_{n,1} = 0.445\sqrt{k/m} \quad (4a)$$

$$\omega_{n,2} = 1.247\sqrt{k/m} \quad (4b)$$

$$\omega_{n,3} = 1.802\sqrt{k/m} \quad (4c)$$

Using Eq. 4 it is possible to obtain a value of the stiffness $k = 2.925e+06$ N/m by averaging the obtained results. Assuming a Rayleight proportional damping ($\alpha = 0$ and $\beta = 0.01$) for the FE simulations, the FRF of the model can be computed and used to identity a suitable damping factor c via iterative procedure. A final value of $c = 2.925e + 04$ Ns/m is identified. At this point, it is possible to compare the FRF of the FE model and the identified ROM: the comparison is reported in Fig. 3 and shows that the reduced-model is able to reproduce the dynamic behaviour of the full model for the first three modes. Using the knowledge of the identified underlying linear model, it is possible to compute the nonlinear restoring forces for each DOF as follows:

$$NLF_1 = -m\ddot{x}_1 - c(\dot{x}_1 - \dot{y}) - k(x_1 - y) + c(\dot{x}_2 - \dot{x}_1) + k(x_2 - x_1) \quad (5a)$$

$$NLF_2 = -m\ddot{x}_2 - c(\dot{x}_2 - \dot{x}_1) - k(x_2 - x_1) + c(\dot{x}_3 - \dot{x}_2) + k(x_3 - x_2) \quad (5b)$$

$$NLF_3 = -m\ddot{x}_3 - c(\dot{x}_3 - \dot{x}_2) - k(x_3 - x_2) \quad (5c)$$

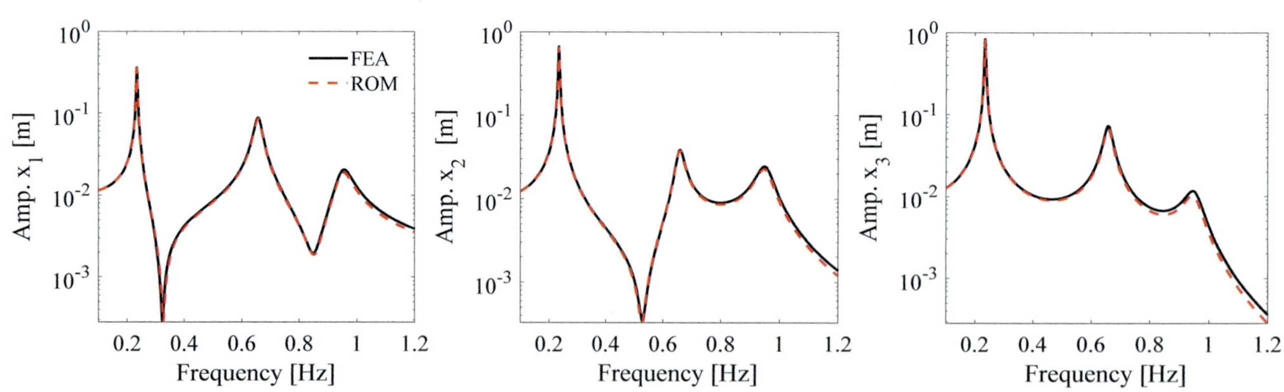

Fig. 3 Comparison between linear Frequency Response Functions (FRF) of the FE model and the selected ROM.

The nonlinear restoring forces are plotted to identify the nonlinear characteristics of the system. To this end, nonlinear transient analyses are performed in ANSYS. For the sake of brevity, only the nonlinear restoring force associated with the

[1] the mass of the beams is considered for 1/3

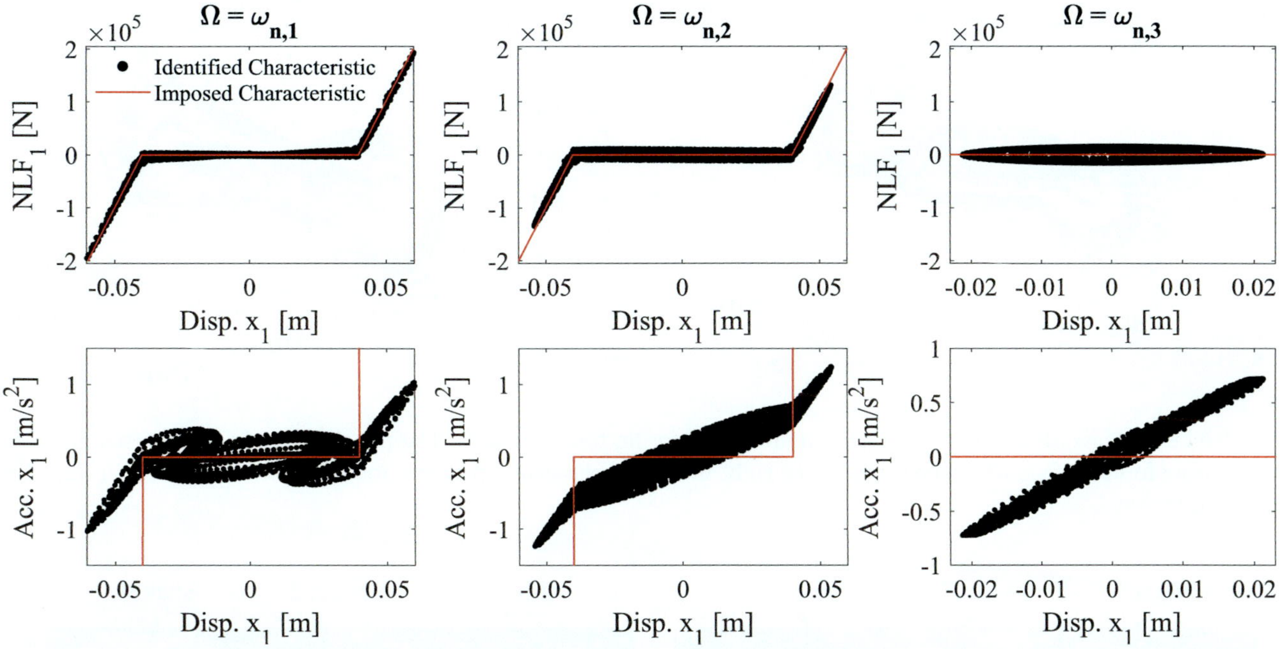

Fig. 4 Lateral view of the identified linear/nonlinear restoring force surfaces (black dots) and associated nonlinear characteristics imposed in ANSYS (red line). The panels in the top row show the nonlinear restoring force surfaces of the first DOF as defined in Eq. 5a, while the bottom panels shows the results obtained with the classical restoring force surface method. From left to right, the panels in the first, second, and third column shows the results obtained using the transient response of the system when, respectively, the first, second, and third natural frequency are used as excitation source.

first DOF of the structure are plotted, for three different excitation frequencies, namely the three natural frequencies $\omega_{n,1}$, $\omega_{n,2}$, and $\omega_{n,3}$. The nonlinear analysis shows that, when the excitation frequency Ω is equal to the first ($\omega_{n,1}$) or the second ($\omega_{n,2}$) natural frequency of the system, contact between the first storey and the limit motion constraint occurs; instead, when the frequency of excitation is increased and matches $\omega_{n,3}$ no contact occurs in the system. At this stage it is possible to plot the nonlinear restoring forces in a subspace of coordinate which is chosen to be x_1 and $\dot{x}_1$. The results are reported in Fig. 4 where the later view of surface (black dots) is shown and compared against the characteristic to be identified (red line). In order to make a comparison with the classical restoring force surface method the acceleration of the mass, typically used in base excited structures (see for example [11]), is plotted in the second row of Fig. 4. The results demonstrates that at the first two excitation levels (i.e. $\Omega = \omega_{n,1}$ and $\Omega = \omega_{n,2}$), the NLRF surface is able to reconstruct the nonlinear piecewise characteristic applied to the system. When the excitation Ω is equalt to $\omega_{n,3}$ contact does not occur and thus the NLRF surface shows a flat line close to zero. On the contrary, the classical approach perceives the present of non-smooth characteristic but it does not allow the accurate identification of the nonlinearity. In addition, the presence of the linear contributions associated to all the DOFs of the system, leads to the generation of an hypersurface rather than simple 3D surface. This concept is clearly highlighted by Fig. 5 where the surfaces associated to $\Omega = \omega_{n,1}$ are plotted in isometric view.

Experimental Validation: Identification of Smooth and Non-Smooth Nonlinearities

The proposed method is utilised to identify the nonlinear characteristics of two similar test-rigs: the first test rig is characterised by smooth geometric nonlinearities while the second one presents an additional localised non-smooth characteristic. The two test rigs are shown in Fig. 6: the first test rig (left panel of Fig. 6) is mounted on a vibrating table whose amplitude is controlled by a feedback sensor (optical laser). The second test rig (right panel of Fig. 6), instead, is excited by the motion of single support whose motion is controlled by the voltage that is imposed on the shaker. Firstly, the linear dynamic behaviour of the two experimental model need to be identified. In this paper, the NLRF method is utilised to identify the full nonlinear behaviour of the first test rig and the the localised nonlinearity of the second test rig. This affect the definition of the ROMs that are used to identify the sought nonlinear characteristics.

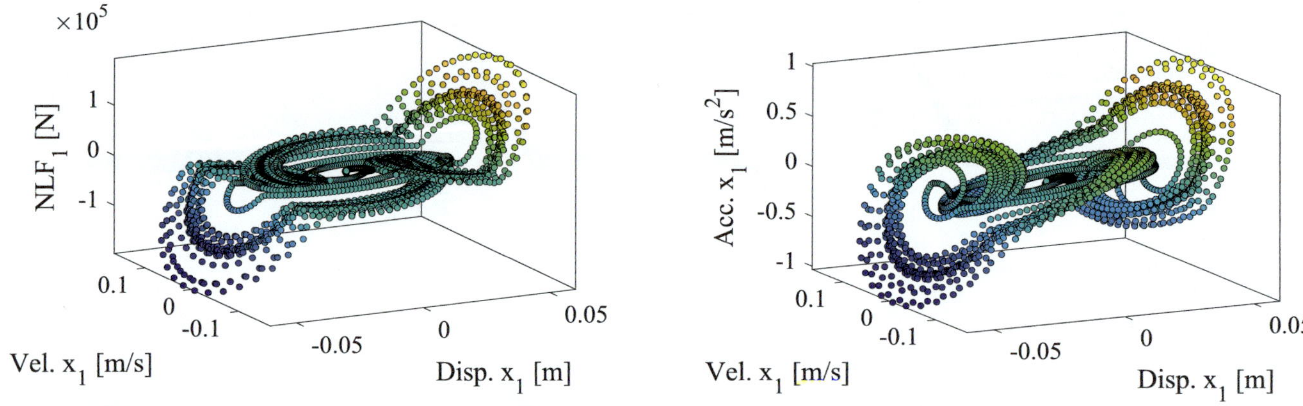

Fig. 5 Isometric view of the identified linear/nonlinear restoring force surfaces. The panel on the left represents the nonlinear restoring force surface while the panel on the right shows the classical representation of the restoring force surface for base-excited systems.

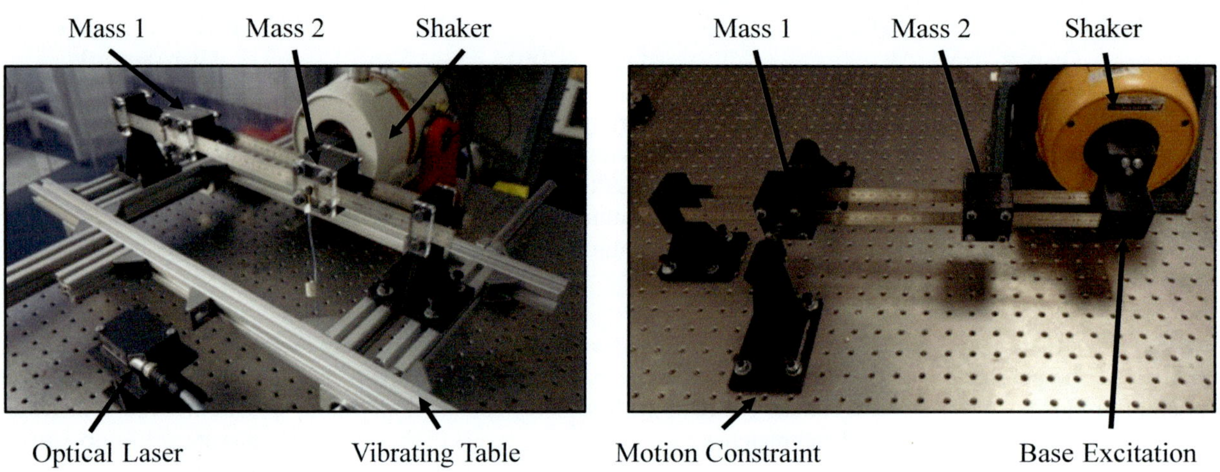

Fig. 6 Experimental test rigs: two-mass system characterised by smooth (left panel) and non-smooth (right panel) nonlinearities. The first experimental model is excited by a vibrating table while the second one is excited by the motion of single support.

The linear behaviour of the first test rig is identified using a two-DOF model which is described by the following set of ODEs:

$$m\ddot{z}_1 + c_1\dot{z}_1 + c_2(\dot{z}_1 - \dot{z}_2) + k_1 z_1 + k_2(z_1 - z_2) + NLF_1(z_1, z_2, \dot{z}_1, \dot{z}_2) = -m\ddot{y} \tag{6a}$$

$$m\ddot{z}_2 + c_3\dot{z}_2 - c_2(\dot{z}_1 - \dot{z}_2) + k_3 z_2 - k_2(z_1 - z_2) + NLF_2(z_1, z_2, \dot{z}_1, \dot{z}_2) = -m\ddot{y} \tag{6b}$$

where y represents the base displacement, $z = x - y$ indicates the relative displacement between the base and the masses, and NLF_1 and NLF_2 are the nonlinear restoring terms associated with the first and second DOF of the model. The linear coefficients are identified via circle fitting method and are reported in Tab. 1. The interested reader can find additional details about the identification of the underlying linear system of the first test rig at the following reference [15].

The linear and nonlinear smooth behaviour of the second test rig is described by the following ROM:

$$m\ddot{x}_1 + c_1\dot{x}_1 + k_1 x_1 + c_2(\dot{x}_1 - \dot{x}_2) + k_2(x_1 - x_2) + \mu_1 x_1^3 + \mu_2 x_1^3 - \mu_{2,b} x_1^2 x_2 + \mu_{2,b} x_1 x_2^2 - \mu_2 x_2^3 + NLF_1 = 0 \tag{7a}$$

$$m\ddot{x}_2 + c_3(\dot{x}_2 - \dot{y}) + k_3(x_2 - y) - c_2(\dot{x}_1 - \dot{x}_2) - k_2(x_1 - x_2) + \mu_3(x_2 - y)^3 - \mu_2 x_1^3 + \mu_{2,b} x_1^2 x_2 - \mu_{2,b} x_1 x_2^2 + \mu_2 x_2^3 = 0 \tag{7b}$$

where x represents the absolute displacement of the two masses, y indicates the base displacement, k_1, k_2, and k_3 are the linear stiffness coefficients, c_1, c_2, and c_3 denote the linear damping coefficients, and μ_1, μ_2, $\mu_{2,b}$, and μ_3 are the nonlinear stiffness coefficients. This ROM has been identified and validated in [18] and this information are utilised to identify localised nonlinearities in the second test rig, using the proposed method. The linear parameters of the system are once again identified via circle fitting method and are reported in Tab. 1. The remaining smooth nonlinear parameters have been identified via meta-heuristic optimisation and, for completeness, are reported in Tab. 2. Details of the identification and validation procedures can be found at the following references [18, 16].

At this stage it is possible to proceed with the identification of the nonlinear characteristics of both rest rigs. Firstly, the NLRF surfaces must be plotted in a reduced sub-space of coordinates: the results are reported in Fig. 7 where the NLRF surfaces of the first DOF for the first and second test rig are shown. It is worth mentioning, that the two test rigs are excited via sinusoidal frequency sweeps in two different ways: the vibrating table of the first test rig is controlled in amplitude via closed-loop control. This allows imposing a certain amplitude of the base excitation. On the contrary, the second test rig does not present a closed-loop control, and frequency sweeps across the first two linear resonances are performed by imposing the amplitude of the sinusoidal voltage signal the is passed to the shaker. Further details about the experimental analysis of the two test rigs can be found in [19, 15].

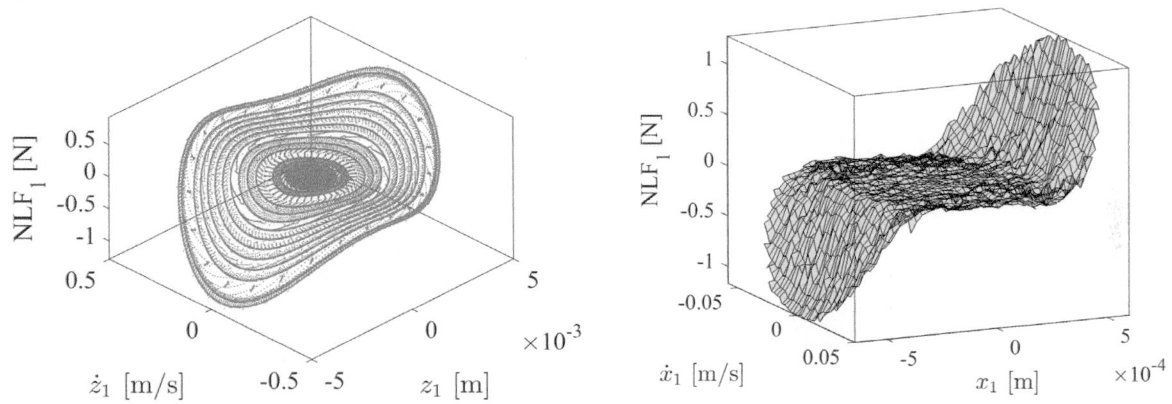

Fig. 7 Experimental NLRF surfaces of the first test rig (left panel) and second test rig (right panel). The surfaces are obtained by imposing, respectively, a base amplitude of 0.05 mm and voltage input of 0.3 V. The coordinates of the first DOF, i.e. $z_1 - \dot{z}_1$ and $x_1 - \dot{x}_1$, are used to plot the NLRF surfaces associated with the first DOF of the selected ROM.

In both cases, the coordinates of the local DOF, e.g. x_1 and $\dot{x}_1$ for the first DOF, are used as reduced sub-space to plot the surfaces. The NLRF surface of the first test rig considers all the nonlinear dynamics of the system and it can be represented as a 3D function in the considered sub-space of coordinates. Similarly the NLRF surface of the second DOF of the first test rig (here not represented) shows the same property. This result is suggesting that the selected coordinates, i.e. $z_1 - \dot{z}_1$ for the first DOF and $z_2 - \dot{z}_2$ for the sscond DOF, are sufficient to describe the nonlinear characteristics of the first test rigs. This information is then utilised to mathematically model the sought nonlinear characteristic which results in the following expression:

$$\begin{cases} NLF_1 = a_1 z_1^2 + a_2 z_1 \dot{z}_1 + a_3 \dot{z}_1^2 + a_4 z_1^3 + a_5 \dot{z}_1^3 + a_6 z_1 \dot{z}_1^2 + a_7 z_1^2 \dot{z}_1; \\ NLF_2 = b_1 z_2^2 + b_2 z_2 \dot{z}_2 + b_3 \dot{z}_2^2 + b_4 z_2^3 + b_5 \dot{z}_2^5; \end{cases} \tag{8}$$

Eq. 8 represents the form of the nonlinear characteristic of the first test rigs and its polynomial degree can be obtain thorugh the iterative procedure described in [15].

Table 1 Linear coefficients identified from the experimental linear transfer functions in Ref. [16, 15]. The first line indicates linear coefficients of the first test rig while the second line shown the linear coefficients identified from the experimental data of the second test rig.

Coefficients	m [kg]	c_1 [Ns/m]	c_2 [Ns/m]	c_3 [Ns/m]	k_1 [N/m]	k_2 [N/m]	k_3 [N/m]
Test rig 1	0.1130	0.0582	0.0205	0.0600	1232.9	225.1	1261.9
Test rig 2	0.1365	0.0454	0.0075	0.0366	714.9	89.6	624.9

Table 2 Nonlinear smooth parameters identified in [18]. The parameters are identified via meta-heuristic optimisation.

μ_1 [N/m^3]	μ_2 [N/m^3]	$\mu_{2,b}$ [N/m^3]	μ_3 [N/m^3]
2.732e+06	1.324e+06	8.850e+06	7.069e+06

Similarly, from inspection of the NLRF surface of the second test rig, a nonlinear function representing the type of nonlinearity is identified of the second experimental model. In this case, a piecewise function is considered to be a valid representation of the nonlinearity. The sought nonlinear characteristic is mathematically represented by the following expression:

$$NLF = \begin{cases} k_p(x_1 - a_2) & \text{if } x_1 > a_2 \\ 0 & \text{if } -a_1 < x_1 < a_2 \\ k_p(x_1 + a_1) & \text{if } x_1 < -a_1 \end{cases} \tag{9}$$

where k_p is the piecewise stiffness and, a_1 and a_2 are the lower and upper limits of the semi free-play gap. At this stage, it is possible to identify the unknown coefficients by imposing and solving a minimisation problem. The results are reported in Tab. 3 for the first test rig and in Tab. 4 for the second test rig.

Table 3 Experimental nonlinear coefficients, identified from the nonlinear restoring force surfaces in [15].

Monomial (M_1)	Coefficient	Value	Monomial (M_2)	Coefficient	Value
z_1^2	a_1	-8.4608×10^3 N/m^2	z_2^2	b_1	-8.3989×10^3 N/m^2
$z_1\dot{z}_1$	a_2	5.3189 Ns/m^2	$z_2\dot{z}_2$	b_2	-4.8546 Ns/m^2
$\dot{z}_1^2$	a_3	0.9055 Ns^2/m^2	$\dot{z}_2^2$	b_3	0.9047 Ns^2/m^2
z_1^3	a_4	1.1237×10^7 N/m^3	z_2^3	b_4	1.1231×10^7 N/m^3
$\dot{z}_1^3$	a_5	1.9777 Ns^3/m^3	$\dot{z}_2^5$	b_5	-8.0214 Ns^5/m^5
$z_1\dot{z}_1^2$	a_6	-1.3876×10^2 Ns^2/m^3	–	–	–
$z_1^2\dot{z}_1$	a_7	1.0495×10^4 Ns/m^3	–	–	–

Table 4 Parameters of the piecewise function identified in [16].

a_1 [m]	a_2 [m]	k_p [N/m]
4.091×10^{-4}	3.995×10^{-4}	5.472×10^3

Once the nonlinear characteristics are identified, the dynamics of the numerical models and experimental test rigs are compared. The comparison is carried out using dynamic responses of the experimental test rigs that were not used during the identification procedure: specifically, the NLRF surfaces of the first and second test rigs were obtained by imposing an amplitude of excitation of, respectively, 0.05 mm and 0.3 V. The dynamics of the first test rig is investigated also at amplitude of the vibration table equal to 0.04 mm and 0.03 mm; these additional data are utilised to carry out the comparison between the numerical and experimental FRCs in the case of the first test rig. The first numerical model is solved using the toolbox COCO [20] via numerical continuation and the results are reported in Fig. 8 where the FRCs are shown in terms of amplitude of the first DOF. The figure shows that the numerical model is able to capture the dynamics of the experimental test rig in terms of amplitude of response and jumps from high to low amplitude solution, even when the experimental and numerical models are compared in excitation conditions that are different from the one utilised during the identification phase. The results of Fig. 8 are satisfactory in terms of prediction capabilities of the model therefore the identified numerical model is considered to be validated.

A similar validation procedures is repeated for the numerical model representing the second test rig: in this case, time-histories and phase-portraits at different frequencies of excitation are used to carry the comparison. This choice is motivated by the very rich and complicate dynamic response of this class system which include grazing bifurcations, multi-periodic isolas, quasi-periodic responses, and chaos. The reader interested in the numerical and experimental dynamics of this class of system is invited to read the following references [19, 21]. The NLRF surface of the second test rig is obtained using data associated to an amplitude of excitation of 0.3 V. The test rigs is also tested using a voltage input of 0.35 V, thus the dynamics responses associated to this condition are utilised to carry out the validation process. Specifically the following conditions are considered: a single period attractor, identified at 11.9 Hz, and a multi-periodic attractor, found when the excitation frequency is equal to 12.1 Hz. The results of the comparison are shown in Fig. 9 where it is demonstrated that the identified nonlinear model is able to capture the dynamics of the experimental test rig from a quantitative and qualitative

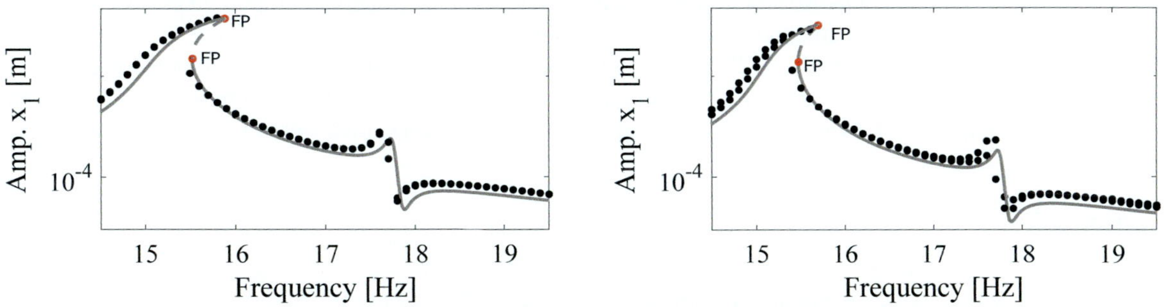

Fig. 8 Comparison between experimental (dots) and numerical FRC (continuous line) in terms of amplitude of the first DOF. The panel on the left shows the comparison between FRCs when an amplitude of excitation equal 0.04 mm is utilised. The panel on the right shows the same comparison when an amplitude of 0.03 mm is imposed to the base.

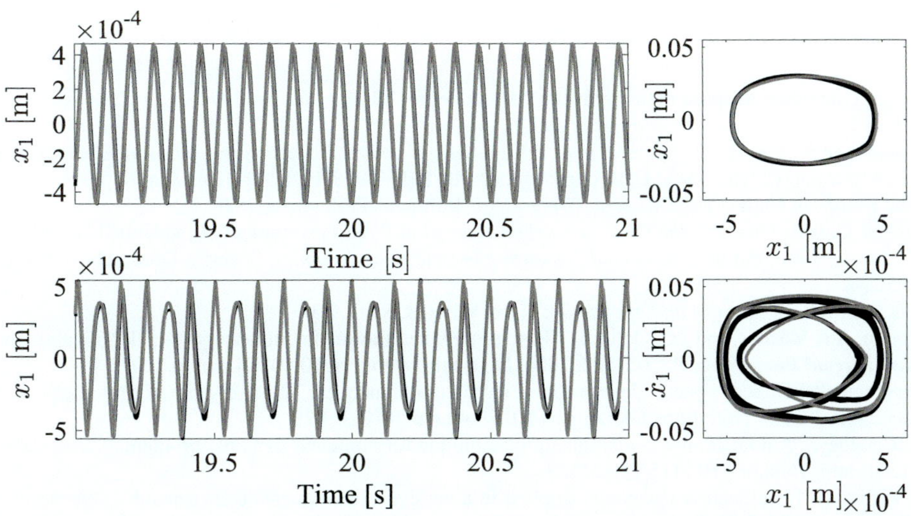

Fig. 9 Comparison between experimental (black continuous line) and numerical results (gray continuous line). The panels on the left show the time-histories while the panels on the right show the orbits. Different attractors are considered, namely, a single period attractor at 0.35 V (panels in the top row) and multiperiodic attractor at 0.35 V (panels in the bottom row).

point of view, in term of time histories and type of attractor. These results confirms the prediction capability of the identified model which is considered to be validated.

Conclusions

This paper presented an overview of novel identification method named the Nonlinear Restoring Force (NLRF) method. The method builds on the separation of linear and nonlinear components of the restoring force, allowing to be interfaced with the current model identification practices utilised in industry. By obtaining a graphical representation of the pure nonlinear response in a reduced sub-space of coordinates, the user can obtain additional insights on the nonlinear dynamics of the system. This enables a quick and easy identification of localised nonlinearities and helps to understand if the selected set of coordinates can fully describe the nonlinear restoring force of the system.

Firstly, the mathematical background, the methodology, and the limitations of the proposed method are presented. Then a numerical example of a nonlinear three storeys building is proposed. To this end, a numerical FE model of the building is created and Ansys and it is utilised to obtain the dynamic response of the structure. Then the NLRF method is utilised to reconstruct the nonlinear restoring force surfaces and identify the nonlinearity. The NFRM method is then used to identify the nonlinear characteristics of two experimental test rigs characterised by smooth and non-smooth nonlinearities. Once again, the NLRF surfaces are obtained using the proposed methodology, identifying the associated nonlinear characteristics. The identified models are then validated against experimental data; to this end, experimental dynamic responses at different amplitudes of excitation are used. Specifically, the validation process is done using data associated to excitation conditions that were not used during the identification process. This guarantees an higher degree of reliability of the identified model, especially outside the ranges of experimental analysis. The comparison shows that the identified numerical models are able capture the dynamics of the experimental test rigs under several excitation conditions, thus validating and concluding the study.

Data Availability

The experimental data utilised in this paper are available at the following DOI [22, 23].

Acknowledgments The authors would like to acknowledge the following NERC and EPSRC grants: GALLANT, Glasgow as a Living Lab Accelerating Novel Transformation (No. NE/W005042/1), RELIANT, Risk EvaLuatIon fAst iNtelligent Tool for COVID19 (No. EP/V036777/1).

References

1. F.A. Prince. "Weight and the future of space flight hardware cost modeling". In *International Society of Parametric Analysis/Society of Cost Estimating and Analysis 2003 International Conference*, 2003.
2. R.N. Yancey. *Challenges, opportunities, and perspectives on lightweight composite structures: aerospace versus automotive.* Lightweight composite structures in transport (2016): 35-52.DOI: https://doi.org/10.1016/B978-1-78242-325-6.00002-5
3. Official Journal of the European Union. *The European Green Deal.* European Parliament, 2020.
4. Department of Energy & Climate Change. *The UK Low Carbon Transition Plan.* Government of the United Kingdom, 2009.
5. M. Friswell, and J.E. Mottershead, *Finite element model updating in structural dynamics.* Springer Dordrecht, 1995. DOI: https://doi.org/10.1007/978-94-015-8508-8
6. D.J. Ewins. *Modal testing: theory, practice and application.* John Wiley & Sons, 2009.
7. G. Kerschen, K. Worden, A.F. Vakakis, and J.C. Golinval. "Past, present and future of nonlinear system identification in structural dynamics". *Mechanical System and Signal Processing*, 20:505–592, 2006. DOI: https://doi.org/10.1016/j.ymssp.2005.04.008
8. J.P. Noël and G. Kerschen. "Nonlinear system identification in structural dynamics: 10 more years of progress". *Mechanical Systems and Signal Processing*, 83:2–35, 1 2017. DOI: https://doi.org/10.1016/j.ymssp.2016.07.020
9. S.F. Masri and T.K. Caughey. "A nonparametric identification technique for nonlinear dynamic problems". *Journal of Applied Mechanics*, 46:433–447, 1979. DOI: https://doi.org/10.1115/1.3424568
10. E. Bonisoli and A. Vigliani. "Identification techniques applied to a passive elasto-magnetic suspension". *Mechanical Systems and Signal Processing*, 21:1479–1488, 4 2007. DOI: https://doi.org/10.1016/j.ymssp.2006.05.009
11. A. Cammarano, S. G. Burrow, and D. A.W. Barton. "Modelling and experimental characterization of an energy harvester with bi-stable compliance characteristics". *Proceedings of the Institution of Mechanical Engineers. Part I: Journal of Systems and Control Engineering*, 225:475–484, 6 2011. DOI: https://doi.org/10.1177/0959651811403093
12. G. Kerschen, M. Peeters, J.C. Golinval, and C. Stéphan. "Nonlinear modal analysis of a full-scale aircraft". *Journal of Aircraft*, 50:1409–1419, 2013. DOI: https://doi.org/10.2514/1.C031918
13. J.P. Noël, L. Renson, and G. Kerschen. "Complex dynamics of a nonlinear aerospace structure: Experimental identification and modal interactions". *Journal of Sound and Vibration*, 333:2588–2607, 2014. DOI: https://doi.org/10.1016/j.jsv.2014.01.024
14. S.F. Masri, F. Tasbihgoo, J.P. Caffrey, A.W. Smyth, and A. G. Chassiakos. "Data-based model-free representation of complex hysteretic mdof systems". *Structural Control and Health Monitoring*, 13:365–387, 1 2006. DOI: https://doi.org/10.1002/stc.147
15. C. Martinelli, A. Coraddu, and A. Cammarano. "Strongly nonlinear multi-degree of freedom systems: Experimental analysis and model identification". *Mechanical Systems and Signal Processing*, 218:111532, 2024. DOI: https://doi.org/10.1016/j.ymssp.2024.111532
16. C. Martinelli, A. Coraddu, and A. Cammarano. "Identifying localised nonlinearities: Nonlinear restoring force surface in piecewise multi-degree-of-freedom systems". In *Society for Experimental Mechanics Annual Conference and Exposition.* pages 19-26. Springer, 2024, DOI: https://doi.org/10.1007/978-3-031-69409-7_4
17. K. Worden, and G.R. Tomlinson. *Nonlinearity in structural dynamics: detection, identification and modelling.* CRC Press, 2001.
18. C. Martinelli, A. Coraddu, and A. Cammarano. "Experimental parameter identification of nonlinear mechanical systems via meta-heuristic optimisation methods". In *Society for Experimental Mechanics Annual Conference and Exposition*, pages 215–223. Springer, 2023. DOI: https://doi.org/10.1007/978-3-031-36999-5_28
19. C. Martinelli, A. Coraddu, and A. Cammarano. "Experimental analysis of a nonlinear piecewise multi-degrees of freedom system". In *International Conference on Nonlinear Dynamics and Applications,* pages 665-675. Springer International Publishing, 2023, DOI: https://doi.org/10.1007/978-3-031-50631-4_56
20. H. Dankowicz and F. Schilder. "Recipes for continuation". *SIAM*, 2013.
21. C. Martinelli, A. Coraddu, and A. Cammarano. "Approximating piecewise nonlinearities in dynamic systems with sigmoid functions: advantages and limitations". *Nonlinear Dynamics*, 111, 8545–8569, 2023. DOI: https://doi.org/10.1007/s11071-023-08293-1
22. C. Martinelli, A. Coraddu, and A. Cammarano. "Strongly nonlinear multi-degree of freedom systems: Experimental analysis and model identification [data collection]". *Enlighten Research Data*, 2023. DOI: 10.5525/gla.researchdata.1621
23. C. Martinelli, A. Coraddu, and A. Cammarano. "Experimental analysis of a nonlinear piecewise multi-degrees of freedom system [data collection]". *Enlighten Research Data*, 2023. DOI: 10.5525/gla.researchdata.1471

Chapter 17
Forced Response of Nonlinear Systems Under Multi-Frequency Excitations by Using Harmonic Balance Method

Dersu Celiksoz and Ender Cigeroglu

Abstract Even though most of the engineering systems falls under the category of single-frequency excitation, there exist several systems, such as double rotor systems, turbomachinery airfoils, gear boxes, micromotors etc., where the excitation is composed of multiple frequencies. Therefore, this study explores the influence of multi-frequency excitation on the vibration behavior of a nonlinear dynamic system. A system with gap nonlinearity is examined in order to better understand the behavior of the discontinuous nonlinear elements under harmonic forces that are exhibiting different frequency characteristics. The multi-frequency excitation is modeled to include two frequencies that are non-integer multiples of a fundamental frequency. Using harmonic balance method, parametrical modal domain analysis is conducted to present the behavior of the system when multiple excitation frequencies are either close to each other or distinct. The findings also illustrate the relationships between excitation sources and the resulting vibration responses of each multi-frequency harmonics. Based on the observed relations, contributions of the harmonics are evaluated and prediction of the vibration sources within a specific frequency range is facilitated.

Keywords Multi-Frequency Excitations · Nonlinear Dynamic Systems · Nonlinear Vibrations · Gap Nonlinearity · Harmonic Balance Method

Introduction

In dynamic systems, mechanical vibrations caused by external sources are considered during the design, analysis, and maintenance phases to ensure the stability and reliability of operating components. Various excitation sources exist for such systems, most of which oscillate at a single frequency. However, some engineering systems, particularly those operating in complex environments, are excited by forces with multiple frequencies. Examples of such systems include double rotor systems, turbomachinery airfoils, gearboxes, and micromotors. Understanding the dynamics and vibration behavior of engineering systems excited by multiple frequencies is crucial for predicting loading conditions and the response of mechanical components.

Unlike single-frequency excited engineering problems, multi-frequency cases require special attention to solve the forced response of the systems, since the harmonics are not integer multiples of each other. Moreover, the effect of multi-frequency phenomena becomes more complicated in the presence of nonlinear elements that influence the frequency response significantly. One of the most common types of nonlinear elements in rotating mechanical systems is gap nonlinearity. Gap nonlinearity arises in systems with contact separations or clearances where mechanical connections or joints exist. These systems exhibit response-dependent system stiffness, since some connections only occur if certain response conditions are met. In other words, the stiffness varies based on the system response, such as displacements or load thresholds are exceeded. This leads to nonlinear behavior, and modeling such systems requires attention to these threshold values. Several studies from different application areas have been conducted on gap nonlinearity in mechanical elements. For instance, Okuizumi [1] investigated the influence of gap size on the nonlinear vibration characteristics of a satellite truss structure with a gap element. Wang et al. [2] examined a non-direct-drive servo motor system with gap nonlinearity to propose a strategy for improving system robustness. Based on such findings, it can be concluded that further investigation of gap nonlinearity across various discontinuous mechanical systems is crucial.

Dersu Celiksoz · Ender Cigeroglu
Department of Mechanical Engineering, Middle East Technical University, Ankara, Türkiye
Center for Gear, Power Transmission Systems and Vibration Research, Middle East Technical University, Ankara, Türkiye
e-mail: dersuceliksoz@gmail.com; ender@metu.edu.tr

Ludovic Renson et al. (eds.), *Nonlinear Structures & Systems, Vol. 1*, Conference Proceedings of the Society for Experimental Mechanics Series, https://doi.org/10.13052/9788743801467-17

For the periodic problems of differential algebraic equations, Harmonic Balance Method (HBM) is used to find a numerical solution. As noted by Jazar [3], substituting an assumed periodic solution using Fourier series into the differential equation generates a residual term. The method requires that the Fourier coefficients of this residual term should vanish, ensuring a balanced solution. Additionally, variants and extensions of the HBM have been applied to diverse types of problems as summarized in [4].

When subjected to multi-frequency excitations, nonlinear systems produce responses whose dynamics are difficult to predict using conventional methods. Thus, different approaches have been studied for such excitation cases. Junge et al. [5] proposed a creative harmonic balance technique based on multidimensional Fourier transformations. This technique extends the standard harmonic balance method by constructing separate time domains for each base frequency. Furthermore, Prabith and Krishna [6] developed a time variational method (TVM) for analyzing mechanical systems experiencing multiple-frequency excitations. This analysis is simplified compared to the multi-harmonic balance method and avoids the need to transform between frequency and time domains. Wu et al. [7] refined the semi-analytical HBM variants by proposing a new formulation of the Alternating Frequency-Time method and introduced a new arc-length continuation condition to expand the capabilities for solving quasi-periodic excitations. Guskov and Thouverez [8] developed a multidimensional HBM and an adjusted HBM to address the nonlinear response and stability of a duffing oscillator subject to bi-periodic excitation. Also, Huang and Zhu [9] proposed a novel incremental harmonic balance method that employs two timescales to handle quasiperiodic motion in nonlinear systems, allowing precise identification of quasiperiodic behavior.

Considering the findings of prior research, this study focuses on the influence of multi-frequency excitation on the vibration behavior of a nonlinear dynamic system with a gap element. Excitation forces are modeled by considering two excitation frequencies that are non-integer multiples of a fundamental frequency. This type of forcing represents a common scenario in rotating systems with two excitation sources. To solve the equation of motion, HBM is extended to balance the harmonic components including different combinations of multiple excitation frequencies and to cover contribution of each harmonic term. The study also examines system response when the excitation frequencies are either close to each other or relatively distinct. Overall, this research provides valuable insights into the design and analysis of dynamic systems subjected to multi-frequency excitations.

Mathematical Modeling

As shown in Fig. 1, single degree of freedom nonlinear system with gap nonlinearity under the action of multi-frequency excitation is considered in this study. The quasi-periodic external forcing and system response are assumed to be functions of two excitation frequencies which are non-integer multiples of a fundamental frequency.

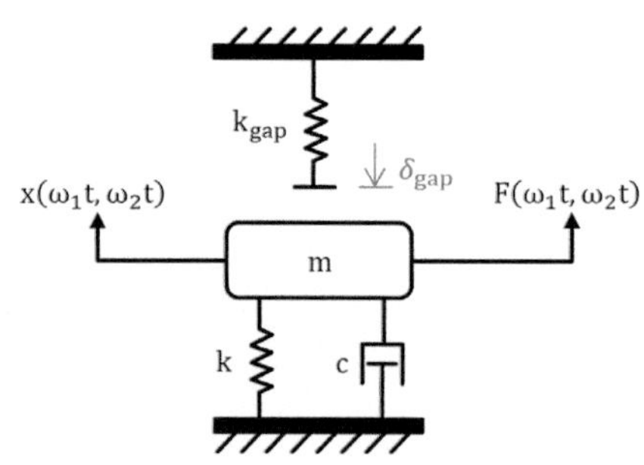

Fig. 1 Nonlinear dynamic system model with gap element.

The equation of motion of the nonlinear system can be written as follows.

$$m\ddot{x}(t) + c\dot{x}(t) + kx(t) + k_{gap}(x(t) - \delta_{gap})\,g(x) = F(t), \tag{1}$$

$$g(x) = \begin{cases} 1, & x \geq \delta_{gap} \\ 0, & x < \delta_{gap} \end{cases}. \tag{2}$$

where, x is the displacement of the system. m, k, c and F are the mass, stiffness, viscous damping in the system and external forcing acting on the system, respectively. k_{gap} and δ_{gap} are the gap stiffness and gap value of the gap nonlinearity. Furthermore, $g(x)$ is used to define when the gap is closed and stiffness is in contact with the system. Plot of the nonlinear force for the gap element is shown in Fig. 2.

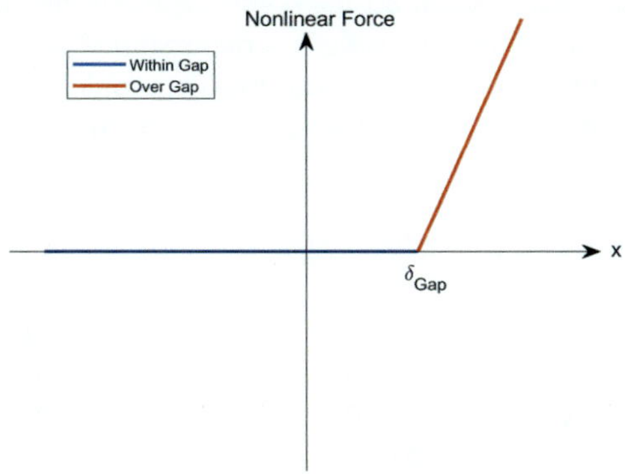

Fig. 2 Variation of nonlinear force.

The quasi-periodic response and forcing are assumed to be functions of two excitation frequencies. Similar to the multi-dimensional representation described by Guskov [8], the response and forcing are defined in Eqs. (3) and (4) by considering the contribution of different harmonic variations of the excitation frequencies.

$$x(t) = x_0 + \sum_{(k_1,k_2)\in Z^2} \left[x_{s(k_1,k_2)}\sin(k_1\omega_1 t + k_2\omega_2 t) + x_{c(k_1,k_2)}\cos(k_1\omega_1 t + k_2\omega_2 t)\right], \quad (3)$$

$$F(t) = f_0 + \sum_{(k_1,k_2)\in Z^2} \left[f_{s(k_1,k_2)}\sin(k_1\omega_1 t + k_2\omega_2 t) + f_{c(k_1,k_2)}\cos(k_1\omega_1 t + k_2\omega_2 t)\right]. \quad (4)$$

Note that, k_1 and k_2 in the equations represent harmonic multipliers within the integer-couple domain Z^2, which includes various combinations of multipliers up to a specified harmonic number. Also, x_0, $x_{s(k_1,k_2)}$, $x_{c(k_1,k_2)}$, f_0, $f_{s(k_1,k_2)}$ and $f_{c(k_1,k_2)}$ are used for the Fourier coefficients of corresponding terms.

Furthermore, the excitation frequencies are defined in the Eq. (5) as non-integer multiples of a fundamental frequency of ω_f, with the multipliers of β_1 and β_2.

$$\omega_1 = \beta_1\omega_f,\ \omega_2 = \beta_2\omega_f. \quad (5)$$

Harmonic Balance Method For Multi-Frequency Excitation

This work analyzes the system response in the frequency domain using an enhanced harmonic balance method. Following the summary by Sert and Cigeroglu [10], the Multi-Harmonic Balance Method is utilized to address systems defined by a finite number of harmonics. The residual equation, including selected harmonics, is derived by combining both linear and nonlinear terms into a matrix form for systems excited by single-frequency as follows.

$$\boldsymbol{r}(\omega) = \begin{bmatrix} k & \mathbf{0} & \mathbf{0} & \dots & \mathbf{0} \\ \mathbf{0} & \boldsymbol{\Lambda}(\omega) & \mathbf{0} & \dots & \mathbf{0} \\ \mathbf{0} & \mathbf{0} & \boldsymbol{\Lambda}(2\omega) & \dots & \mathbf{0} \\ \dots & \dots & \dots & \dots & \dots \\ \mathbf{0} & \mathbf{0} & \mathbf{0} & \mathbf{0} & \boldsymbol{\Lambda}(n_H\omega) \end{bmatrix} \begin{Bmatrix} x_0 \\ x_{s1} \\ x_{c1} \\ x_{s2} \\ x_{c2} \\ \dots \\ x_{sn_H} \\ x_{cn_H} \end{Bmatrix} + \begin{Bmatrix} f_{N0} \\ f_{Ns1} \\ f_{Nc1} \\ f_{Ns2} \\ f_{Nc2} \\ \dots \\ f_{Nsn_H} \\ f_{NcN_H} \end{Bmatrix} - \begin{Bmatrix} F_0 \\ F_{s1} \\ F_{c1} \\ F_{s2} \\ F_{c2} \\ \dots \\ F_{sn_H} \\ F_{cN_H} \end{Bmatrix} = 0. \quad (6)$$

In Eq. (6), $\boldsymbol{\Lambda}(\omega)$ represents the dynamic stiffness of each harmonic up to the harmonic number of n_H, and defines the linear portion of the residual as follows.

$$\boldsymbol{\Lambda}(\omega) = \begin{bmatrix} k - \omega^2 m & -\omega c \\ \omega c & k - \omega^2 m \end{bmatrix}. \quad (7)$$

Moreover, f_{N_r} are used to denote the bias, sine or cosine components of the nonlinear internal forcing. Building upon these studies on the Multi-Harmonic Balance Method, as well as the assumed solutions and force representations provided by Eqs. (3) and (4), the HBM is extended to address a nonlinear problem under multi-frequency excitation. As a result, the modified residual equation and corresponding dynamic stiffness are rederived as follows.

$$\boldsymbol{r}(\omega_1,\omega_2) = \begin{bmatrix} k & \mathbf{0} & \mathbf{0} & \mathbf{0} & \cdots & \mathbf{0} \\ \mathbf{0} & \boldsymbol{\Lambda}(\omega_1,0) & \mathbf{0} & \mathbf{0} & \cdots & \mathbf{0} \\ \mathbf{0} & \mathbf{0} & \boldsymbol{\Lambda}(0,\omega_2) & \mathbf{0} & \cdots & \mathbf{0} \\ \mathbf{0} & \mathbf{0} & \mathbf{0} & \boldsymbol{\Lambda}(\omega_1,\omega_2) & \cdots & \mathbf{0} \\ \vdots & \vdots & \vdots & \vdots & \ddots & \vdots \\ \mathbf{0} & \mathbf{0} & \mathbf{0} & \mathbf{0} & \mathbf{0} & \boldsymbol{\Lambda}(n_{H1}\omega_1, n_{H2}\omega_2) \end{bmatrix} \begin{Bmatrix} x_{0(0,0)} \\ x_{s(1,0)} \\ x_{c(1,0)} \\ x_{s(0,1)} \\ x_{c(0,1)} \\ x_{s(1,1)} \\ x_{c(1,1)} \\ \vdots \\ x_{s(n_{H1},n_{H2})} \\ x_{c(n_{H1},n_{H2})} \end{Bmatrix} +$$

$$\begin{Bmatrix} f_{N(0,0)} \\ f_{N\,s(1,0)} \\ f_{N\,c(1,0)} \\ f_{N\,s(0,1)} \\ f_{N\,c(0,1)} \\ f_{N\,s(1,1)} \\ f_{N\,c(1,1)} \\ \vdots \\ f_{N\,s(n_{H1},n_{H2})} \\ f_{N\,c(n_{H1},n_{H2})} \end{Bmatrix} - \begin{Bmatrix} F_{(0,0)} \\ F_{s(1,0)} \\ F_{c(1,0)} \\ F_{s(0,1)} \\ F_{c(0,1)} \\ F_{s(1,1)} \\ F_{c(1,1)} \\ \vdots \\ F_{s(n_{H1},n_{H2})} \\ F_{c(n_{H1},n_{H2})} \end{Bmatrix} = 0 \tag{8}$$

$$\boldsymbol{\Lambda}(\omega_1,\omega_2) = \begin{bmatrix} k-(\omega_1+\omega_2)^2 m & -(\omega_1+\omega_2)c \\ (\omega_1+\omega_2)c & k-(\omega_1+\omega_2)^2 m \end{bmatrix} \tag{9}$$

Note that the terms of n_{H1} and n_{H2} in Eqs. (8) and (9) represents the number of considered harmonics for each excitation frequency. Additionally, the nonlinear force resulted from nonlinear gap element is denoted by $f_{N\,(k_1,k_2)}$ and it is calculated using Fourier integrals for each harmonic combination. This integration involves calculating the nonlinear force contributions over a common period to capture the nonlinear behavior across the multi-frequency range in which the gap threshold is exceeded.

Following the selection of the contributing harmonics, the residual equation is solved using the arc length continuation method, which is effective in handling nonlinearities through a path-following approach. This solution process involves predicting a future solution that lies on a circle being centered on the current solution and having a predefined radius in the solution domain. The predicted solution is then iteratively corrected employing the Newton-Raphson method until the convergence threshold is met. Details of Newton's method with arc-length continuation can be found in [11–12].

Results and Discussion

The analysis approach described in the previous section is conducted for various cases where different multipliers are used for the excitation frequencies. This parametric study aims to understand the behavior of the system when multiple excitation frequencies are either close to each other or distinct, as well as the influence of harmonic pairs on the solution. First, the harmonic pairs present in the system are specified and their multipliers used in the study are given in Table 1. The quasi-periodic response of the system is obtained by substitution of those harmonic pairs into the multi-frequency response given by Eq. (3). Furthermore, external forcing is selected to excite the system with two different frequencies, as shown in Eq. (10).

$$F(t) = f_{s(1,0)} \sin(\omega_1 t) + f_{s(0,1)} \sin(\omega_2 t + \phi_{(0,1)}) \,. \tag{10}$$

Table 1 Multipliers of selected harmonic couples

Harmonic Couple	1	2	3	4	5	6
k_1	0	1	1	1	0	-1
k_2	0	0	1	-1	1	1

Subsequently, a system model and corresponding residual equation are created using the parameters provided in Table 2. The frequency response of the system is obtained employing Newton's Method with the arc length continuation method. For clarity, phase angle is assumed to be 0 in the analysis and its effect is investigated separately.

Table 2 System Parameters.

Parameter	Unit	Value
Mass, m	kg	2.00
Stiffness, k	N/m	300
Damping, c	Ns/m	0.30
Gap Stiffness, k_{gap}	N/m	200
Gap Distance, δ_{gap}	m	0.03
External Forcing, $f_{s(1,0)}$	N	0.50
External Forcing, $f_{s(0,1)}$	N	0.35
Phase Angle, $\phi_{(0,1)}$	rad	0

To better understand the contributions of multi-frequency harmonics, the frequency response results for non-integer frequency multipliers of 0.75 and 1.20, as defined in Eq. (5), are obtained, and shown in Figs. 3 and 4 both. The results include the total solution along with the harmonic solution components. The linear response under multi-frequency excitation is also included in the figures in order to highlight the influence of nonlinearity.

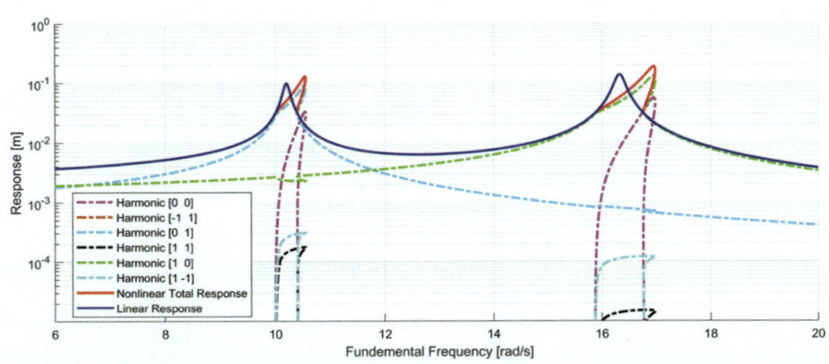

Fig. 3 Frequency response analysis results in logarithmic scale for $\beta_1 = 0.75$ and $\beta_2 = 1.20$ within fundamental frequency range.

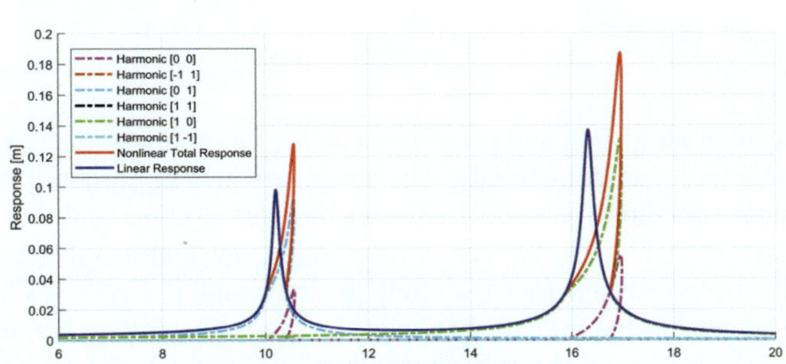

Fig. 4 Frequency response analysis results in linear scale for $\beta_1 = 0.75$ and $\beta_2 = 1.20$ within fundamental frequency range.

By investigating the resonance frequencies given in Figs. 3 and 4, it can be observed that resonance occurs at the specified non-integer multiples of the linear natural frequency. These resonance points are primarily due to the contributions of the fundamental harmonic of each excitation component. For instance, the first resonance occurs around 10.20 rad/s, which, when multiplied by the harmonic multiplier of 1.2, results in a natural frequency of 12.25 rad/s. Bias term is smaller than the fundamental harmonics but it is significantly larger than the combination harmonics considered in the study.

The analysis discussed above is repeated for different non-integer multiples of the fundamental frequency to examine the effect of the closeness of the multipliers. A range of multiplier pairs, starting from the same excitation frequency to excitation frequencies away from each other, are considered and given in Table 3.

Table 3 Non-integer multiplier pairs for the parametrical analysis.

Multiplier Pair	1	2	3	4
β_1	0.75	0.75	0.75	0.75
β_2	0.75	0.80	1.2	1.45

For each set of multipliers, the analysis is conducted, and the total responses of each case are given as a single plot, as shown in Fig. 5. The results indicate that as the multipliers become closer, the maximum response amplitude increases. This is because the close resonances of the corresponding harmonics result in more interaction, leading to a higher overall response. Conversely, when the multipliers are apart from each other, the distinct harmonics behave more independently and reduce their interaction, which leads to smaller response amplitudes. This is an expected result, since in the case of distant natural frequencies, energy given by the excitation forces are applied at different frequencies: whereas, for close frequencies energy of the excitations are localized in a narrow frequency range resulting in the increase of the response amplitude.

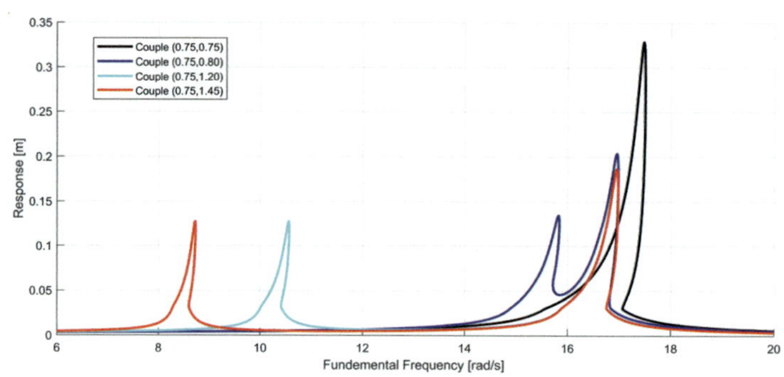

Fig. 5 Total frequency responses for selected multiplier pairs within fundamental frequency range.

In addition to the study of the effect of multiplier pairs on the response, the influence of the phase angle between harmonic components of the forcing is investigated. By keeping the multiplier pairs of 0.75 and 1.20 constant, analysis is repeated for different phase angles of 0, $\pi/3$ and $4\pi/3$ radians. It is observed that results are affected from the phase angle variation negligibly; hence, for brevity these results are not given here.

Conclusion

This research emphasizes the impact of multi-frequency excitation on the vibrational behavior of nonlinear dynamic systems by examining a mechanical system with gap nonlinearity. The study provides important findings by modeling the system using an extended harmonic balance method for multi-frequency excitation cases and solving the forced response with the arc-length continuation method. Firstly, it investigates the harmonic contributions that are non-integer multiples of the fundamental frequency. Secondly, it conducts a parametric study to understand the system's behavior as the frequency multiples get closer. The results indicate that as the excitation frequencies come closer together, the resultant vibrations become stronger and require careful design considerations for these systems. Furthermore, the research highlights the potential for identifying unexpected vibration sources by analyzing the system's multi-frequency response. In conclusion, understanding multi-frequency excitations and their effects is crucial for enhancing the reliability and performance of mechanical systems.

Future research extending this analysis to more complex mechanical systems could lead to a better understanding of various engineering systems under multi-frequency excitations.

References

1. Okuizumi, N. and Natoriy, M.C., "Nonlinear Vibrations of a Satellite Truss Structure with Gaps" *45th AIAA/ASME/ASCE/AHS/ASC Structures, Structural Dynamics & Materials Conference*, pp. 19–22, 2004.
2. Wang, B., Ji, R., Zhou, C., Sohel, R.M., Liu, K., Hua, W., and Ye, H., "A Gap Nonlinearity Compensation Strategy for Non-Direct-Drive Servo Systems" *Energies*, vol. 17, p. 3582, 2024.
3. Jazar, R.N., Perturbation Methods in Science and Engineering, *Springer*, pp. 262–277, 2021.
4. Yan, Z., Dai, H., Wang, Q., and Atluri, S.N., "Harmonic Balance Methods: A Review and Recent Developments" *Computer Modeling in Engineering & Sciences*, vol. 137(2), pp. 1419–1459, 2023.
5. Junge, L., Frey, C., Ashcroft, G., and Kugeler, E., "A New Harmonic Balance Approach Using Multidimensional Time" *Journal of Engineering for Gas Turbines and Power*, vol. 143, 2021.
6. Prabith, K. and Krishna, I.R.P., "A Time Variational Method for the Approximate Solution of Nonlinear Systems Undergoing Multiple-Frequency Excitations" *Journal of Computational and Nonlinear Dynamics*, vol. 15(3) 2020.
7. Wu, J., Hong, L., and Jiang, J., "A Comparative Study on Multi- and Variable-Coefficient Harmonic Balance Methods for Quasi-Periodic Solutions" *Mechanical Systems and Signal Processing*, vol. 187, 2023.
8. Guskov, M. and Thouverez, F., "Harmonic Balance-Based Approach for Quasi-Periodic Motions and Stability Analysis" *Journal of Vibration and Acoustics*, vol. 134, pp. 031003-1–031003-11, 2012.
9. Huang, J.L. and Zhu, W.D., "An Incremental Harmonic Balance Method with Two Timescales for Quasiperiodic Motion of Nonlinear Systems Whose Spectrum Contains Uniformly Spaced Sideband Frequencies" *Nonlinear Dyn*, vol. 90(2), pp. 1015–1033, 2017.
10. Sert, O. and Cigeroglu, E., "Adaptive Harmonic Balance Methods, A Comparison" *Special Topics in Structural Dynamics: Conference Proceedings of the Society for Experimental Mechanics Series, The Society for Experimental Mechanics*, vol. 6, pp. 279–289, 2016.
11. Von Groll, G., and Ewins, D. J., "The Harmonic Balance Method with Arc-Length Continuation in Rotor/Stator Contact Problems" *Journal of Sound and Vibration*, vol. 241(2), pp. 223–233, 2001.
12. Dimitriadis, G., "Continuation of higher-order harmonic balance solutions for nonlinear aeroelastic systems" *Journal of Aircraft*, vol. 45(2), pp. 523–537, 2008.

Chapter 18
Finding the Limits of Beam Elements: Modeling Bolted Joints as an Effective Joint Region

Nicholas Pomianek, Casey Whitworth, Enrique Gutierrez-Wing, Trevor Jerome, and J. Gregory McDaniel

Abstract Interrupting a beam with a bolted joint introduces dynamic nonlinearity and time dependent variability arising from complex interfacial physics. High fidelity finite element simulation of large jointed structures is therefore expensive and has diminishing returns when considering only global behavior. We propose a simplified method to model bolted joints where a characteristic set of dynamic properties is determined for the jointed regions of a structure. Euler-Bernoulli beam elements are used to represent joints as continuous solid regions with tunable damping, stiffness, and density properties. These properties are updated by an optimization process based on experimental data using classical and recent frequency response similarity metrics. Assembly and reassembly effects are accounted for experimentally through variation of bolt tightening orders. The Brake-Reuß standardized beam is used as a test case for this work; results show robust agreement between the updated model and modal test data. This work develops a practical and effective framework for joint modeling that is readily applicable in any finite element software.

Keywords Joint Mechanics· Model Updating· Similarity Metrics· Frequency Response· Finite Elements

Introduction

It is easy to find structural design cases where an estimation of global response is required at the modeling stage but including nonlinearity is not strictly necessary. These are typically large structures which often contain many instances of the same joint and loading scenarios significantly below expected service levels. If the common joints in these designs are sufficiently standardized and prevalent, then prototypes should be available and easy to test *ex situ*. Methods that leverage the available information under these circumstances to produce a high fidelity model that is inexpensive enough for structural design model updating are of increasing interest to the joint modeling community.

Modeling nonlinear systems requires a choice between fidelity and computation time. The highest-fidelity techniques involve detailed scanning of the joint interface surfaces and modeling asperity interactions [1]. For modeling scenarios where joint surfaces are not available a priori, statistical descriptions of surface asperities are often employed alongside a combination of linear modal extraction and a quasi-static approach [2]. If the structure isn't expected to encounter significant macroslip, then a Iwan or Masing contact element formulation is typically used [3]. More recent techniques for this application include the definition of Contact-Zone elements [4], or the replacement of bolts with springs [5]. Commercial finite element packages offer joint elements that entirely replace joints but they require empirical stiffness data. Finally, sufficiently linear systems are adequately modeled using friction contact conditions such as the penalty method that are available in commercial finite element software. All these modeling techniques require validation and calibration with experimental data for best results. If these data are available and joints are expected to be in the linear regime, then a simplified and calibrated linear model that can predict the most significant dynamics of joints is a powerful tool for structural design.

Many joint modeling techniques have been developed to better predict the effects of friction across a wide range of loading scenarios but all require model updating and most require element types that are unavailable to finite element practitioners. Furthermore, the nonlinear approaches that many of these modeling techniques use prevent the use of frequency

Nicholas Pomianek · Casey Whitworth · Enrique Gutierrez-Wing · J. Gregory McDaniel
Department of Mechanical Engineering, Boston University, Boston, MA 02215
e-mail: pomianek@bu.edu; caseyw12@bu.edu; esgw@bu.edu; jgm@bu.edu

Trevor Jerome
Naval Surface Warfare Center, Carderock Division, West Bethesda, MD 20817
e-mail: trevor.w.jerome.civ@us.navy.mil

Ludovic Renson et al. (eds.), *Nonlinear Structures & Systems, Vol. 1*, Conference Proceedings of the Society for Experimental Mechanics Series, https://doi.org/10.13052/9788743801467-18

domain eigenvalue solvers. These limitations have inspired the recent work of Black et. al. [6], which addresses them by modeling joints as tied contact regions between solid elements. By updating the size of this contact region along with the stiffness and density of the surrounding elements, the global response of a jointed structure was predicted accurately. The present study also uses the idea of defining a 'virtual material' [6] but instead uses an approach tailored to situations where a structural prototype is unavailable.

This work seeks to develop an accessible and effective framework for joint modeling for use with standard FEA tools that does not require special contact elements or parameters. This work seeks to demonstrate that updating the damping, stiffness, and density characteristics of the elements within an effective joint region (EJR) can create a computationally efficient model that preserves the most important dynamic response features of a joint.

Experimental setup

The Brake-Reuß standardized test beam (BRB) [7] was used as a case study for this work because of the extensive modal testing data available in the existing literature. The prismatic geometry of the BRB also lends itself well as a best-case-scenario for the modeling approach developed in the following sections. Pictured in Figure 1, the test specimen was manufactured according to the machining specifications outlined in [8] but was made from ASTM A283 steel for material availability reasons. The procedure for assembling the test structure from [8] was followed, where both sides of the BRB are clamped to a flat surface, a shim is inserted in the interface parallel to the bolt axis, and an instrumented torque wrench is used to tighten the bolts to 20 Nm. This assembly procedure is followed during each re-assembly of the system throughout experimentation. The assembled system is suspended using monofilament fishing line to approximate free-free boundary conditions.

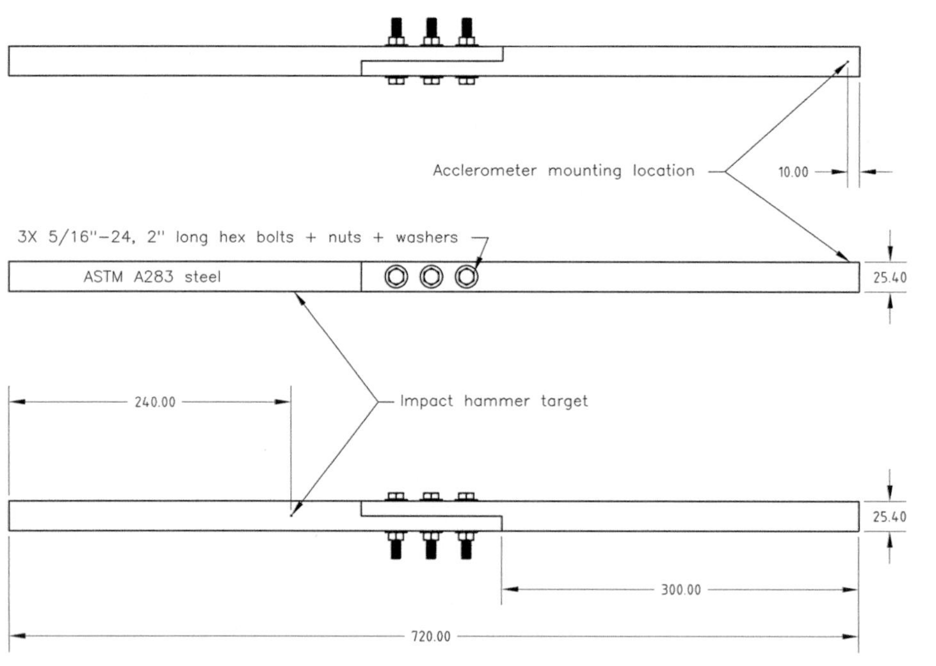

Fig. 1 Brake-Reuß beam test specimen, dimensions in mm.

Modal testing was performed using an aluminum-tipped Brüel & Kjær model 8206-002 impact hammer and a Brüel & Kjær model 4534-B accelerometer, the placement of the accelerometer and impact hammer target location are shown in Figure 1. Both the accelerometer and impact hammer signals are fed through PCB model 482A21 signal conditioners into a Digilent Analog Discovery 2 digital oscilloscope which records data to a computer running WaveForms software. The system is oriented such that the impact direction is perpendicular to the suspension lines so that they do not add stiffness to the beam. The experimental instrumentation setup is shown in detail in Figure 2.

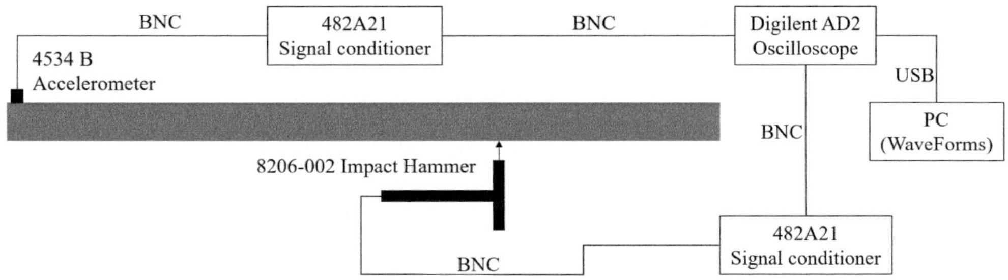

Fig. 2 Experimental instrumentation setup used for impact hammer modal testing.

Bolted joints, specifically the BRB, have been demonstrated to exhibit non-repeatability and sensitivity to bolt tightening order [8]. To account for this, six bolt tightening orders were tested, ten impacts were recorded for each order and the average frequency response function (FRF) was computed for each order in units of accelerance a as

$$A^{(n)}(f) = \tilde{a}^{(n)}(f)/\tilde{F}^{(n)}(f) \tag{1}$$

where $\tilde{a}$ is the frequency domain acceleration, $\tilde{F}$ is the frequency domain force, f is the frequency, and n is the hit number index. The average FRF is thus

$$A_{ave}(f) = \frac{1}{10}\sum_{n=1}^{10} A^{(n)}(f). \tag{2}$$

Finite element model updating

This work uses the Euler-Bernoulli beam element as a test case for the EJR method. Despite its lower performance compared to other finite element formulations for beams, the Euler-Bernoulli element is computationally inexpensive and widely accessible. The properties chosen for updating are the damping loss factor η, the Young's modulus E, and the density ρ. These properties were selected because joints have been observed to influence stiffness and damping, and the addition of bolts, nuts, washers and holes necessarily changes the density of an otherwise homogeneous region. Using this formulation, the dynamic response of an element at a given frequency ω as a function of these properties is

$$\vec{U}(\eta, E, \rho) = \left[(1+\eta i)\mathbf{K}(E) - \omega^2\mathbf{M}(\rho)\right]^{-1}\vec{F} \tag{3}$$

where $\vec{F}$ are the nodal applied forces, $\vec{U}$ are the nodal displacements, $\mathbf{K}$ is the stiffness matrix, and $\mathbf{M}$ is the mass matrix. In order to compare to modal test data, the response is converted to accelerance $\vec{A}$.

$$\vec{A} = (i\omega^2)\vec{U} \oslash \vec{F} \tag{4}$$

A unit force is applied at the instrumented hammer impact location by finding it's equivalent nodal location and substituting the corresponding element of $\vec{F}$ in equation (3). The frequency response function for comparison to modal data is calculated by finding the equivalent nodal location of the accelerometer and extracting the corresponding element from the accelerance vector in equation (4). All models evaluated in this work use free-free boundary conditions and neglect the added mass of the accelerometer. Mesh convergence studies were performed and an element length of 1 cm was chosen for model updating.

The EJR method separately defines properties for the joint region as η_j, E_j, and ρ_j, allowing them to differ from the known properties for the monolithic sections of the structure as shown in Figure 3. In the non-joint region, these material properties are assigned according to test specimen material data. EJR properties are assigned by finding the equivalent nodal locations of the joint boundaries and replacing η, E, and ρ with η_j, E_j, and ρ_j for the elements within these boundaries. The resulting model is that of a solid beam with properties η, E, and ρ varying as a function of beam length. The known material properties and bounds used for optimization of EJR properties are summarized in Table 1.

The finite element model used in this work was developed in MATLAB and benchmarked against Abaqus. Firstly, a set of frequencies is extracted from modal test data when the experimental frequency response function is calculated. Next, the finite element model is evaluated at each frequency in this set to construct a frequency response function that can be compared directly with the experimental modal test data. The degree of similarity between the experimental FRF and the

Table 1 Model properties and bounds set during optimization.

Property	Lower bound	Upper bound	Units
η	0.005	0.005	
η_j	0	0.03	
E	200	200	GPa
E_j	0	20000	GPa
ρ	7800	7800	kg/m^3
ρ_j	7000	11000	kg/m^3

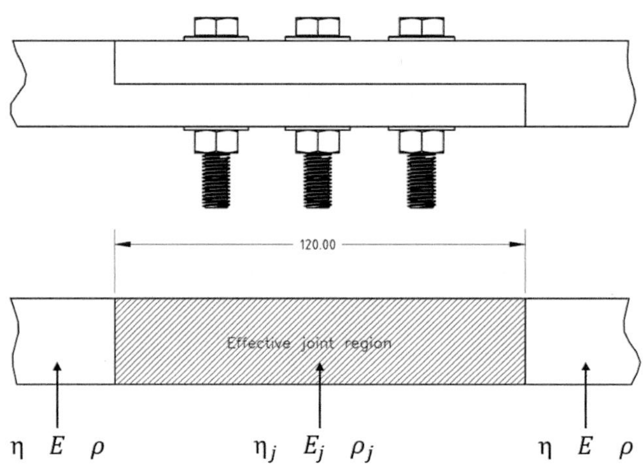

Fig. 3 The effective joint region is defined as any part of the structure that includes a frictional interface, dimension in mm.

finite element FRF is used as the cost function for a particle swarm optimization process. The EJR properties η_j, E_j, and ρ_j are the decision variables for this optimization process. Upon each iteration of the optimization, the EJR element properties are updated, the FEA model is evaluated, and the degree of similarity to the experimental dataset is calculated. The degree of similarity was calculated using a handful of different FRF similarity metrics from the literature.

The most straightforward metric tested is the root mean squared percent error (RMSPE) between the two FRFs being compared [9]. This is a simplistic approach that does not leverage the modal information contained in FRFs but is valuable as a point to compare against more sophisticated techniques. The R^2 value of a linear fit (ILR2) between the imaginary parts of each FRF serves as another metric focused on phase [9]. The FRF scaling factor (FRFSF) criterion is another alternative focused on the magnitude differences between FRFs.

$$\mathrm{FRFSF} = \frac{\sum_{f=f_1}^{f_2} |X(f)|}{\sum_{f=f_1}^{f_2} |Y(f)|} \tag{5}$$

In this equation, X is the reference FRF, Y is the comparison FRF, and f_1 and f_2 are the lower and upper bounds of the frequency range of interest. The frequency response assurance criterion (FRAC) is a well established inner-product approach that was used as a baseline to compare to more recent approaches [10].

$$\mathrm{FRAC} = \frac{\left|\sum_{f=f_1}^{f_2} X(f)Y^*(f)\right|^2}{\sum_{f=f_1}^{f_2} X(f)X^*(f)\ \sum_{f=f_1}^{f_2} Y(f)Y^*(f)} \tag{6}$$

Both FRAC and the modified frequency response assurance criterion (MFRAC), which scales FRAC by the ratio of the minimum to maximum powers of each FRF, are investigated. As an alternative to FRAC, the mean cross signature scale factor (CSF) was also applied due to its improved sensitivity to damping. [9].

$$\mathrm{CSF} = \left(\sum_{f=f_1}^{f^2} \frac{2\,|X^*(f)Y(f)|}{(X^*(f)X(f)) + (Y^*(f)Y(f))}\right)\frac{1}{N_f} \tag{7}$$

This equation includes the total number of frequencies of interest N_f. Finally the mean magnitude similarity function (MSF) and shape similarity function (SSF) criteria are the most recent and highest performing metrics used [11]. The similarity of magnitude and shape are calculated separately to increase the amount of information provided by each metric. For magnitude the metric is calculated as

$$|H(f)| = \left|\frac{Y(f)}{X(f)}\right| \tag{8}$$

$$M_\alpha(f) = 1 - \tanh\left(\frac{\ln 3}{2}\frac{|20\log_{10}|H(f)||}{\alpha}\right) \tag{9}$$

$$M \text{ index} = \frac{1}{f_2 - f_1}\int_{f_1}^{f_2} M_\alpha(f)df \tag{10}$$

where α is a scaling factor set to 20 in this work and M index is the average MSF value across the frequency range of interest. The phase similarity is calculated as

$$argH(f) = \phi_y(f) - \phi_x(f) \tag{11}$$

$$\tau(f) = \frac{1}{2\pi}\frac{d\, argH(f)}{df} \tag{12}$$

$$S_\beta(f) = 1 - \tanh\left(\frac{\ln 3}{2}\frac{|\tau(f)|}{\beta}\right) \tag{13}$$

$$S \text{ index} = \frac{1}{f_2 - f_1}\int_{f_1}^{f_2} S_\beta(f)df \tag{14}$$

where ϕ_x and ϕ_y are the phase of the FRFs X and Y and β is a scaling factor set to 1 for this work. The combined mean magnitude and shape similarity metric MSSF is

$$MS \text{ index} = w_m(M \text{ index}) + w_s(S \text{ index}) \tag{15}$$

where w_m and w_s are weighting factors such that $w_m + w_s = 1$. Only equal weighting is analyzed in this work.

The frequency window being used has been shown to influence model updating results when using single-number FRF similarity metrics [12]. In light of this concern, a windowed approach was developed where the frequency range of interest was subdivided into between one and ten sections which are each analyzed separately by the optimization process. Before optimization the experimental data is decimated and filtered using the Savitzky–Golay technique. The result of this numerical analysis setup is a set of EJR material properties that vary as a function of frequency which can be easily added to any commercial finite element software to model joints using beam elements.

Results

The aim of this work is to develop a computationally inexpensive joint modeling technique that yields acceptable results for applications where joints are not subject to high enough loads to induce highly nonlinear behavior. For comparison, a Euler-Bernoulli element model, a Timoshenko element model, and a solid element model were evaluated using known material properties. The properties applicable to the test specimen used are $\eta = 0.005$, $E = 200E9$ Pa, and $\rho = 7800$ kg/m^3, these properties are also assigned to the non-jointed region during EJR optimization. The Euler-Bernoulli and Timoshenko models are that of a monolithic beam with geometry equivalent to the BRB. The solid element model was developed in Abaqus following standard practices for joint modeling outlined in [13]. All geometry of the beams and bolts are modeled explicitly, the surface contact between beams and between washers is modeled explicitly, loading is applied internal to the bolts, and friction is modeled using a penalty formulation. As seen in the upper plot in Figure 4, agreement of all nominal models is poor when compared to experimental data and this comes at a computational cost of 7.83 minutes for the solid element model, 10 seconds for the Timoshenko model, and 0.055 seconds for the Euler-Bernoulli model. The EJR method, pictured in the lower plot of Figure 4, shows notably better agreement with the significant features of the experimental FRF. Because the optimized EJR model uses only Euler-Bernoulli elements, it also takes 0.055 seconds to run. The EJR optimization process that updated the nominal Euler-Bernoulli model takes 13.52 seconds.

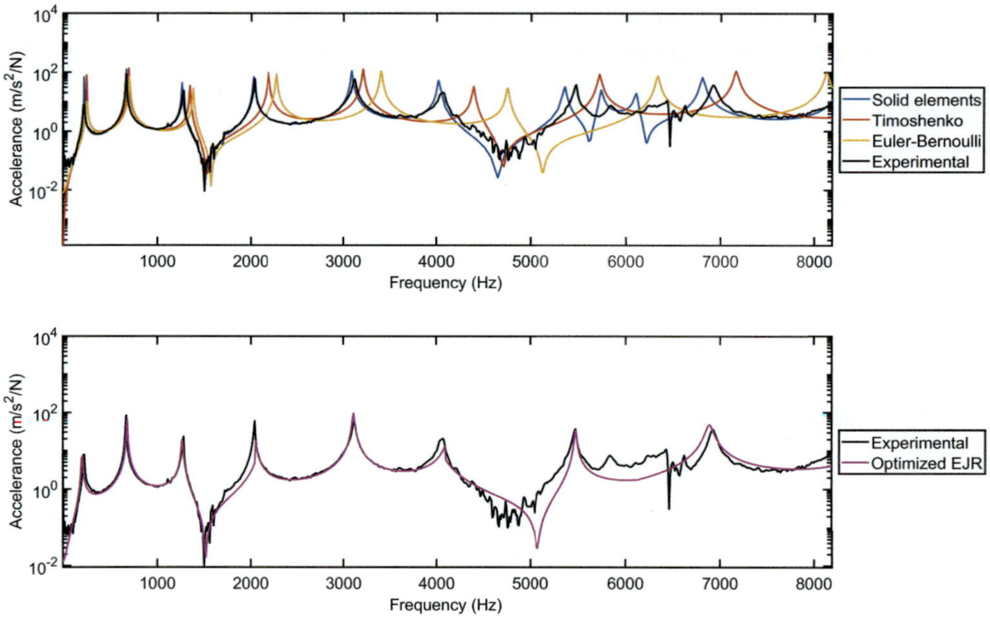

Fig. 4 Comparison of nominal FE models with EJR method optimization results. Solid element simulation run in Abaqus using explicitly modeled joint geometry, contact interactions, and bolt forces. MSSF similarity metric, bolt tightening order 3-2-1.

The effect of different FRF similarity metrics was also considered. With the exception of the RMSPE and FRFSF metrics, all considered similarity metrics performed well. Figure 5 shows the optimized EJR models generated using the eight similarity metrics analyzed in this work. The optimal number of frequency window divisions depended on the similarity metric but for the ILR2, FRAC, MFRAC, CSF, MSF, and MSSF optimized models the highest performing number of windows was four. The highest performing similarity metric overall was found to be MSSF.

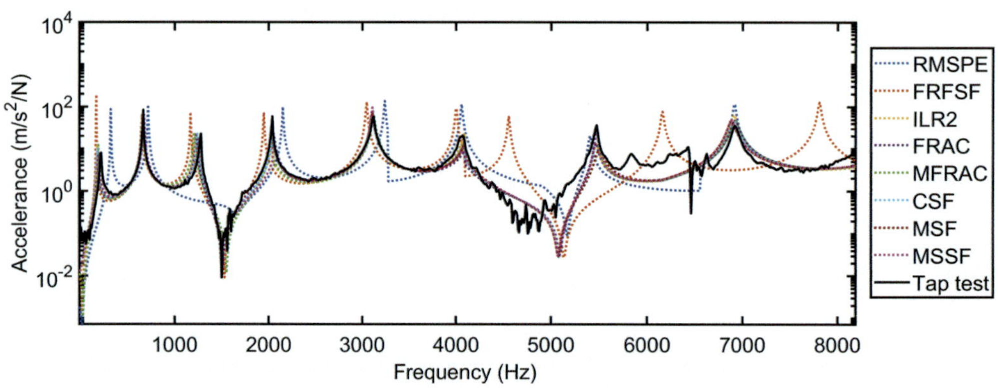

Fig. 5 Performance comparison of FRF similarity metrics, bolt tightening order 3-2-1.

A repeatability study was performed to test the sensitivity of the EJR method to system variability due to assembly conditions. Experimental data was taken on the same BRB assembled with six bolt tightening orders, as shown in the upper plot of Figure 6. An EJR optimized model was then generated for each data set, showing minor sensitivity to joint repeatability. The lower plot of Figure 6 shows the EJR model results corresponding to each bolt tightening order.

Because four windows were used to update the models shown in the previous figures, the EJR properties vary as a function of frequency. Figure 7 shows the variation in properties in the different frequency windows used for optimization. The upper and lower bounds of each variable available to the particle swarm optimization algorithm are shown by the vertical axis limits of each plot. While many of these results are non-physical, the resulting FRF is in good agreement with experimental data. The MSSF similarity metric between the EJR models and experimental data in Figure 6 ranges from 0.924 to 0.967 (its maximum value is 1). One of the inherent limitations of beam elements is their lower performance at

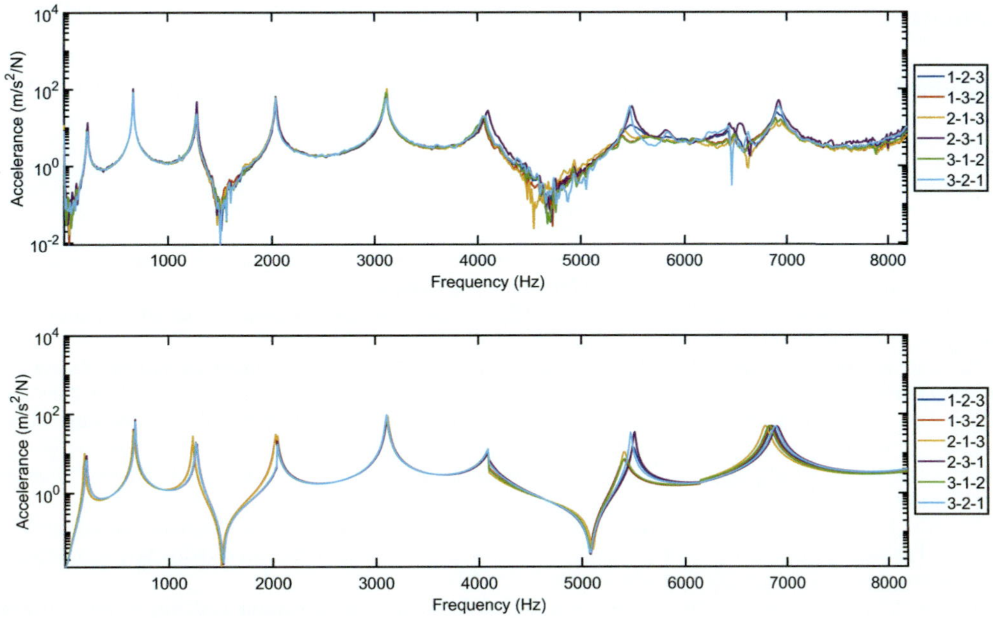

Fig. 6 Comparison of experimental data and optimization results for six bolt tightening orders, MSSF similarity metric.

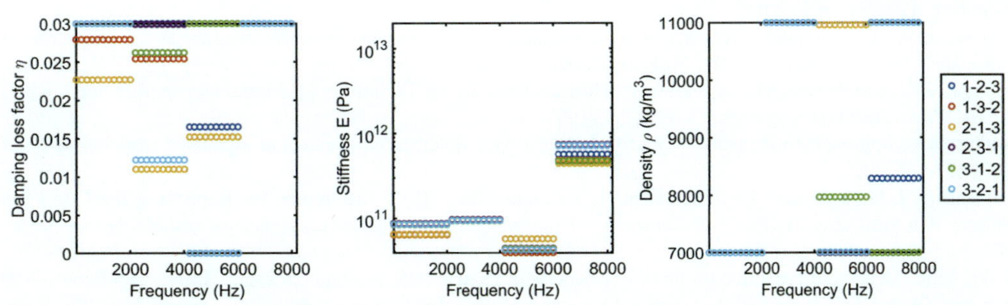

Fig. 7 Comparison of EJR properties across six bolt tightening orders, frequency range divided into four windows, MSSF similarity metric.

higher frequencies, the optimization process was required to compensate for this which likely contributed to the prediction of non-physical properties.

Conclusion

A method for modeling joints using inexpensive element formulations was developed and demonstrated to have good performance in the case of the BRB benchmark structure. A 99.988% reduction in computation time was achieved when compared to standard practice solid element joint modeling with better agreement to experimental data. The EJR method was shown to have low sensitivity to joint non-repeatability and similarity metric cost function formulation for optimization. The performance of various FRF similarity metrics for model updating in joints was also explored. In comparison to previous model updating techniques that parameterize joints, the EJR method can be applied in any commercial finite element modeling package which gives it the potential to be a powerful and accessible tool for modeling the dynamic response of jointed structures. Further research into the methods developed by this work must be done to validate EJR models against a structure containing many joints. Improvements in property physicality can likely be made by further dividing the EJR elements into bolt regions and non-bolt regions. Additionally, generalization of the EJR method to more than just Euler-Bernoulli beam elements is a necessary step in creating a complete method of analysis for real-world structures.

Acknowledgement

This work was supported through the Navel Engineering Education Consortium, Grant Number N00174-22-1-0015.

References

1. Wang, X., An, B., Xu, Y., and Jackson, R.L. "The effect of resolution on the deterministic finite element elastic-plastic rough surface contact under combined normal and tangential loading". *Tribology International*, 144:106141 (2020)
2. Porter, J.H. and Brake, M.R. "Towards a predictive, physics-based friction model for the dynamics of jointed structures". *Mechanical Systems and Signal Processing*, 192:110210 (2023)
3. Brake, M.R. "An overview of the analysis of jointed structures". In *The Mechanics of Jointed Structures: Recent Research and Open Challenges for Developing Predictive Models for Structural Dynamics*, pages 11–23. Springer International Publishing (2017)
4. Shokrollahi, S. and Adel, F. "Finite element model updating of bolted lap joints implementing identification of joint affected region parameters". *Journal of Theoretical and Applied Vibration and Acoustics*, 2(1):65–78 (2016)
5. Gant, F., Rouch, P., Louf, F., and Champaney, L. "Definition and updating of simplified models of joint stiffness". *International Journal of Solids and Structures*, 48(5):775–784 (2011)
6. Black, J.K., Callis, S.J., Feizy, A., Johnson, C.L., Lieven, N.A., and Vega, M.A. "A Simplified Finite Element Joint Model Updated with Experimental Modal Features". In *Conference Proceedings of the Society for Experimental Mechanics Series*, volume 5, pages 107–123. Springer (2024)
7. Brake, M.R. and Reuß, P. "The Brake-Reuß beams: A system designed for the measurements and modeling of variability and repeatability of jointed structures with frictional interfaces". In *The Mechanics of Jointed Structures: Recent Research and Open Challenges for Developing Predictive Models for Structural Dynamics*, pages 99–107. Springer International Publishing (2017)
8. Brake, M.R., Schwingshackl, C.W., and Reuß, P. "Observations of variability and repeatability in jointed structures". *Mechanical Systems and Signal Processing*, 129:282–307 (2019)
9. Marinone, T. and Moya, A. "Comparison of frf correlation techniques". In *Conference Proceedings of the Society for Experimental Mechanics Series*, volume 3, pages 299–309. Springer New York LLC (2015)
10. Pascual, R., Golinval, J.C., and Razeto, M. "A Frequency Domain Correlation Technique for Model Correlation and Updating". *Proceedings of the International Modal Analysis Conference - IMAC*, 1 (1997)
11. Shin, K. "An alternative approach to measure similarity between two deterministic transient signals". *Journal of Sound and Vibration*, 371:434–445 (2016)
12. Kramer, H.R., Manring, L.H., Schultze, J.F., Zimmerman, S.J., and Mann, B.P. "Strategies for Improving the Comparison of Frequency Response Functions with Similarity Metrics". In *Conference Proceedings of the Society for Experimental Mechanics Series*, pages 101–109. Springer (2024)
13. Groß, J. and Brake, M.R. "A standard practice for modeling bolted joints in a finite element package". In *The Mechanics of Jointed Structures: Recent Research and Open Challenges for Developing Predictive Models for Structural Dynamics*, pages 415–426. Springer International Publishing (2017)

Chapter 19
Design of a Tuned Vibration Absorber with Friction Contact for High Location Ratio Application

Matthew W. Hancock, Nathan P. Lovell, Yusuf R. Shehata, Keegan J. Moore, Bogdan I. Epureanu, Thomas N. Thompson, and Sean T. Kelly

Abstract Tuned vibration absorbers (TVAs) are passive devices commonly used to reduce the vibration amplitude of a host structure subject to harmonic excitations. These devices are designed so that vibration energy flows from the host structure to the TVA, which dissipates that energy using linear damping mechanisms. However, devices using nonlinear energy dissipation methods, such as friction-enhanced TVAs, have recently been explored. Additionally, it is desirable to place a TVA on the host structure where large vibration amplitudes occur, but this is often not feasible in practice. The ratio of the amplitudes of a high-vibration-amplitude area to a low-vibration-amplitude area is referred to as the location ratio (LR). Careful tuning of the TVA is critical to ensure effective energy absorption, especially when the TVA is placed in a high-LR area. This paper presents a case study of a novel TVA design that is optimized for high-LR applications and utilizes nonlinear friction contacts to dissipate energy. The host structure analyzed is the Box Assembly (BA), an experimental structure previously analyzed for vibration testing. A finite element model of the BA, validated by experimental ping tests, is considered, and a target mode of the BA is selected for TVA design. Simulations are performed to illustrate the effectiveness of the TVA in a high-LR application by varying linear system parameters. The effects of TVA nonlinear friction damping on the vibration amplitude reduction are also investigated. From these simulations, TVA design considerations are suggested.

Keywords Tuned Vibration Absorber · Friction · Location Ratio · Nonlinear Dynamics · Vibrations

Introduction

Large vibration amplitudes in structures subject to harmonic excitations can increase risk of high cycle fatigue and failure, especially when the structure's operating and natural frequencies coincide. For such systems, vibration suppression is critical

Matthew W. Hancock
Department of Mechanical Engineering, Texas A&M University
e-mail: matthewhancock@tamu.edu

Nathan P. Lovell
Department of Civil and Environmental Engineering, The Catholic University of America
e-mail: nathanplovell@gmail.com

Yusuf R. Shehata
Department of Mechanical and Aerospace Engineering, University of California, San Diego
e-mail: yusufrshehata@gmail.com

Keegan J. Moore
Daniel Guggenheim School of Aerospace Engineering, Georgia Institute of Technology
e-mail: kmoore@gatech.edu

Bogdan I. Epureanu
Department of Mechanical Engineering, University of Michigan, Ann Arbor
e-mail: epureanu@umich.edu

Thomas N. Thompson · Sean T. Kelly
Los Alamos National Laboratory
e-mail: tnthompson@lanl.gov; stkelly@lanl.gov

Ludovic Renson et al. (eds.), *Nonlinear Structures & Systems, Vol. 1*, Conference Proceedings of the Society for Experimental Mechanics Series, https://doi.org/10.13052/9788743801467-19

to the health of structural components that experience cyclical loading. A common solution for vibration reduction is to introduce a tuned vibration absorber (TVA), a linear, passive device designed to reduce the amplitude of vibration of a host structure subject to harmonic excitation. The TVA (also referred to as a tuned mass damper, or TMD) is tuned to target a specific frequency of interest, usually a natural frequency at which unwanted vibration amplitudes are experienced by the host. TVAs are typically less than 5% of the mass of the host structure [1], and can vary based on the application, placement, and host structure geometry, among other factors. A typical TVA absorbs the host's vibration energy by transferring that energy to an external mass, causing it to oscillate. As a result of this energy transfer, the TVA mass can have a relatively high response amplitude while the host's vibration is reduced.

Because of their ability to effectively mitigate unwanted vibrations, there are a myriad of applications in which TVAs have been applied across civil [1, 2], mechanical [2], and aerospace fields [3–6]. For example, skyscrapers using TVAs include the John Hancock Tower in Boston and the Citicorp Center in Manhattan [1]. In these structures, one or more TVAs are installed on the upper floors of the building to dissipate energy and reduce building response due to wind loads. An example of a mechanical application of TVAs is in pipelines above the arctic circle, in which wind loads and the associated vortex shedding can excite vibration modes of the pipelines [2]. In this example, TVAs have been installed on each span of the pipe to prevent fatigue cracking at the pipe joints. Another example is that presented by Aksoy et al. in which a TVA was designed to reduce vibrations of an air vehicle missile launcher [7]. Being so common, the linear dynamics of TVA-host systems have been studied extensively, and TVA theory is well known [1].

Recently, work has been conducted to increase the effectiveness of TVAs by incorporating nonlinear energy dissipation methods, such as friction [3, 8–12] and impacts [3, 8, 9, 12]. These methods have been shown to provide significant vibration reduction; however, introducing nonlinearity presents additional challenges and computational effort with respect to modeling and tuning to the resonant frequency of interest. In particular, frictional nonlinearities arise from transitions between a "stuck" state, in which there is no relative motion between friction contacts, and a "slip" or "sliding" state, in which contacts are moving freely with respect to each other. As a result, relative to linear TVA dynamics, predicting the response of such designs is more challenging. Many "friction-enhanced" TVA designs have been developed [10–14], with their associated modeling and experimental difficulties having been studied extensively in literature [15, 16]. Despite the additional modeling challenges, introducing nonlinear energy dissipation methods can provide additional vibration reduction in to-be-discussed scenarios where a linear TVA is insufficient.

Another important factor that affects TVA effectiveness is the location at which the TVA is placed on the host structure. It is typically desirable to place a TVA at the location of highest vibration amplitude on the host; this positioning can best facilitate energy flow from the host structure into the TVA to be dissipated. However, in many cases, it is not feasible to place the TVA at these high-amplitude locations, so the TVA must be designed to be effective at a low-amplitude location on the host structure. Location ratio (LR) is a metric that quantifies this amplitude difference, defined as the ratio of the highest vibration amplitude on the host structure to the amplitude at the mounting location of the TVA. This metric has been considered in previous work by Lupini et al. [5] and Mitra et al. [17]. At the location of highest amplitude, the LR is equal to one, and increases as the TVA is moved to areas of lower vibration amplitude. High-LR areas on a structure are areas of relatively low vibration amplitude, meaning less kinetic energy can be transferred to a TVA at this location. An example of a high-LR application is a ring damper designed for a turbomachinery blisk, like that in Refs. [3, 18, 19]. Blisks are rotating, fanlike structures used in modern turbomachinery whose service environment and inherent mistuning (i.e., deviations from nominal geometry and material properties) can create damaging vibration-induced stresses that may lead to high cycle fatigue failure. It is not feasible to mount a TVA at the tips of the blades where the highest vibration amplitude often occurs; doing so would interfere with the fluid flow, and thus lower engine performance. Therefore, it is necessary to mount the TVA in the disk portion of the blisk, where the LR is high. TVAs for this application have been designed and simulated by Lupini et al. [4, 19] and Cimpuieru et al. [3, 8, 9] which have yielded promising results.

Although effective TVA designs have been produced for high-LR applications, even those which include nonlinear energy dissipation methods, these designs primarily target a niche application in their respective field. Given that these designs have proven promising, a flexible, generalized set of design guidelines would be useful in helping designers develop effective friction-enhanced TVAs for other structures. As such, this paper presents a set of design considerations for friction-enhanced TVAs for high-LR placement. This paper describes the design of a friction-enhanced TVA for application on a simple host structure, known as the Box Assembly (BA). First, a brief overview of TVA theory is presented. Second, the process of creating and validating a finite element (FE) model of the BA is described. Third, a directional study considering a single-degree-of-freedom (SDoF) TVA is conducted using a reduced-order model (ROM) of the BA. Fourth, the TVA design process is discussed, and the final TVA design used for the BA is presented. Fifth, multiple BA-TVA systems are simulated with and without including nonlinear friction contacts considering various TVA attachment locations. The paper concludes with a discussion of the simulation results and design considerations that can be drawn from this study.

Methods

This section begins with a discussion of relevant theory concerning host-TVA systems, followed by the methods used for TVA design. First, an FE model of the host structure, i.e., the BA, was created and calibrated using experimental frequency response functions (FRFs) extracted from ping testing. Second, a reduced-order model (ROM) of the BA was created and used for an initial SDoF TVA directional analysis. Third, the TVA was designed and combined with the BA to form multiple BA-TVA FE models considering various TVA attachment locations. Fourth, the BA-TVA combined systems were simulated with and without friction. The results of these simulations informed the tuning of the TVA, making this step iterative. Finally, the results of the simulations with the tuned TVA led to the results and conclusions. This TVA design workflow is portrayed in Fig. 1.

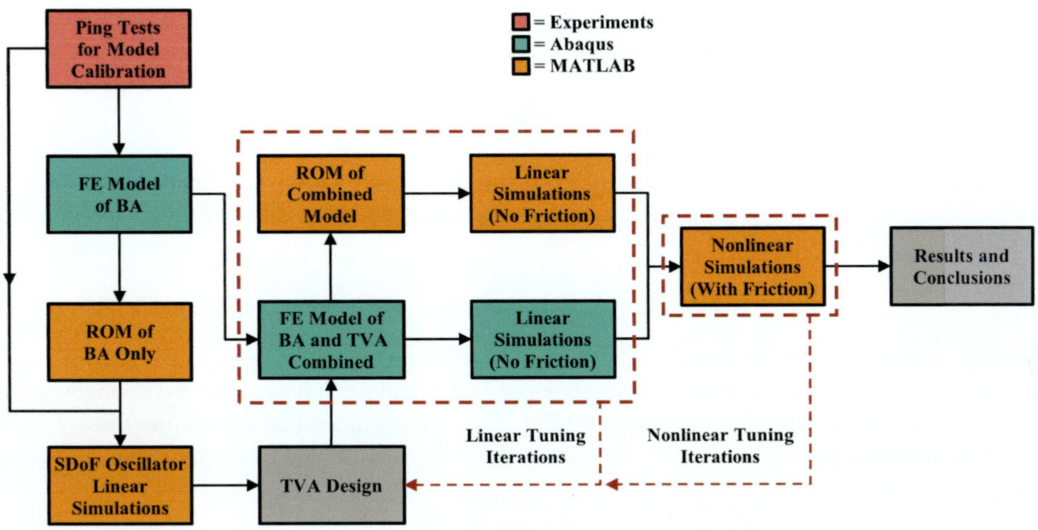

Fig. 1 TVA design workflow.

Theory

The linear equations of motion (EoMs) for a host structure combined with a TVA are given as:

$$\boldsymbol{M}\ddot{\boldsymbol{x}} + \boldsymbol{C}\dot{\boldsymbol{x}} + \boldsymbol{K}\boldsymbol{x} = \boldsymbol{f}_E \tag{1}$$

where $\boldsymbol{f}_E$ is the external forcing vector, $\ddot{\boldsymbol{x}}$, $\dot{\boldsymbol{x}}$, and $\boldsymbol{x}$ are the system response vectors in terms of physical acceleration, velocity, and displacement, respectively, and $\boldsymbol{M}$, $\boldsymbol{C}$, and $\boldsymbol{K}$ are the system mass, damping, and stiffness matrices, respectively. For an elastically coupled host-TVA system, $\boldsymbol{M}$ and $\boldsymbol{K}$ have the general form:

$$\boldsymbol{M} = \begin{bmatrix} \boldsymbol{M}_{host} & 0 \\ 0 & \boldsymbol{M}_{TVA} \end{bmatrix}; \; \boldsymbol{K} = \begin{bmatrix} \boldsymbol{K}_{host} & \boldsymbol{K}_c \\ \boldsymbol{K}_c^T & \boldsymbol{K}_{TVA} \end{bmatrix} \tag{2}$$

where $\boldsymbol{M}_{host}$ and $\boldsymbol{K}_{host}$ correspond to the host structure alone, $\boldsymbol{M}_{TVA}$ and $\boldsymbol{K}_{TVA}$ the TVA alone, and $\boldsymbol{K}_c$ the linear elastic coupling between host and TVA. For the applications considered in this work, the external forcing is assumed to be harmonic, and can be defined as $\boldsymbol{f}_E = \widehat{\boldsymbol{f}}_E e^{i\omega t}$, where $\widehat{\boldsymbol{f}}_E$ is the complex forcing in the frequency domain, $i = \sqrt{-1}$, ω is the forcing frequency in rad/s, and t is time. Accordingly, for the linear EoMs in Eq. (1), a harmonic input force yields a harmonic response of the form $\boldsymbol{x} = \widehat{\boldsymbol{x}} e^{i\omega t}$, where $\widehat{\boldsymbol{x}}$ is the complex response in the frequency domain. Substituting these terms into Eq. (1) yields the frequency-domain EoMs:

$$\left[-\omega^2 \boldsymbol{M} + i\omega \boldsymbol{C} + \boldsymbol{K}\right] \widehat{\boldsymbol{x}} = \widehat{\boldsymbol{f}}_E \tag{3}$$

For the host structure considered in this work, while the model fidelity is such that Eq. (3) could be solved directly in system coordinates without significant computational cost, when nonlinearities are introduced, these equations become prohibitive to solve. Thus, to reduce computational cost, the system matrices are reduced using Craig-Chang component mode synthesis (CC-CMS) [20] with a Galerkin projection scheme. Reduced mass, damping, and stiffness matrices are denoted as $\widetilde{\boldsymbol{M}}$, $\widetilde{\boldsymbol{C}}$, and $\widetilde{\boldsymbol{K}}$, respectively, along with reduced frequency-domain forcing $\widehat{\widetilde{\boldsymbol{f}}}_E$ and responses $\widehat{\widetilde{\boldsymbol{x}}}$, where $\widehat{\widetilde{\boldsymbol{x}}}$ contains master (i.e., retained) degrees of freedom (DoFs), $\widehat{\widetilde{\boldsymbol{x}}}_m$, and secondary DoFs, $\widehat{\widetilde{\boldsymbol{x}}}_s$. In addition, damping for this system is approximated using Rayleigh damping, where the damping matrix $\widetilde{\boldsymbol{C}}$ (and $\boldsymbol{C}$) is assumed proportional to the mass matrix such that $\widetilde{\boldsymbol{C}} = \alpha\widetilde{\boldsymbol{M}}$. The Rayleigh damping coefficient α is related to the modal damping ratio, ζ, by Eq. (4):

$$\zeta = \frac{\alpha}{2\omega} \tag{4}$$

For this work, α is assumed to be constant for all frequencies. Using the natural frequency for the host structure's targeted mode (here, the fundamental mode) ω_1 and experimentally measured damping, ζ_{exp}, yields Eq. (5):

$$\alpha = 2\zeta_{exp}\omega_1 \tag{5}$$

Using this definition of $\widetilde{\boldsymbol{C}}$, Eq. (3) can be rewritten in terms of reduced quantities as:

$$\left[\left(-\omega^2 + i\omega\alpha\right)\widetilde{\boldsymbol{M}} + \widetilde{\boldsymbol{K}}\right]\widehat{\widetilde{\boldsymbol{x}}} = \widehat{\widetilde{\boldsymbol{f}}}_E \tag{6}$$

All linear simulations to follow involve solving Eq. (6) (using MATLAB's mldivide function) considering ω values in a frequency range of interest. When a nonlinear force such as friction is added to the system, an additional term, $\boldsymbol{f}_{NL}$, is included in the system-level EoMs in Eq. (1) to give:

$$\boldsymbol{M}\ddot{\boldsymbol{x}} + \boldsymbol{C}\dot{\boldsymbol{x}} + \boldsymbol{K}\boldsymbol{x} = \boldsymbol{f}_E + \boldsymbol{f}_{NL}(\boldsymbol{x}, \dot{\boldsymbol{x}}) \tag{7}$$

Again, the damping matrix is assumed as $\boldsymbol{C} = \alpha\boldsymbol{M}$. Additionally, CC-CMS is used, where all DoFs at nodes that: 1) have non-zero nonlinear friction contact forces, 2) are response DoFs, and/or 3) have non-zero applied external forces are included in $\boldsymbol{x}_m$. For this work and adopting notation from Ref. [21], it is assumed that the friction force $\boldsymbol{f}_{NL}$ is represented by dry Coulomb friction characterized by a signum function, which is approximated with a hyperbolic tangent function as:

$$\boldsymbol{f}_{NL}\left(\boldsymbol{x}_m, \dot{\boldsymbol{x}}_m\right) \approx \sum_e \boldsymbol{w}_e \mu N \tanh\left(\frac{\dot{x}_{m,e}}{\epsilon}\right) \tag{8}$$

where e is a summation index/identifier for a nonlinear DoF contact pair, $\boldsymbol{w}_e$ is a vector with a single $+1$ and -1 at indices of the contact pair DoFs for nonlinear element e and zeros elsewhere, $\dot{x}_{m,e}$ is the relative velocity between the contact pair e DoFs (always a scalar), and ϵ is a regularization coefficient dictating the sharpness of the hyperbolic tangent function's transition. ϵ can be modified to aid in computation and soften the set of equations, with smaller ϵ yielding closer approximations to the signum function [21]. While μ, N, and ϵ can be chosen to be dependent on e, for this work in which only a single friction contact pair is considered for all nonlinear simulations (i.e., max $e = 1$), this was not applicable. Substituting Eq. (8) into Eq. (7) and using CC-CMS reduced-order quantities yields:

$$\widetilde{\boldsymbol{M}}\ddot{\widetilde{\boldsymbol{x}}} + \alpha\widetilde{\boldsymbol{M}}\dot{\widetilde{\boldsymbol{x}}} + \widetilde{\boldsymbol{K}}\widetilde{\boldsymbol{x}} = \widetilde{\boldsymbol{f}}_E + \sum_e \widetilde{\boldsymbol{w}}_e \mu N \tanh\left(\frac{\dot{\widetilde{x}}_{m,e}}{\epsilon}\right) \tag{9}$$

Note that any non-zero entries in $\widetilde{\boldsymbol{w}}_e$ are associated with master DoFs, with zeros for all secondary DoFs for all e. Similarly, retained DoFs are chosen such that the relative velocity $\dot{\widetilde{x}}_{m,e}$ will only be between master DoFs for all e. The nonlinear simulations to follow involve solving Eq. (9) in the time domain using the first-order, state space form given by:

$$\begin{bmatrix} \dot{\boldsymbol{x}} \\ \dot{\boldsymbol{y}} \end{bmatrix} = \begin{bmatrix} \boldsymbol{y} \\ \widetilde{\boldsymbol{M}}^{-1}\left(\widetilde{\boldsymbol{f}}_E - \widetilde{\boldsymbol{K}}\boldsymbol{x} - \alpha\widetilde{\boldsymbol{M}}\boldsymbol{y} + \sum_e \widetilde{\boldsymbol{w}}_e \mu N \tanh\left(\frac{y_{m,e}}{\epsilon}\right)\right) \end{bmatrix} \tag{10}$$

where $\boldsymbol{x}$ and $\boldsymbol{y}$ in Eq. (10) are general state variables still considering the reduced-order coordinates, and $y_{m,e}$ still represents the relative velocity between the retained DoFs of contact pair e. Because steady-state response amplitudes are sought for all nonlinear FRFs, Eq. (10) is simulated for enough time steps to ensure convergence to a steady-state response for a given ω.

Model Creation and Validation

Before beginning the TVA design process, a model of the host structure, i.e., the BA, was needed. Though previous models of the BA have been created using nominal dimensions [22], it was found that the measured dimensions of the BA considered in this study varied from the nominal. Therefore, the FE model created for this study considered measured dimensions, with significant dimensions shown in Fig. 2. For this work, all high-fidelity models and ROM matrices were created using Abaqus/CAE 2023 software. A mesh convergence study was completed to: 1) ensure modal analysis results (i.e., natural frequencies) were converged, and 2) choose a practical seed size to reduce the computational effort of any high-fidelity simulations and ROM matrix generation. It was determined that a 3 mm seed size using quadratic elements yielding a mode 1 natural frequency of 82.38 Hz was sufficient. Note that this was only for initial model development.

Because TVA effectiveness is highly dependent on the target mode, particularly for a structure with low inherent damping such as the BA, it is critical that the FE model accurately represent the as-manufactured BA with respect to: 1) the target mode's natural frequency and damping, and 2) the FRF response amplitude at and around the targeted mode. For reasons discussed in later sections, mode 1 was chosen as the target mode for this study. As such, identification of the natural frequency and damping of mode 1 was of greatest interest.

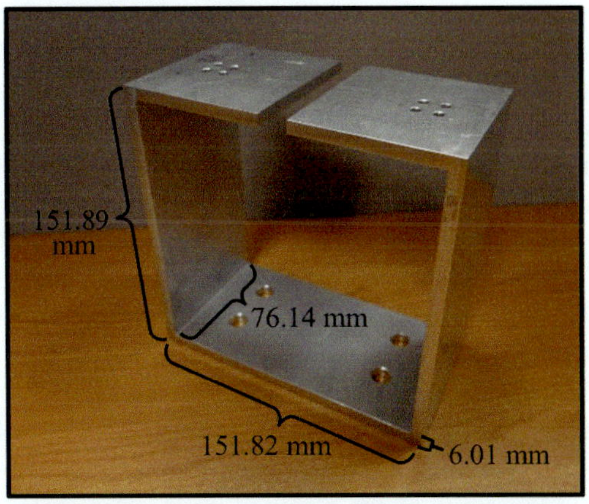

Fig. 2 BA with measured dimensions shown.

A series of ping tests (i.e., impact or hammer tests) were conducted on the BA to experimentally determine the mode 1 damping ratio and first four natural frequencies. For this study, free-free boundaries were considered. Thus, to mitigate any external interference and approximate a free-free boundary condition, the BA was suspended in such a way as to reduce interference and friction at attachment points. As shown in Fig. 3, this low-interference suspension system was achieved using an aluminum box frame, from which fishing line (South Bend M1420 Monofilament Line) connected to rubber bands (Pale Crepe Gold) held the BA aloft in the center of the frame. Fishing line was used to reduce friction at suspension points, and the flexibility of the rubber bands ensured that the tensile forces from the fishing line had a negligible effect on the dynamics of the system.

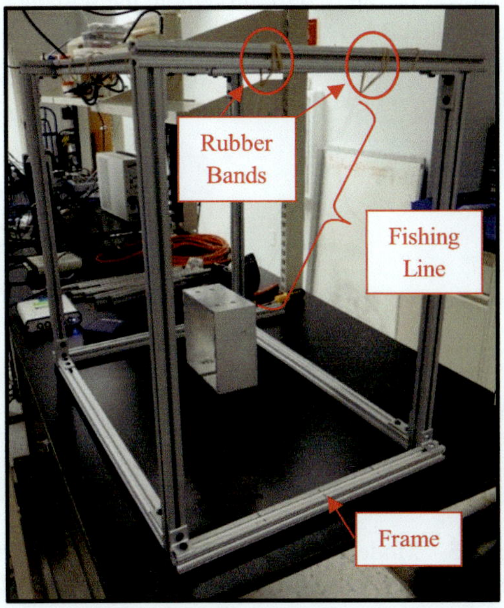

Fig. 3 Ping test experimental setup with BA suspended using rubber bands, fishing line, and an aluminum frame. Accelerometer not shown.

The hammer impact location and the placement location and orientation of a single-axis accelerometer were varied across seven ping tests to test the sensitivity of the BA's response to these variables. Additionally, the angle at which the BA was suspended from the fishing line attachments was varied for this same purpose. It was found that the measured natural frequencies varied negligibly across all hammer impact and accelerometer locations and suspension configurations. The average measured natural frequency of mode 1 was 83.25 Hz, and all measurements were within 0.1% of that value.

Initially, the average measured mode 1 natural frequency differed from the predicted mode 1 natural frequency from the FE model (82.38 Hz) by 0.87 Hz, a 1.05% error. Although this error is appreciably low, due to the BA's low inherent damping, an error of less than 0.50% was desired to consider the model acceptable for this study. Therefore, further model refinement was necessary. Three parameters – Young's modulus, mass density, and geometry – were considered for the purpose of refining the FE model to more closely match the experimentally measured mode 1 natural frequency. Because the mass density and geometry were based on experimental measurements, Young's modulus was modified uniformly across the BA. It was found that increasing the Young's modulus by 1.42% (from 69×10^9 Pa to 70×10^9 Pa) yielded a mode 1 natural frequency of 82.96 Hz, reducing the error to an acceptable 0.35%.

With the model geometry and targeted mode 1 natural frequency calibrated, the final model calibration step was to match the experimental and BA ROM FRFs (ROM details are discussed shortly), specifically around the mode 1 natural frequency. In this work, FRFs are defined as the displacement magnitude of the x-DoF at node 1 divided by the external forcing magnitude (here, the terms magnitude and amplitude are used interchangeably, both referring to the peak-to-peak amplitude divided by 2). Calibrating the ROM FRF was achieved by modifying the mass proportional Rayleigh damping coefficient α, initially estimated as 0.4 rad/s for $\zeta_{exp} \approx 0.0004$ obtained for mode 1 using the half-power method. After reducing α to 0.31 rad/s, the BA ROM FRF magnitude was approximately 0.015 m/N at the mode 1 resonance, corresponding to $\zeta \approx 0.0003$

(or 0.03%). Figure 4 shows strong agreement around the mode 1 natural frequency between the calibrated BA ROM FRF and a representative experimental FRF. While experimental ζ estimates can provide initial estimates of α using Eq. (5), for an isolated target mode like that considered here and for host structures with low inherent damping, small modifications like those done in this work are still likely necessary. Figure 5 shows the shape of mode 1 from the calibrated FE model, which exhibits an opening-closing motion with much of the displacement occurring in the x-y plane.

In summary, after calibration, the three quantities of interest – natural frequency, damping, and FRF magnitude – were in agreement between the experiment and BA ROM for mode 1. It is important to note that while this FE model of the BA may not be valid for other studies, applications, or higher-order modes, this study is only concerned with the behavior of the BA at frequencies at or near the natural frequency of mode 1. Thus, this FE model was considered valid for the purposes of this study.

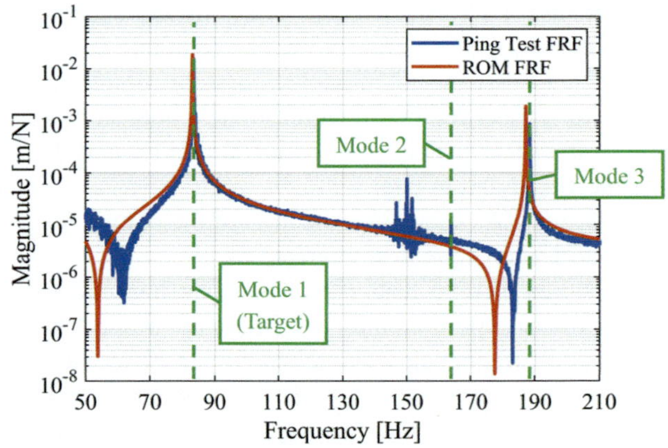

Fig. 4 Comparison of FRFs from the calibrated BA ROM and a representative ping test with first three modes indicated.

For all high-fidelity FE models (with or without a TVA), CC-CMS reduced-order mass and stiffness matrices were generated within Abaqus/CAE and exported to be processed in MATLAB. For the BA alone, master DoFs included all DoFs at all considered TVA placement locations (see Fig. 6). This ROM was used for the model calibration and included 5 retained nodes with 3 translational DoFs each as master DoFs, and 14 secondary (modal) DoFs. The locations of the retained nodes were selected to represent a range of LRs for mode 1 and are shown in Fig. 6. These nodes (node 1 – node 5) are the considered TVA attachment locations and will be referred to throughout the remainder of this work. For combined BA-TVA systems, master DoFs included all DoFs at forcing and response nodes and all DoFs for nodes included in the friction contact.

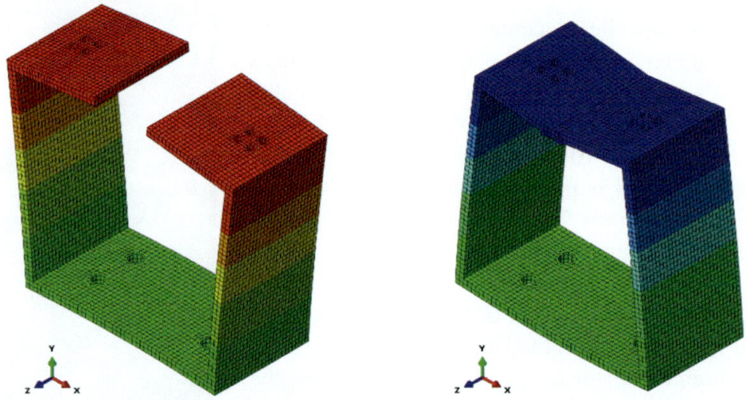

Fig. 5 Mode 1 of calibrated BA FE model. Green represents locations with little to no displacement magnitude, while red and blue represent large positive and negative displacement magnitudes, respectively.

Location Ratio

At a given excitation frequency, the LR for node i is defined in Eq. (11), where the total displacement magnitude at node 1 is divided by the displacement magnitude at node i. Note that the LR can also be calculated using the mode shape alone, which, because mode 1 is significantly isolated from all other modes, is almost identical.

$$\mathrm{LR}_i = \frac{\|\boldsymbol{x}_1\|_2}{\|\boldsymbol{x}_i\|_2} \tag{11}$$

Node 1 was chosen as the reference node for calculating LR because that is where the maximum deflection was observed for the considered harmonic forcing (applied in the x-direction at node 1). The LR of each considered node is shown in Fig. 6.

SDoF TVA Directional Analysis

Before performing any high-fidelity TVA design, a series of linear simulations were conducted to determine the effect of the oscillation direction on TVA effectiveness. Knowing which TVA oscillation direction (at each node) has the greatest effect on host response amplitude can aid design efforts and inform understanding of the energy transfer from the host structure. For these simulations, the reduced-order mass and stiffness matrices were augmented to be of the form shown in Eq. (2) by independently including a SDoF oscillator along each DoF for all nodes shown in Fig. 6, yielding 15 BA-SDoF-TVA systems (5 nodes with 3 DoFs each). These augmented mass and stiffness matrices were used to compute linear FRFs for each BA-SDoF-TVA system. This allowed for the study of linear TVA effectiveness across different TVA placements and orientations with a simple, computationally efficient system.

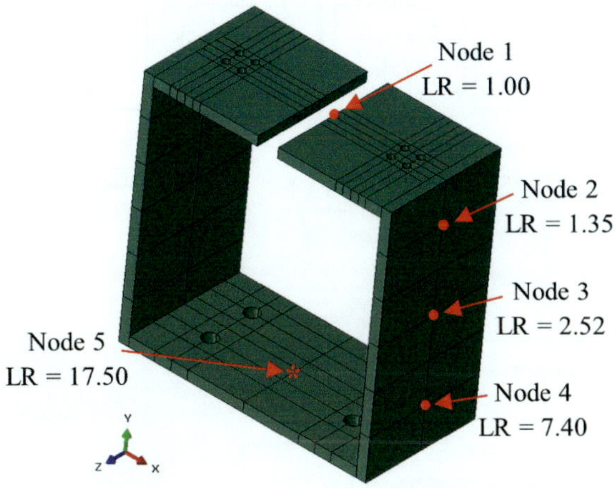

Fig. 6 The 5 nodes considered for TVA placement with respective location ratios (* indicates node on backside relative to view shown).

Because the TVA is modeled as a base-excited SDoF oscillator, it is most excited when the base vibrates in the same direction as the oscillator (i.e., when the SDoF TVA direction and predominant host motion at the attachment location are aligned). Considering a harmonic forcing excitation in the x-direction at node 1, Fig. 7 shows the effect of changing the SDoF TVA's oscillation direction on the peak magnitude reduction of the BA FRF when targeting mode 1 (response reduction is considered for the x-DoF FRF at node 1). Peak magnitude reduction refers to the percent decrease in maximum magnitude across all frequencies in the FRF. The frequency range considered for calculating quantities in Fig. 7 was 65 to 105 Hz. It can be seen in Fig. 7 that a TVA aligned in the x-direction has the greatest reduction at nodes 1 – 4, but alignment in the y-direction has the greatest reduction at node 5. Based on the mode shape in Fig. 5 (and because this mode is isolated as shown in Fig. 4), these results are intuitive given that the forced response deflection shape in this frequency range will be significantly dominated by mode 1. However, for a host structure with a more complex dynamic response, this may not be as evident, and performing this initial study provides a quantitative estimate of TVA directional effectiveness to inform the design process. Additionally, one may consider a local coordinate system that better aligns with dominate motions at candidate nodes.

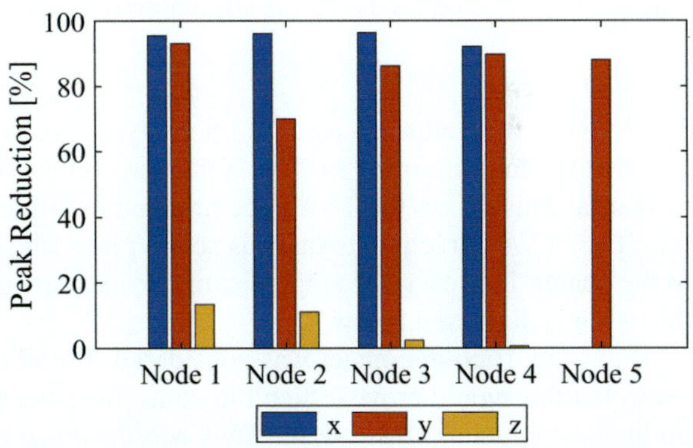

Fig. 7 Effect of SDoF TVA oscillation direction at each considered node on BA peak reduction.

One can also include a friction contact between the BA and SDoF TVA to study the effect of friction on SDoF TVA performance as a function of location and orientation. Though nonlinear analyses for BA-SDoF-TVA systems are not shown for brevity, such simulations can be useful to provide an estimate of nonlinear effectiveness.

TVA Design

To conduct the presented linear and nonlinear parametric studies, a novel TVA for the BA was designed. The TVA had to meet several criteria to be applicable for this study including: 1) a dominant oscillation direction, 2) predictable oscillatory behavior, 3) ease of manufacturing and repairability, 4) a predictable point of friction contact, and 5) tunability with respect to linear and nonlinear behavior. This section discusses the latter portions of the workflow presented in Fig. 1, particularly the iterative TVA design process. Included are notes of design concepts that were not selected for the final model but may be of interest when designing for different host structures and applications. For reference, Fig. 8a displays the detailed model of the TVA design. For the purpose of focusing this study's scope to nonlinear friction contact effects, this model was defeatured, as shown in Fig. 8b, and potential complexities due to nonlinear geometric and fastener physics ignored.

The first design aspect considered was the mechanical apparatus used for oscillation. Apparatuses considered include strings, a cylindrical cantilever beam, a mass-loaded spring, and a rectangular fixed-fixed flexure. The rectangular fixed-fixed flexure was ultimately chosen due to: 1) potential for high discrepancy in moments of inertia between its primary dimensions, 2) approximately linear dynamics assuming fixed bolted connections and relatively small displacements, and

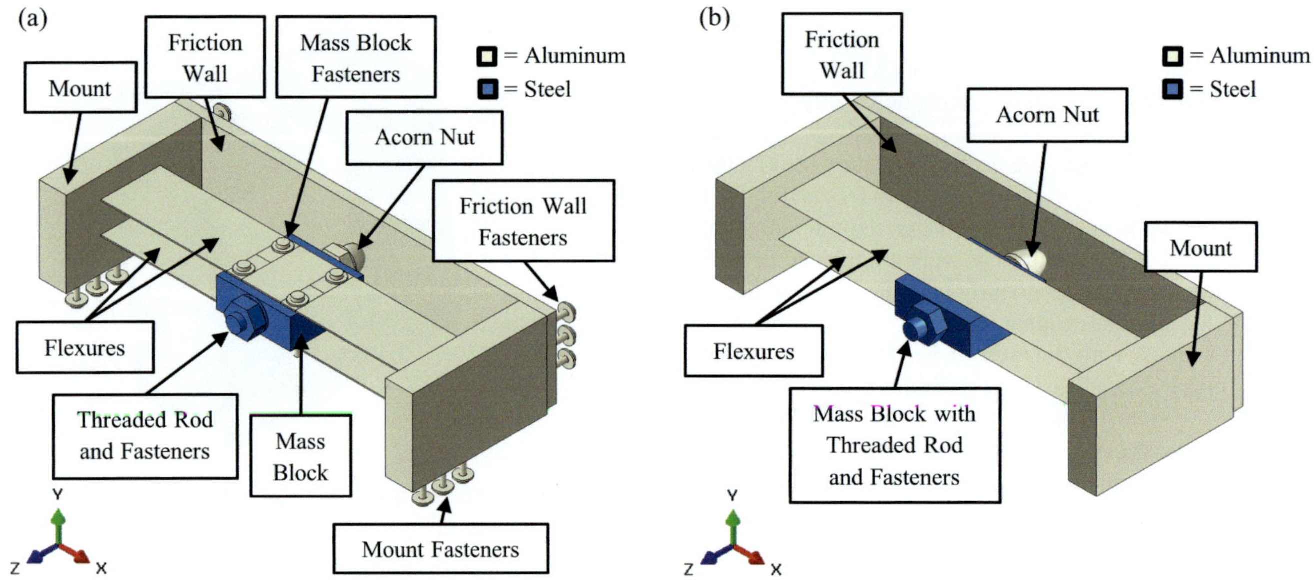

Fig. 8 (a) Detailed model and (b) defeatured model of TVA design.

3) relative manufacturing ease. Additionally, a fixed-fixed flexure, along with the threaded rod, eliminates any reliance on gravity to induce a normal load for a friction contact and can minimize any gravitational effects with well-chosen moments of inertia. This allows the TVA to be oriented in different directions, a need clearly demonstrated by the directional study in the SDoF TVA Directional Analysis section (see Fig. 7). Originally, the TVA was designed as a single flexure with a mass in the center. However, the single flexure design saw undesired torsional modes that were too easily excited; the solution to this issue is discussed below.

Next, the friction contact was considered. Initial design concepts included friction contact with the surface of the BA itself, but this would prove difficult in controlling the normal force while also potentially being an undesired source of wear on the host structure. Instead, the TVA was designed to include a built-in "friction wall" (see Fig. 8) to serve as the contact surface for nonlinear energy dissipation, making the friction contact surfaces independent of the host structure attachment. Degradation of the TVA is preferable to damage of the host structure, as oftentimes replacing the TVA is much less costly than repairing or replacing the host structure. Furthermore, the nature of the contact required careful handling in terms of simulation and practical considerations; the options lied between emulating surface-to-surface contact or point-to-surface contact (edge-to-surface contact, though plausible, was not considered in this design). Surface-to-surface contact emulation (e.g., a flat face rubbing on the friction wall) came with the benefit of providing a counteracting moment force (between the mass and friction wall) to prevent the excitation of torsional modes. However, it is this reactionary moment that raised a concern of uneven pressure distribution at the point of contact. Furthermore, true surface-to-surface contact is often not realized in practice due to inherent variabilities in the contact surfaces. To mitigate this issue, a second fixed-fixed flexure of equal length and thickness was introduced similar to that presented in [23]. This had the effect of greatly increasing the torsional stiffness to minimize the potential for undesired torsional oscillation. From here, the design settled on a point-to-surface contact which was emulated through contact between a rounded end and the friction wall. For the rounded end, several configurations were considered, including a carriage bolt head (like in Lupini et al. [5]), a round-end ball screw, and acorn nut attached to a threaded rod. All of these components are readily available for purchase from retail providers, an important aspect for minimizing manufacturing costs. The acorn nut and threaded rod were ultimately chosen due to their tunability and repairability with respect to both linear and nonlinear system properties.

The last consideration in the design process was the tunability of the TVA. This refers to the ease in which the TVA parameters can be adjusted to affect its target natural frequency and response amplitude. For the TVA in Fig. 8, the main factors affecting the natural frequency are the flexure stiffness, mass properties, and geometry. The factors affecting the frictional force are the normal force applied at the point of contact and the friction coefficient between the surfaces. Designing the TVA to include adjustable flexure stiffness was deemed impractical. Instead, the flexures' length and thickness were chosen to result in a TVA natural frequency that was reasonably close to the target mode; these parameters remained constant. Additionally, the threaded rod comes with the capability to thread on additional mass at the end not in contact with the friction wall, as well as allow for an adjustable applied normal force. Lastly, the acorn nut is not only replaceable, but its material can be changed to modify the friction coefficient between itself and the friction wall. Were this design to be used in practice, it is recommended that that the acorn nut be the softer of the two materials so that it experiences the majority of the wear, as an acorn nut is more easily replaced.

Another important aspect of the TVA is its mass distribution, namely between the "oscillating mass" and the "dead mass" (see Fig. 9). The term "dead mass" is used to define the mass of the TVA which is not contributing to energy dissipation and primarily moves as a rigid body with the host structure. The dead mass typically consists of the hardware used to attach the TVA to the host, though in this study also includes the friction wall and mounts. The mounts consist of blocks with enough bending stiffness to act as fixed ends for the flexures, and in a physical realization of the TVA would be bolted to the host structure. Conversely, oscillating mass is mass whose motion is caused by absorption of energy from the host. Together these two masses make up the total mass of the TVA.

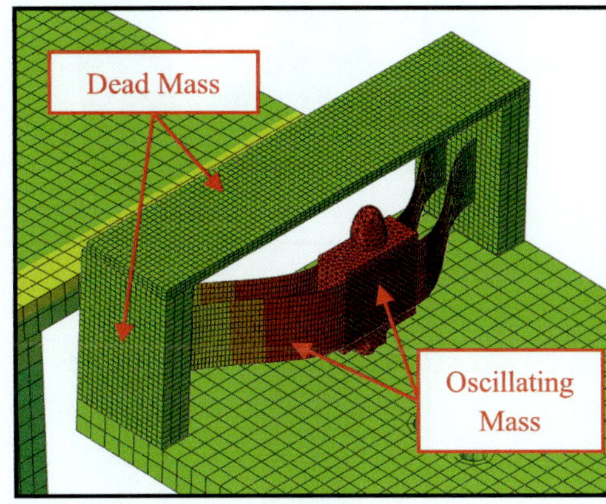

Fig. 9 "Dead mass" vs. "oscillating mass".

This design can be analytically approximated as two fixed-fixed beams acting in parallel. Being able to predict the natural frequency of one's TVA is important in expediting the design process and verifying computational results. The analytical approximation of this TVA's fundamental frequency f_1, while not exact due to assumptions and simplifications needed for analytical formulation, is given in Eq. (12) as:

$$f_1 = \frac{1}{2\pi}\sqrt{\frac{384EI}{L^3 m}} \tag{12}$$

with E, I, and L referring to Young's modulus, area moment of inertia (for configuration shown in Fig. 9, about axis in same direction as the global y-axis), and length (not including the portion attached to the oscillating block) of a single flexure, respectively, and m referring to the mass of the oscillating block. Using parameter values from one of the studied TVA configurations (e.g., $E = 69\times10^9$ Pa, $I = 1.56\times10^{-14}$ m^4, $L = 0.05$ m, and $m = 0.0115$ kg), Eq. (12) gives an estimated frequency of 85.34 Hz. The derivation of Eq. (12) is provided in the Appendix. For designs that closely resemble simple beam structures with or without attachments that approximate point masses, analogous analytical equations are recommended to be used initially to provide estimates informing configuration feasibility and design parameters.

As mentioned previously, the design process was iterative. While the goal was to create a TVA that was applicable to the BA, the design process was also intended to include steps that could be applied more generally for TVA design. Important considerations realized throughout the process are noted and discussed in the Discussion section.

Combined BA-TVA Systems and TVA Tuning

Once the TVA design was satisfactory, five separate FE models of BA-TVA systems (henceforth referred to as "combined systems") were created. In each model, the TVA was mounted at one of the five locations of interest (corresponding to the five nodes in Fig. 6). Figure 10 shows the five FE models, illustrating the five TVA attachment locations.

Initial linear simulations of the combined-system ROMs showed that the TVA split the resonant frequency into an upper peak and a lower peak, as is expected of a well-designed TVA [1]. However, the TVA's effectiveness varies with the tuning of the linear parameters, and thus an optimal tuning had to be found for each attachment location. "Optimal tuning" varies greatly by application; some designers may value how far the new FRF peaks are moved from the original resonance, while others might place greater weight on amplitude reduction.

For this study, two metrics were defined to quantify the performance of the TVA. The first metric, called Peak Split Factor (PSF), quantifies how symmetrically split the two peaks are around the original natural frequency. PSF is defined as:

$$\text{PSF} = \frac{\omega_0 - \omega_{lower}}{\omega_{upper} - \omega_{lower}} \tag{13}$$

where ω_0 is the original target natural frequency of the BA alone, and ω_{upper} and ω_{lower} refer to the combined system's upper and lower natural frequencies, respectively. A PSF value can range from 0 to 1, where a value of 0.5 indicates a perfectly symmetric split. Values lower than 0.5 indicate that the lower peak is closer to the original resonance than the upper peak, and values higher than 0.5 indicate the opposite. Values out of this 0 to 1 range indicate that both peaks are either above or below the original target resonant frequency, making the metric inapplicable. For some applications, it is desirable to have the two peaks as far from the target resonance as possible (i.e., equidistant from the target resonance). For

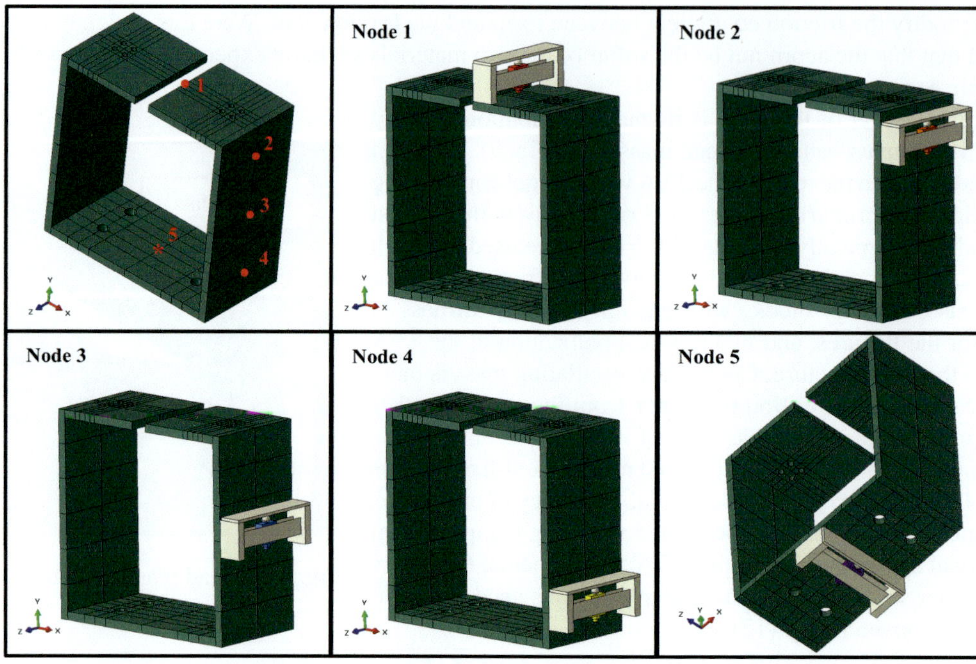

Fig. 10 From top left – BA model with five nodes of interest (* indicates node on backside relative to view shown), the combined model with the TVA at nodes 1 and 2. From bottom left – the combined models with the TVA at nodes 3, 4, and 5. Different colored TVA mass blocks represent different material densities.

example, consider a scenario where the operational environment is predicted to induce a narrowband harmonic forcing with a frequency of ± 1.0 Hz around a host structure's original natural frequency, and a TVA can create a maximum of a 3.0 Hz split between the new natural frequencies. In this case, an effective peak split would result in the new natural frequencies being symmetrically spaced to provide a 0.5 Hz tolerance on both sides of the forcing frequency band. In this example, a symmetric split leads to the lowest risk of resonating at the new natural frequencies.

In this study, it was initially assumed that an optimally tuned TVA was one which prioritized PSF. For this work, if a PSF value between 0.45 and 0.55 could be achieved, the TVA designed was considered tuned for that specific placement location. As will be discussed in the Results section, this tuning scheme was found to be adequate when the TVA was placed at nodes 1, 2, and 3 (i.e., low LRs). However, this was not the case for nodes 4 and 5 (i.e., high LRs). For an application where the applied harmonic forcing has a frequency band greater than the split achievable by the TVA, the host structure could still resonate at one of the new natural frequencies, potentially damaging the structure. As a result, a second tuning iteration was conducted for nodes 4 and 5 using a new performance metric, called Peak Reduction (PR). PR is defined as:

$$\mathrm{PR} = \frac{\max\left(Y_0\left(\omega\right)\right) - \max\left(Y\left(\omega\right)\right)}{\max\left(Y_0\left(\omega\right)\right)} \cdot 100 \tag{14}$$

where $\max\left(Y_0\left(\omega\right)\right)$ is the maximum response amplitude of the BA alone and $\max\left(Y\left(\omega\right)\right)$ the maximum response amplitude of the combined system, both over the entire frequency range of interest. PR quantifies the response amplitude reduction due to the addition of the linear TVA, regardless of the shift in natural frequencies of the new resonant peaks. Higher values of PR correspond to greater amplitude reductions for the host structure, and therefore indicate greater TVA effectiveness. A PR greater than 45% was considered adequate for this tuning iteration.

As mentioned previously, linear tuning of this TVA design is achieved through the addition of mass. In practice, this would be achieved be screwing an additional mass onto the end of the threaded rod going through the oscillating mass. For simulation purposes, the density of the mass in the TVA model was adjusted to emulate this additional (or detracted) mass, rather than adding new geometry to the model. This allowed for mesh consistency across all simulations while performing linear tuning iterations, in addition to avoiding: 1) potential computational discrepancies due to mesh regeneration, and 2) time needed for remeshing. Table 1 shows the mass densities of the oscillating mass used at each considered node/attachment location for both tuning iterations along with their resulting PSF and PR values.

In total, seven combined system FE models were created (five tuned for PSF, and two tuned for PR considering node 4 and 5 attachment locations specifically), and CC-CMS reduced-order mass and stiffness matrices were extracted for each model.

Table 1 Densities of oscillating mass and equivalent added mass after tuning the TVA for each node/attachment location, with resulting PSF and PR values.

Target Tuning Metric	Node	Density of Oscillating Mass, [kg/m^3]	Equivalent Added Mass, [%]	Resulting PSF	Resulting PR, [%]
PSF	1	7800*	–	0.477	51.4
	2	8000	2.3	0.452	50.2
	3	8400	6.9	0.456	53.1
	4	8860	12.2	0.487	20.2
	5	9000	13.8	0.458	3.8
PR	4	8400	6.9	0.033	50.8
	5	8220	4.8	–	48.8

*Nominal steel density

For each ROM, three master nodes (i.e., 9 retained master DoFs) were retained: 1) a response node, which is coincident with node 1, and 2) a contact pair of nodes – one on the friction wall and one on the tip of the acorn nut. Because the harmonic forcing was applied at the response node in the x-direction (see Fig. 6) for all cases, this minimal set of retained nodes allows for: 1) applying the external harmonic force and nonlinear forces directly in the reduced-order space, and 2) calculating desired responses without necessitating a projection back to the full-order system coordinates. All linear and nonlinear MATLAB simulations of the combined systems presented in the Results section considered these reduced-order matrices for the respective TVA attachment locations and tunings indicated.

Results

Linear Simulations

Linear simulations of the combined systems were performed using each of the seven ROMs, and the resulting FRFs are shown in Fig. 11 along with the FRF of the BA alone (i.e., no TVA) for comparison. As with the SDoF TVA simulations, the harmonic excitation was applied in the x-direction at node 1, and the presented FRFs are for the x-DoF at node 1.

With the linear TVA tuned with respect to PSF, a nearly even split is seen for all five nodes. As the LR increases from node 1 to node 5, the resulting natural frequencies converge toward that of the BA alone with no TVA, though the two resulting peaks remain approximately equidistant from the original resonance. For the cases with a TVA attached at nodes 1, 2, and 3, a PR above 50% is achieved. However, for the cases with a TVA attached at nodes 4 and 5 and PSF-based tuning, peak magnitudes are similar to that of the BA alone, having PRs of 20.2% and 3.8%, respectively. When the TVAs for these placements were retuned with respect to PR, their resulting FRF peaks were reduced yielding PR values above 48%. However, their respective PSF values after tuning for PR shifted out of the desired range (0.45 – 0.55) as shown in Table 1. Because both resulting natural frequencies of the combined system with the TVA placed at node 5 were above the original target natural frequency ω_0, the PSF was not applicable and is thus not reported in Table 1. This phenomenon is discussed further in the Design Considerations section.

Nonlinear Simulations

The same reduced-order matrices used for linear simulations were also used for all nonlinear simulations, with friction contact now included between the contact pair DoFs. For all presented nonlinear simulations, the same harmonic excitation and response DoF as that used for linear simulations is considered. Time series simulations were computed using Eq. (10) and ode45 in MATLAB for individual excitation frequencies within the frequency range of interest. Steady-state response amplitudes (peak-to-peak amplitude divided by 2) were calculated for each excitation frequency once response amplitudes were no longer changing in time, and the results compiled to produce a nonlinear FRF.

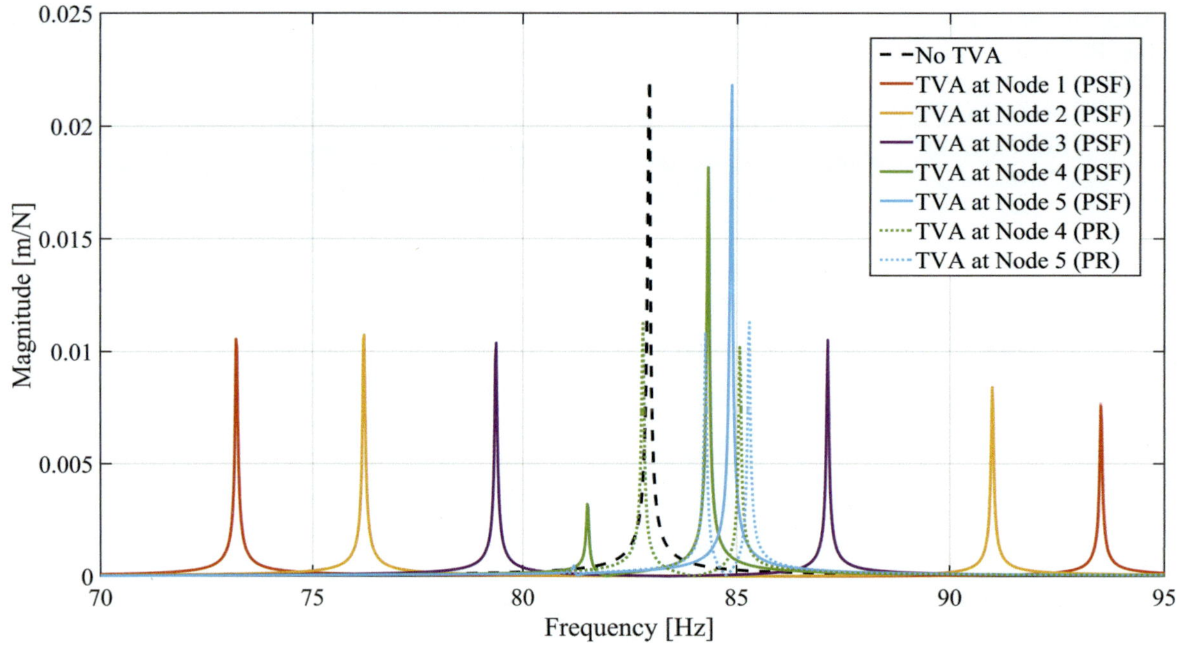

Fig. 11 Linear FRFs of the BA alone and combined systems with TVA attached at the five nodes of interest. PSF-based tunings are shown for attachment at nodes 1, 2, and 3, and both PSF-based and PR-based tunings are shown for attachment at nodes 4 and 5.

From linear simulations, the highest achievable peak reduction across all attachment locations was 48.8%; with the addition of friction, the goal was to attain over 90% PR for each TVA attachment location. Thus, the PR-based TVA tuning was considered for nonlinear simulations with the TVA attached at nodes 4 and 5. However, as mentioned previously, PSF-based TVA tuning with the TVA attached at nodes 1, 2, and 3 was sufficient, and thus the PSF-tuned versions of the TVA were maintained. Nonlinear simulations consisted of varying the value of the dimensionless microslip parameter $\mu N/F$,

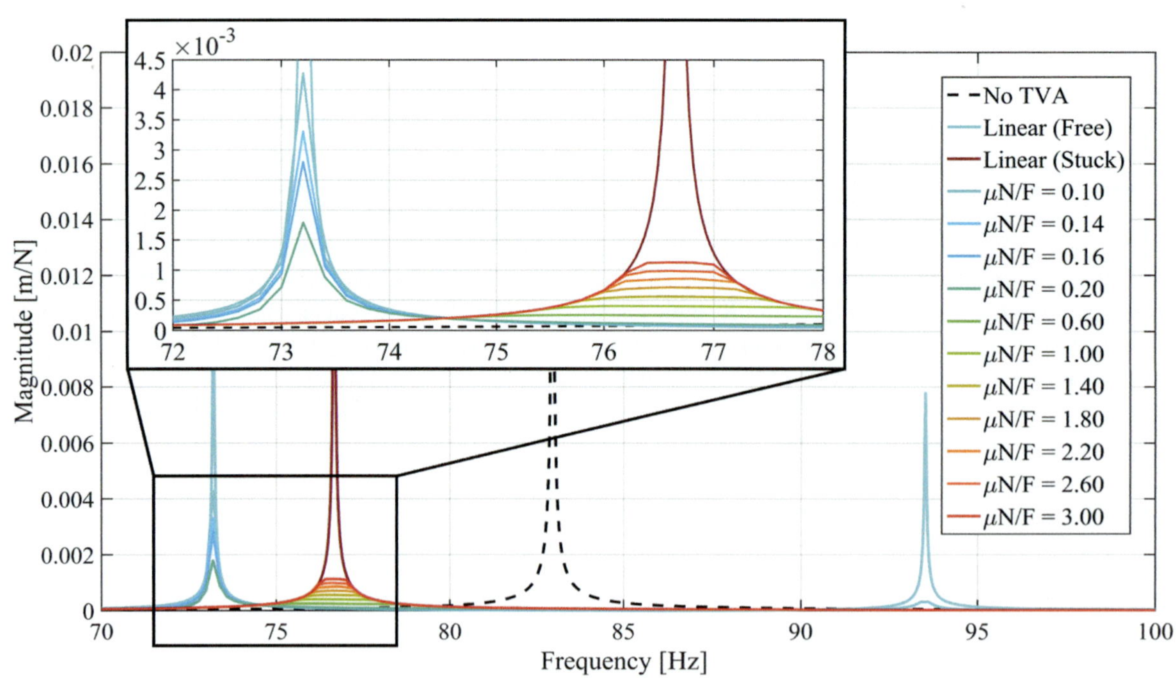

Fig. 12 Nonlinear FRFs for friction-enhanced TVA attached at node 1.

i.e., the ratio of the friction force, μN, to the external forcing amplitude F. For the systems considered here, note that any combinations of μ, N, and F that yield an identical $\mu N/F$ will yield the same nonlinear response, regardless of the specific values of μ, N, and F considered.

Figure 12 shows nonlinear FRFs considering a range of $\mu N/F$ values for the TVA placed at node 1, and includes the linear FRFs of the "No TVA" and linear TVA (labeled "Linear (Free)") shown in Fig. 11 for reference. Figure 12 also includes the linear "stuck" case (labeled "Linear (Stuck)"), in which $\mu N/F$ is increased to the point of having no relative motion between the contact pair (i.e., the entire TVA acts as dead mass). For this TVA placement, a $\mu N/F$ value of 0.30 attains the maximum PR of 99%.

The focus of this study, however, is the effectiveness of the TVA at high-LR areas, i.e., nodes 4 and 5. TVAs placed at nodes 2 and 3, being low-LR areas, yielded similar results to those presented for the TVA at node 1 and are therefore

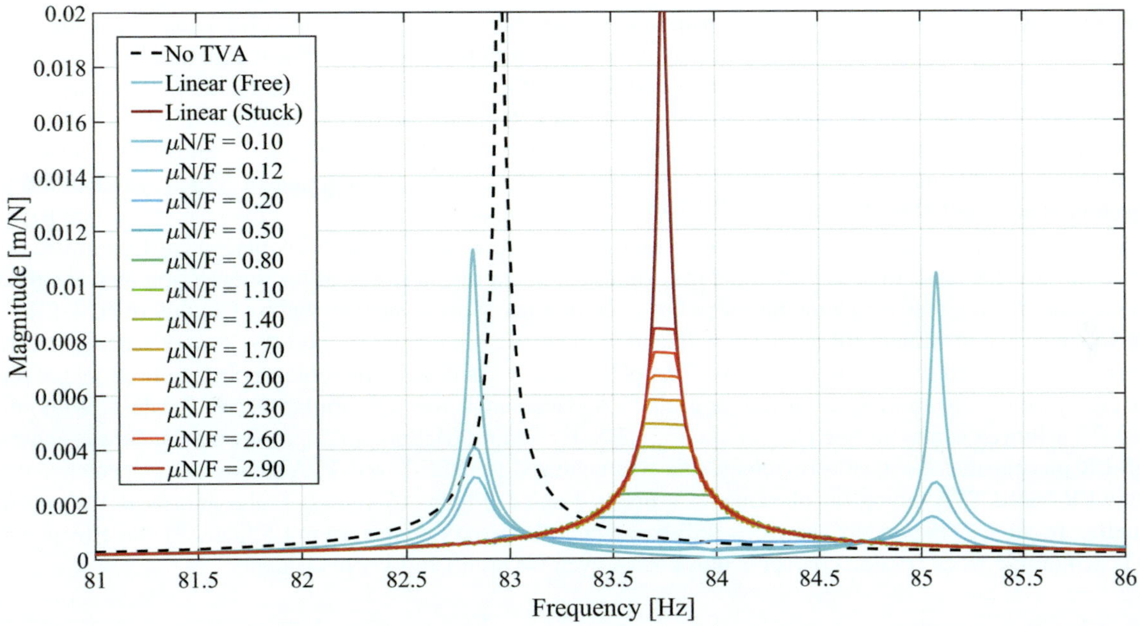

Fig. 13 Nonlinear FRFs for friction-enhanced TVA attached at node 4.

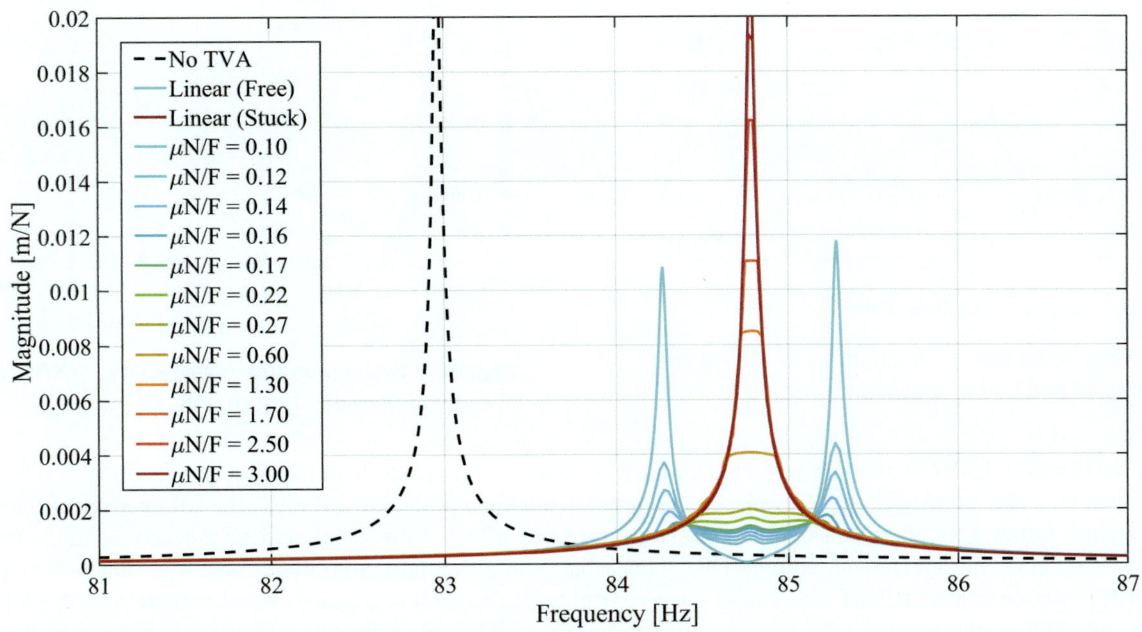

Fig. 14 Nonlinear FRFs for friction-enhanced TVA attached at node 5.

not shown. The nonlinear and reference linear FRFs for TVA placement at nodes 4 and 5 are shown in Figs. 13 and 14, respectively.

When the TVA is placed at node 4, a maximum PR of 96.3% is achieved across all $\mu N/F$ values. When placed at node 5, a maximum PR of 93.3% is achieved. These PR values are achieved for $\mu N/F = 0.20$ and $\mu N/F = 0.18$, respectively.

Discussion

The linear FRFs in Fig. 11 illustrate important trends as the LR increases. For TVA attachment at nodes 1, 2, and 3 with low LRs, tuning for either PSF or PR achieves similar results; however, this is not the case for attachment at nodes 4 and 5 with high LRs. For these TVA attachment locations, there is a dramatic tradeoff between PSF and PR, emphasizing that TVAs attached at high-LR areas can be much more sensitive to tuning than if attached at low-LRs areas. Choosing the optimal tuning scheme is an engineering decision informed by the operational environment and host structure limits.

The nonlinear FRFs considering a friction-enhanced TVA in Figs. 12 – 14 show a dramatic increase in reduction for all attachment locations compared to their respective linear FRFs. For the case of TVA attachment at node 1 (Fig. 12), the nonlinear maximum PR is 99.0%, a vast improvement over that of the linear TVA (51.4%) and comfortably above the desired PR of at least 90%. Similar improvements from 50.8% to 96.3% PR for attachment at node 4 (Fig. 13) and 48.8% to 93.3% PR for attachment at node 5 (Fig. 14) are also achieved. In each case, as $\mu N/F$ increases, the response of the BA begins to leave the linear free FRF and converge to the linear stuck FRF. A trend seen for all TVA placement locations was that the maximum achievable PR tends to occur when the maximum nonlinear response lies at the intersection between the free and stuck curves. This effect agrees with what has been observed in other studies, such as that by Cimpuieru et al. [3].

Figure 15 shows the maximum PR achieved with and without friction for the PR and PSF tuning schemes, with the five data points in each curve corresponding to the five TVA placements considered. For values including friction, the plotted PR values correspond to the greatest PRs achieved for each TVA placement across all simulated $\mu N/F$ values. Without friction, the greatest PR achieved across all five placements is 53.2%. By adding friction, a PR greater than 93% was achieved even at the high-LR placements. As stated previously, this was achieved for PSF-based TVA tuning for placements at nodes 1, 2 and 3, and PR-based TVA tuning for placements at nodes 4 and 5. Figure 15 shows: 1) the improved TVA performance when including friction, 2) the tradeoff between PSF-based and PR-based tuning at high LRs, and 3) that effective vibration reduction is achievable when friction contact is introduced, even for high-LR TVA placements.

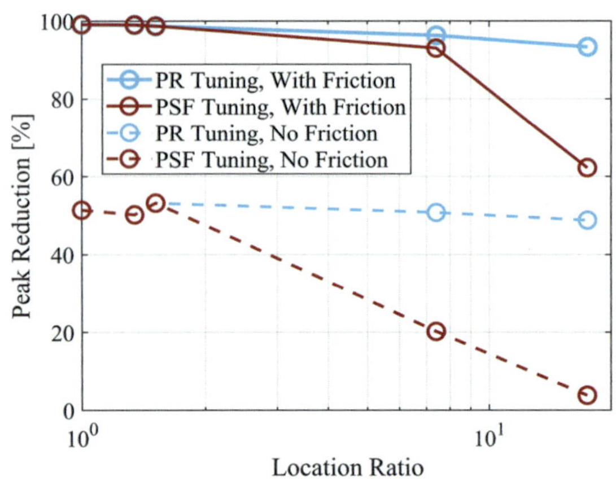

Fig. 15 Peak reduction vs. location ratio for varying TVA placement and tuning scheme.

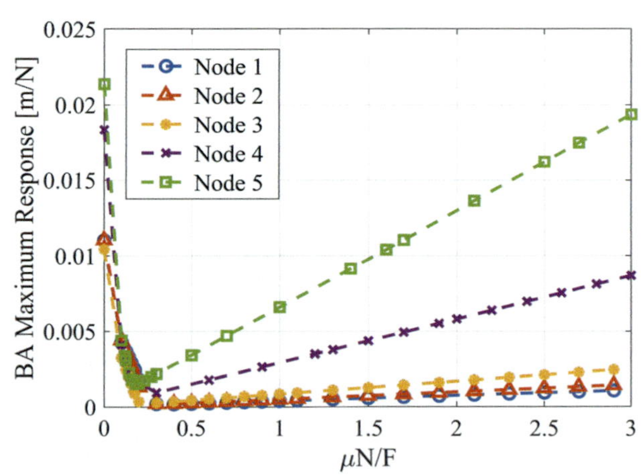

Fig. 16 BA maximum response vs. μN/F for TVA placement at nodes 1 through 5.

Figure 16 displays the BA maximum response for each considered TVA placement location as a function of $\mu N/F$. One can notice that the curves flatten as the TVA placement moves from node 5 to node 1 (i.e., as the LR decreases). That is, for TVA placements at low LRs, there is a significantly larger range of $\mu N/F$ values that achieve a nearly identical reduction as the $\mu N/F$ yielding the greatest reduction. This shows that, while dramatic amplitude response reduction is possible across all LRs, friction contact parameters require more careful tuning to achieve a near-optimal reduction for high-LR TVA placement. In addition, the optimal $\mu N/F$ varies slightly across all TVA placement locations, indicating that the optimal $\mu N/F$ can depend on LR. These results highlight the importance of having a tunable friction contact, as in practice, some iteration is likely needed to achieve optimal, or near optimal, reduction for TVAs placed in a high-LR area.

Design Considerations

Thus far, the presented results have shown agreement with what similar studies in literature have demonstrated – that effective friction-enhanced TVAs can be designed for high-LR applications. However, the design process described here has yielded a number of insights into the nature of friction-enhanced TVA design, which could be used to guide future efforts. These insights have been compiled into the following set of design considerations, with the hope that they will help inform future TVA design efforts and assist in making strategic design choices.

1. Determine potential TVA attachment locations and evaluate feasibility through a location ratio study.
For the host structure considered in this work, it is relatively easy to see where preferable (low-LR) TVA attachment locations exist for the targeted mode through visual examination of the mode shape for mode 1. However, this is may not true for complex structures, and an examination of the displacement amplitudes across the host structure may reveal unexpected variations in LR. After an initial computational modal analysis of the BA, it was found that for some mode/node combinations, the LR was extremely high. For example, the LR at node 5 for mode 2 is approximately 12,751, indicating an almost negligible vibration amplitude at this location. Although mode 2 was not considered in this study, it is reasonable to assume that if a TVA were placed at this location to target mode 2, there would not be enough energy transfer for a TVA to have a significant impact on the BA's response. It is therefore recommended that designers calculate the LR to determine feasible/desirable attachment locations on the host structure for their mode of interest. Doing so allows one to eliminate impractical attachment locations where a TVA would not be able to effectively absorb energy from the host structure.

2. Perform a directional study to determine effective TVA oscillation directions.
The effectiveness of a TVA can vary greatly depending on the direction of oscillation of the TVA mass. In this study, the SDoF TVA directional analyses showed that when targeting mode 1, in most cases, the greatest amplitude reduction was achieved when the TVA mass oscillated in the x-direction (except at node 5), and rarely had any effect when oscillating in the z-direction. Furthermore, a TVA's effectiveness with respect to oscillation direction will differ based on the targeted mode. For the BA, modes 2 and 4 have mode shapes which involve motion in the z-direction, whereas mode 1 did not (see Fig. 17). Therefore, when targeting modes 2 or 4, a TVA oscillating in the z-direction may be more effective.

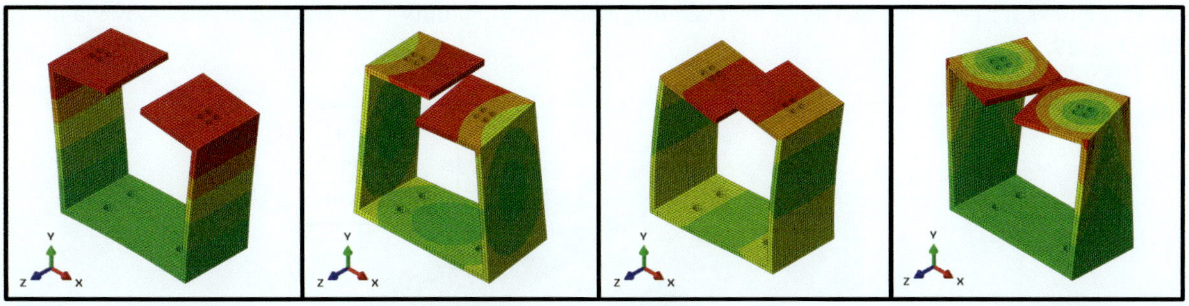

Fig. 17 From left to right, modes 1 through 4 of the BA. The green to red color gradient represents the minimum to maximum displacement gradient, respectively.

For some structures, the choice of oscillation direction is intuitive; if the structure itself oscillates primarily in one direction, that same direction will likely be the optimal choice for the TVA mass' oscillation. This was the case for the BA, where the mode shape associated with mode 1 clearly shows the largest displacement in the x-direction. Consequently, x-direction oscillators were almost always the most effective in this case. However, more complicated host structures may have mode shapes with no obvious primary direction of displacement. Furthermore, there may be oscillation directions that would be preferred, but are not feasible due to geometry or other placement constraints. In these instances, a thorough study of the effectiveness of various oscillation directions is recommended to ensure sufficient energy transfer from the host structure to the TVA.

3. Evaluate the effect of mass loading on the host structure.
In this study, it was found that increasing the amount of dead mass (for the same placement location) decreased the natural frequency of the combined system when the TVA acts a rigidly attached mass (e.g., see linear "stuck" and "No TVA" curves in Fig. 12). For the TVA presented in this work, targeting a lower natural frequency meant increasing the oscillating mass

(see Eq. 12). As such, as the dead mass was increased, the oscillating mass was also increased to compensate for the resulting change in target frequency due to this mass loading effect. Because a heavier TVA is often undesirable for many applications, such as for aerospace structures where mass is typically sought to be minimized, it follows that designers should understand the mass properties of their TVAs and seek to minimize, or utilize, dead mass where possible, and if necessary, account for potential mass loading effects on overall system dynamics.

4. Account for stiffening effects in combined system dynamics due to TVA mounts.
For the high-LR areas in this study (i.e., at nodes 4 and 5), the addition of the TVA increased the combined system natural frequency when the TVA acts as a rigidly attached mass (e.g., see the linear "stuck" and "No TVA" curves in Figs.13 and 14). This is a counterintuitive result, as adding mass is often expected to decrease natural frequency as discussed in design consideration 3. The increase seen for TVA placement at nodes 4 and 5 is due to a stiffening effect caused by the TVA's mounting mechanism as illustrated in Fig. 18.

Considering the area moment of inertia equation for a beam with a rectangular cross-section, $I = bh^3/12$, the BA's area moment of inertia about its bending axis for mode 1 is significantly increased with the introduction of the TVA's mounts. This stiffening effect dominates in comparison to the mass loading effect at high LRs, leading to an overall increase in the combined system's natural frequencies. It follows that designers should understand the stiffening effects of their TVAs and seek to minimize, or utilize, the impact of their mounting mechanisms.

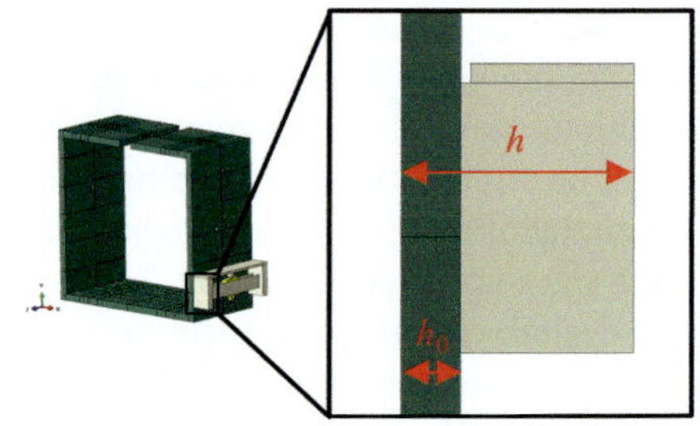

Fig. 18 Structure thickness with, h, and without, h_0, TVA mounts, illustrating effective increase in thickness due to TVA mounts (attachment at node 4 shown).

5. Ensure the TVA is tunable with respect to target natural frequency and friction contact parameters.
TVAs that use friction-based energy dissipation mechanisms are sensitive to many factors, some of which are addressed in this study (i.e., TVA placement, mass, friction coefficient, and normal force). As shown, this is especially true for placements at locations with high LRs (e.g., see Fig. 16). Designing a TVA such that one can adjust these properties based on empirical effectiveness is desirable when optimizing for a selected performance metric. In the case of the presented TVA design, tunability also came with the benefit of repairability due to the replaceable acorn nut. This allows for greater longevity of the overall component. Tunability was achieved in this study via the features portrayed in Fig. 19.

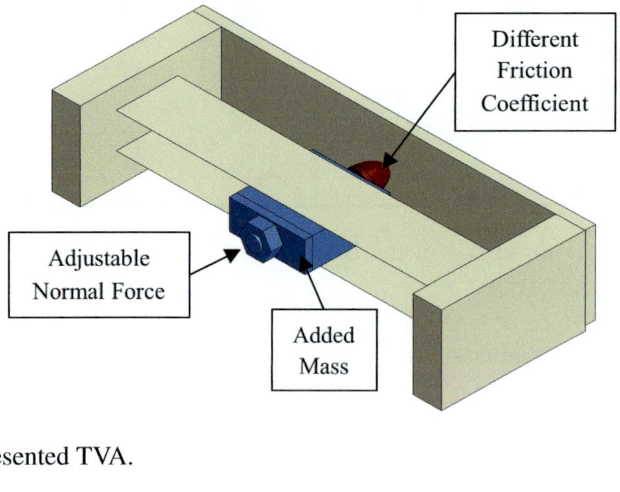

Fig. 19 Tunable features of presented TVA.

6. Tune with respect to the performance metric that best applies to the host structure's operational environment.
Operational environments that can vary dramatically will significantly impact TVA behavior and effectiveness. Understanding the nature of the loads the host structure will experience is crucial to optimizing a TVA's performance. There are several considerations in choosing the best suited performance metric, whether it be from those detailed in this study or one of the designers' choosing. Examples of considerations include, but are not limited to, variety in environmental loading conditions, load frequency band, frequency sweeps encountered during startups and shutdowns, and desired level of vibration absorp-

tion, among others. It is recommended that a metric tailored to the operational environment be chosen, and that the TVA is iteratively tuned as needed to achieve optimal performance.

7. Perform a fatigue analysis to evaluate TVA lifespan within a given operational environment.
Though not shown in this study for brevity, a fatigue analysis is a crucial step that should be performed prior to manufacturing a TVA. An understanding of the anticipated stresses experienced by a TVA during operation will aid designers in predicting the TVA's longevity and scheduling routine repairs/replacement of the TVA or its components.

Future Work

While this study did not consider a physical realization of the TVA, the manufacturing of this TVA would likely reveal insights that could justify (or call into question) the design choices made in this study, and thus give rise to new design considerations. Furthermore, experimental testing of this TVA design would be necessary to corroborate the simulation results and confirm the effectiveness of the TVA. Both steps – the building and testing of the proposed TVA – would further authenticate and inform any augmentations to the design considerations proposed herein.

Acknowledgments This research was funded by Los Alamos National Laboratory (LANL) through the Engineering Institute's Los Alamos Dynamics Summer School. The Engineering Institute is a research and education collaboration between LANL and the University of California San Diego's Jacobs School of Engineering. This collaboration seeks to promote multidisciplinary engineering research that develops and integrates advanced predictive modeling, novel sensing systems, and new developments in information technology to address LANL mission relevant problems. The authors would also like to acknowledge Samantha Ceballes and Julian Davis for contributing their technical expertise to the development of this project.

References

1. Connor, J.J.: Tuned Mass Dampers. In: Introduction to Structural Motion Control. New York, NY, USA. Prentice Hall Pearson Education, Inc., (2003).
2. Sun, J.Q., Jolly, M.R., Norris, M.A.: Passive, Adaptive and Active Tuned Vibration Absorbers—a Survey. Journal of Mechanical Design. (117)B, 234–242 (1995). https://doi.org/10.1115/1.2836462
3. Cimpuieru, M., Kripfgans, A.D., Kelly, S.T., Epureanu, B.I.: Additive Manufacturing of Resonant Vibration Absorbers for Turbomachinery Blisks. Presented at the IMAC XLII, Orlando, Florida, USA, (2024).
4. Lupini, A., Mitra, M., Epureanu, B.I.: Application of Tuned Vibration Absorber Concept to Blisk Ring Dampers: A Nonlinear Study. Journal of Engineering for Gas Turbines and Power. (141)10, 101016 (2019). https://doi.org/10.1115/1.4044684
5. Lupini, A., Shim, J., Epureanu, B.I.: Experimental and Computational Study of a Tuned Damper with Frictional Contacts. AIAA Journal. (58)8, 3607–3613 (2020). https://doi.org/10.2514/1.J059393
6. Dalisay, J.D.E., Moore, K.J., Bergman, L.A., Vakakis, A.F.: Local Nonlinear Stores Induce Global Modal Interactions in the Steady-State Dynamics of a Model Airplane. Journal of Sound and Vibration. (500), 116020 (2021). https://doi.org/10.1016/j.jsv.2021.116020
7. Aksoy, T., Acar, B., Özgen, G.O.: A Novel Method for the Vibration Reduction of an Air Vehicle Missile Launcher. In: AIAA Scitech 2019 Forum. https://doi.org/10.2514/6.2019-0494
8. Cimpuieru, M., Kelly, S.T., Epureanu, B.I.: Computational Analysis of Tuned Vibration Absorbers with Impact and Friction Contacts. In: AIAA Scitech 2023 Forum. https://doi.org/10.2514/6.2023-2075
9. Cimpuieru, M., Kelly, S.T., Epureanu, B.I.: Resonant Vibration Absorbers with Impacts. In: Nonlinear Structures & Systems, Volume 1, Conference Proceedings of the Society for Experimental Mechanics Series. 237–241 (2023). https://doi.org/10.1007/978-3-031-36999-5_31
10. Nasr, A., Mrad, C., Nasri, R.: A Study of Friction Vibration Absorber: Impact of Friction Modeling on the Efficacy of the Absorber and Friction Coefficient Optimization. Mechanics & Industry. (19)6, (2018). https://doi.org/10.1051/meca/2018039
11. Pisal, A.Y., Jangid, R.S.: Dynamic Response of Structure with Tuned Mass Friction Damper. International journal of Advanced Structural Engineering. (8)4, 363–377 (2016). https://doi.org/10.1007/s40091-016-0136-7
12. Wang, C., Brown, J.D., Singh, A., Moore, K.J.: A Two-Dimensional Nonlinear Vibration Absorber Using Elliptical Impacts and Sliding. Mechanical Systems and Signal Processing. (189), 110068 (2023). https://doi.org/10.1016/j.ymssp.2022.110068
13. Wang, Y., Ma, Y., Wang, H., Firrone, C.M., Hong, J.: Dynamic Performance of a Novel Tuned Vibration Absorber with Nonlinear Friction Interfaces. Nonlinear Dynamics. (112)11, 8831–8848 (2024). https://doi.org/10.1007/s11071-024-09347-8
14. Hong, H.K., Liu, C.S.: Coulomb Friction Oscillator: Modelling and Responses to Harmonic Loads and Base Excitations. Journal of Sound and Vibration. (229)5, 1171–1192 (2000). https://doi.org/10.1006/jsvi.1999.2594
15. Wang, X., Wang, R., Huang, B., Mo, J., Ouyang, H.: A Study of Effect of Various Normal Force Loading Forms on Frictional Stick-Slip Vibration. Journal of Dynamics, Monitoring and Diagnostics. (1)1, 46–55 (2021). https://doi.org/10.37965/jdmd.v2i2.48
16. Griffin, J.H., Menq, C.-H.: Friction Damping of Circular Motion and Its Implications to Vibration Control. Journal of Vibration and Acoustics. (113)2, 225–229 (1991). https://doi.org/10.1115/1.2930173

17. Mitra, M., Lupini, A., Epureanu, B.I.: A Study on the Application of Tuned Vibration Absorbers to Nominally Cyclic Structures. In: ASME Turbo Expo 2019: Turbomachinery Technical Conference and Exposition, Volume 7B: Structures and Dynamics, V07BT35A002. Pheonix, Arizona, USA (2019). https://doi.org/10.1115/GT2019-90167
18. Lupini, A., Epureanu, B.I.: A Friction-Enhanced Tuned Ring Damper for Bladed Disks. Journal of Engineering for Gas Turbines and Power. (143)1, 011002 (2021). https://doi.org/10.1115/1.4049203
19. Lupini, A.: Computational Models and Damper Concepts for Vibration Reduction of Turbomachinery Blisks. Ph.D. Thesis. University of Michigan. (2020).
20. Craig, R.R., Chang, C.-J.: Free-Interface Methods of Substructure Coupling for Dynamic Analysis. AIAA Journal. (14)11, 1633–1635 (1976). https://doi.org/10.2514/3.7264
21. Krack, M., Gross, J.: Harmonic Balance for Nonlinear Vibration Problems. Springer Cham, (2019). https://doi.org/10.1007/978-3-030-14023-6
22. Hart, K., Lewis, E., Sinclair, S., Moore, K.J., Flynn, G.S., Haynes, C.: A Characterization of the Uncertainty in Force-Control Testing for Aerospace Applications. Presented at the IMAC XLII, Orlando, Florida, USA, (2024).
23. Singh, A., Moore, K.J.: Identification of Multiple Local Nonlinear Attachments Using a Single Measurement Case. Journal of Sound and Vibration. (513), 116410 (2021). https://doi.org/10.1016/j.jsv.2021.116410

Appendix

This section details the step-by-step derivation of the analytical equation, Eq. (12), used to guide this study's initial TVA design efforts. A sketch of a simplified configuration used to model the presented TVA design is given in Fig. A.1.

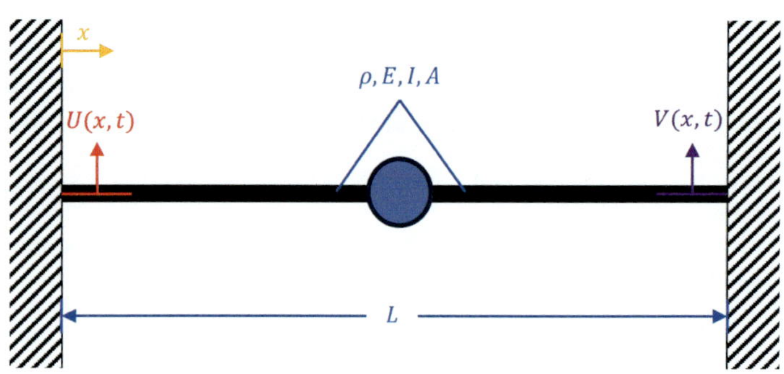

Fig. A.1 Simplified sketch of single flexure with center mass.

The governing equations (considering a local frame) are given by:

$$\rho A U_{tt} + EI U_{xxxx} = 0, \text{ for } 0 \leq x \leq \frac{L}{2} \tag{A.1}$$

$$\rho A V_{tt} + EI V_{xxxx} = 0, \text{ for } \frac{L}{2} \leq x \leq L \tag{A.2}$$

where ρ is the mass density per unit area, A the cross-sectional area, E the Young's modulus, I the area moment of inertia about the out-of-plane axis in Fig. A.1, L the length (not including the portion attached to the oscillating block), m the mass of the oscillating block, and $U(x,t)$ and $V(x,t)$ the respective displacement of the flexures on either side of the oscillating block.

Applying fixed boundary conditions at $x = 0$ and $x = L$ and compatibility conditions at $L/2$ yields the following relations:

$$U(0,t) = U_x(0,t) = 0 \tag{A.3}$$

$$V(L,t) = V_x(L,t) = 0 \tag{A.4}$$

$$U\left(\frac{L}{2},t\right) = V\left(\frac{L}{2},t\right) \tag{A.5}$$

$$U_x\left(\frac{L}{2},t\right) = V_x\left(\frac{L}{2},t\right) \tag{A.6}$$

$$\text{(Assume)}\quad EIU_{xxx}\left(\frac{L}{2},t\right) = P \tag{A.7}$$

$$\text{(Assume)}\quad EIV_{xxx}\left(\frac{L}{2},t\right) = -P \tag{A.8}$$

Considering the static problem (i.e., $U(x)$ and $V(x)$ no longer consider a time dependency) and enforcing defined boundary and compatibility conditions gives:

$$EIU_{xxxx}(x) = 0,\ \text{for } 0 \le x \le \frac{L}{2} \tag{A.9}$$

$$EIV_{xxxx}(x) = 0,\ \text{for } \frac{L}{2} \le x \le L \tag{A.10}$$

$$U(0) = U_x(0) = 0 \tag{A.11}$$

$$V(L) = V_x(L) = 0 \tag{A.12}$$

$$U\left(\frac{L}{2}\right) = V\left(\frac{L}{2}\right) \tag{A.13}$$

$$U_x\left(\frac{L}{2}\right) = V_x\left(\frac{L}{2}\right) = 0 \tag{A.14}$$

$$\text{(Assume)}\quad EIU_{xxx}\left(\frac{L}{2}\right) = P \tag{A.15}$$

$$\text{(Assume)}\quad EIV_{xxx}\left(\frac{L}{2}\right) = -P \tag{A.16}$$

The solution forms for $U(x)$ and $V(x)$ considering integration constants $a, b, c,$ and d for $U(x)$ and $e, f, g,$ and h for $V(x)$, respectively, are:

$$U(x) = \frac{1}{EI}\left(\frac{1}{6}ax^3 + \frac{1}{2}bx^2 + cx + d\right) \tag{A.17}$$

$$V(x) = \frac{1}{EI}\left(\frac{1}{6}ex^3 + \frac{1}{2}fx^2 + gx + h\right) \tag{A.18}$$

Applying boundary conditions in Eqs. (A.11) and (A.12) for Eqs. (A.17) and (A.18) gives:

$$U(0) \to d = 0;\ \ U_x(0) \to c = 0 \tag{A.19}$$

$$V(L) = \frac{1}{EI}\left(\frac{1}{6}eL^3 + \frac{1}{2}fL^2 + gL + h\right) = 0 \tag{A.20}$$

$$V_x(L) = \frac{1}{EI}\left(\frac{1}{2}eL^2 + fL + g\right) = 0 \tag{A.21}$$

Implicitly using Eq. (A.19) and applying the compatibility conditions in Eqs. (A.13), (A.15), and (A.16), respectively, gives:

$$\frac{1}{EI}\left[\frac{1}{6}a\left(\frac{L}{2}\right)^3 + \frac{1}{2}b\left(\frac{L}{2}\right)^2\right] = \frac{1}{EI}\left[\frac{1}{6}e\left(\frac{L}{2}\right)^3 + \frac{1}{2}f\left(\frac{L}{2}\right)^2 + g\left(\frac{L}{2}\right) + h\right] \tag{A.22}$$

$$EIU_{xxx}\left(\frac{L}{2}\right) = a = P \tag{A.23}$$

$$EIV_{xxx}\left(\frac{L}{2}\right) = e = -P \tag{A.24}$$

Similarly, substituting Eqs. (A.17) and (A.18) into Eq. (A.14) respectively gives:

$$U_x\left(\frac{L}{2}\right)=\frac{1}{EI}\left[\frac{1}{2}a\left(\frac{L}{2}\right)^2+b\left(\frac{L}{2}\right)\right]=0 \quad \text{(A.25)}$$

$$V_x\left(\frac{L}{2}\right)=\frac{1}{EI}\left[\frac{1}{2}e\left(\frac{L}{2}\right)^2+f\left(\frac{L}{2}\right)+g\right]=0 \quad \text{(A.26)}$$

Equations (A.20), (A.21), (A.22), (A.25), (A.26), and (A.23) can be written in matrix form (not including $c=0$ and $d=0$ from Eq. (A.19)) as:

$$\begin{bmatrix} 0 & 0 & \frac{L^3}{6EI} & \frac{L^2}{2EI} & \frac{L}{EI} & \frac{1}{EI} \\ 0 & 0 & \frac{L^2}{2EI} & \frac{L}{EI} & \frac{1}{EI} & 0 \\ \frac{-L^3}{48} & \frac{-L^2}{8} & \frac{L^3}{48} & \frac{L^2}{8} & \frac{L}{2} & 1 \\ \frac{L^2}{8EI} & \frac{L}{2EI} & 0 & 0 & 0 & 0 \\ 0 & 0 & \frac{L^2}{8EI} & \frac{L}{2EI} & \frac{1}{EI} & 0 \\ 1 & 0 & 0 & 0 & 0 & 0 \end{bmatrix}\begin{bmatrix} a \\ b \\ e \\ f \\ g \\ h \end{bmatrix}=\begin{bmatrix} 0 \\ 0 \\ 0 \\ 0 \\ 0 \\ P \end{bmatrix} \quad \text{(A.27)}$$

Solving Eq. (A.27) yields the non-zero integration constants:

$$a=P;\ b=\frac{-LP}{4};\ e=-P;\ f=\frac{3LP}{4};\ g=-\frac{L^2P}{4};\ h=\frac{L^3P}{24} \quad \text{(A.28)}$$

where a and e are also in agreement with Eqs. (A.23) and (A.24), respectively. Substituting the integration constants in Eqs. (A.19) and (A.28) into Eqs. (A.17) and (A.18) respectively gives:

$$U(x)=\frac{1}{EI}\left(\frac{P}{6}x^3-\frac{LP}{8}x^2\right) \quad \text{(A.29)}$$

$$V(x)=\frac{1}{EI}\left(\frac{-P}{6}x^3+\frac{3LP}{8}x^2-\frac{L^2P}{4}x+\frac{L^3P}{24}\right) \quad \text{(A.30)}$$

Considering a unit applied load $F=1$ at the mass location $x=L/2$ (corresponding to shear $P=-1/2$) and solving for stiffness $k(x)$ when evaluated at $x=L/2$ yields:

$$k\left(\frac{L}{2}\right)=\frac{F}{U\left(\frac{L}{2}\right)}=\frac{192EI}{L^3} \quad \text{(A.31)}$$

One can note that assuming the flexures are massless and completely dictate the TVA stiffness, Eq. (A.31) is equal to the fixed-fixed beam stiffness at the mass location where max deflection occurs. Lastly, using the expression for $k\left(L/2\right)$ in Eq. (A.31) to calculate the structure's fundamental frequency f_1 (not the integration constant f) gives Eq. (A.32) which is equal to Eq. (12):

$$f_1=\frac{1}{2\pi}\sqrt{\frac{2k}{m}}=\frac{1}{2\pi}\sqrt{\frac{384EI}{L^3m}} \quad \text{(A.32)}$$

where $k=k\left(L/2\right)$ in Eq. (A.31). Note that because there are two flexures acting in parallel for the actual TVA design, an overall stiffness of $2k$ is considered in Eq. (A.32).

Chapter 20
Topology Optimization of Isolated Response Curves in 3D Geometrically-nonlinear Beam

Enora Denimal Goy, Yichang Shen, Samuel Fruchard, Adrien Mélot, and Ludovic Renson

Abstract Topology optimisation is a powerful tool for designing efficient and light structures. However, classical topology optimisation methods (SIMP, LSF), which are gradient-based, are not adapted to deal with nonlinear vibrations in the context of geometrical nonlinearities as the simulation of such systems is computationally expensive, and the strong nonlinear behaviour makes the objective function non-convex with many local minima. The present work investigates the potential of using global optimisation methods to topology optimise those structures. To provide more robust nonlinear features in the optimisation, the bifurcations are directly tracked and optimised. The strategy is applied to a 3D finite element model of a beam.

Keywords Bifurcation · Topology optimisation · Global optimisation · Geometric nonlinearities · Reduced-order Model

Introduction

A major transformation in design methods for mechanical structures (transport, aeronautics, space, etc) is taking place in order to continue to make them lighter and increase their performances while benefiting from the arrival of new technologies such as 3D printing. Optimising the dynamics of these structures is becoming extremely complex due to the intrinsic presence of non-linearities leading to dynamic phenomena rarely observed in industrial mechanical systems. For example, in the presence of non-linearities, the dynamics of structures can exhibit bifurcations.

Previous studies have focused on the structural and topology optimisation of mechanical structures to limit the level of amplitudes or avoid bifurcation behaviour in the context of localised nonlinearities [1–3]. The present work has for objective to expand these works to the context of geometric nonlinearities. Such nonlinearities imply a significant increase in the computational cost, making the use of reduced-order models (ROM) as well as efficient optimisation techniques crucial. In [4], the authors employed gradient-based optimisation for nonlinear resonances mitigation in the context of geometrical nonlinearities, yet the approach is limited to structures composed of beams and can only do shape optimisation.

In this work, a framework is proposed to perform topology optimisation of geometrically nonlinear structures to control the appearance of bifurcations for 3D structures. For each new geometries, a ROM is constructed to solve the nonlinear problem and bifurcations are tracked. To reduce the computational cost of the optimisation, a global optimisation algorithm coupling the CMA-ES to the PSO is proposed. The topology is parametrised with the Moving Morphable Component [5] framework and unconnected geometries are removed in the optimisation loop to reach higher efficiency.

Enora Denimal Goy
Inria, CMAP, Ecole Polytechnique, Palaiseau, France
e-mail: enora.denimal-goy@inria.fr

Yichang Shen · Ludovic Renson
Dynamics Group, Department of Mechanical Engineering, Imperial College London, London, UK
e-mail: yichang.shen@imperial.ac.uk; l.renson@imperial.ac.uk

Samuel Fruchard · Adrien Mélot
Université Gustave Eiffel, Inria, COSYS-SII, I4S, Campus Beaulieu, Rennes, France
e-mail: samuel.fruchard@inria.fr; adrien.melot@inria.fr

Ludovic Renson et al. (eds.), *Nonlinear Structures & Systems, Vol. 1*, Conference Proceedings
of the Society for Experimental Mechanics Series, https://doi.org/10.13052/9788743801467-20

Methodology

Geometric Nonlinearity & Model Reduction

In the design of thin structures, geometric nonlinearity plays an important role because of the large deflections that they exhibit and the resulting complex nonlinear phenomena. Finite element (FE) simulations are frequently employed to predict the nonlinear response. A large number of elements is usuallu needed to obtain a converged discretization. As a consequence, full-order nonlinear analysis of complex structures featuring important nonlinearity is computationally expensive, setting a clear limitation in terms of analysis and parametric design. To address this challenge, one can transform the full-order model by reduction methods, leading to a Reduced-order model (ROM) with less computational complexity. An ideal ROM should accurately capture the nonlinear behavior and characteristics of the full-order model with minimal approximation error while remaining computationally efficient.

Invariant-manifold-based techniques for model reduction of nonlinear mechanical systems have been explored in recent years [6]. These methods ensure the ROM's trajectories are consistent with those of the full-order system due to the invariance property, and thus provide mathematically rigorous model reduction for nonlinear systems [7]. In this work, the Direct Normal Form (DNF) approach [8] was applied to obtain the ROM of the FE models. The approach allows direct computation of the nonlinear mapping enabling it to pass from the physical space (dofs of the FE mesh) to lower-dimensional invariant manifolds. The nonlinear mapping between the physical and reduced coordinates derived by the DNF methods reads:

$$\mathbf{q} = \sum_{i}^{n} \phi_i X_i + \sum_{i=1}^{n}\sum_{j=1}^{n} \bar{\mathbf{a}}_{ij} X_i X_j + \sum_{i=1}^{n}\sum_{j=1}^{n} \bar{\mathbf{b}}_{ij} \dot{X}_i \dot{X}_j + \sum_{i=1}^{n}\sum_{j=1}^{n} \bar{\mathbf{c}}_{ij} X_i \dot{X}_j, \tag{1}$$

where n is the number of master modes used to build the ROM, vector $\mathbf{q}$ is generalized displacements at the nodes, and X_i together with its velocity $\dot{X}_i$ are the coordinates used to span the i-th invariant manifold, ϕ_i is the eigenvector and $\bar{\mathbf{a}}_{ij}$, $\bar{\mathbf{b}}_{ij}$ and $\bar{\mathbf{c}}_{ij}$ are second-order tensors. The reduced dynamics is then given by the normal form of the problem, up to the second-order, by taking into account the Rayleigh damping $\mathbf{C} = \zeta_M \mathbf{M} + \zeta_K \mathbf{K}$ with an assumption of small damping ratios on the master modes. It reads, $\forall p = 1, ...m$:

$$\ddot{X}_p + (\zeta_M + \zeta_K \omega_p^2)\dot{X}_p + \omega_p^2 X_p + \sum_{i=1}^{n}\sum_{j=1}^{n}\sum_{k=1}^{n} (A^p_{ijk} X_i X_j X_k + B^p_{ijk} X_i \dot{X}_j \dot{X}_k + C^p_{ijk} X_i X_j \dot{X}_k) = F_p, \tag{2}$$

where the coefficients $A^p_{ijk}, B^p_{ijk}, C^p_{ijk}$ arise from the cancellation of non-resonant quadratic terms. The full expressions of the tensors and coefficients in Eqs. (1, 2), which are not given here for the sake of conciseness, can be found in [8]. Figure 1 shows an example of reduced-order modeling of a clamped beam discretized with 3D elements and the corresponding backbone curves computed by using ROM.

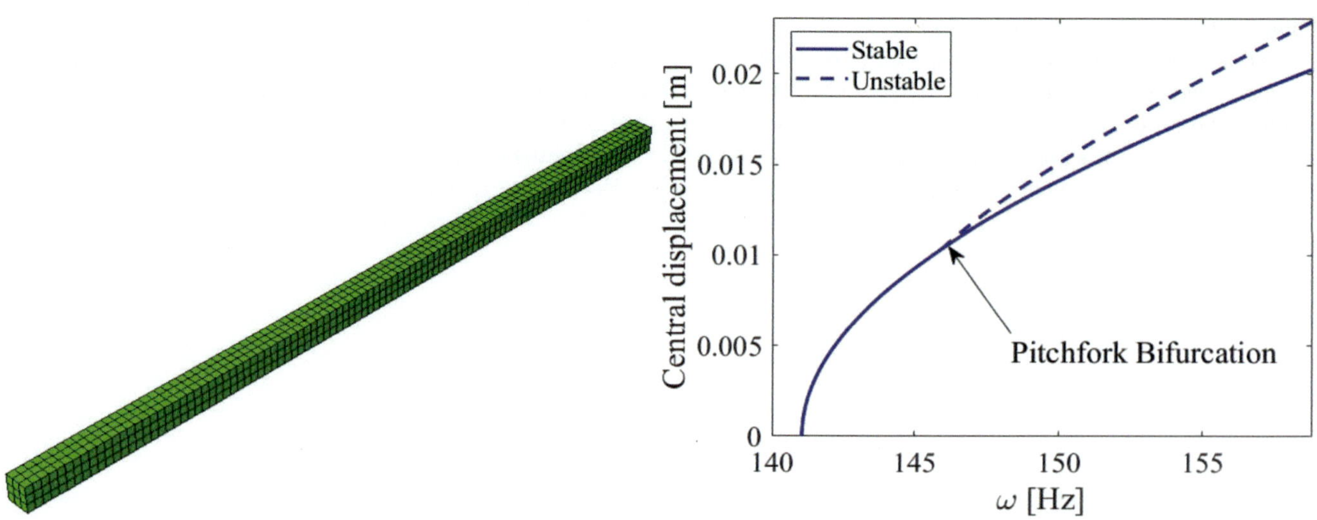

Fig. 1 Left: The clamped beam discretized with 3D elements. Right: Backbone curves of the first mode computed by the ROM

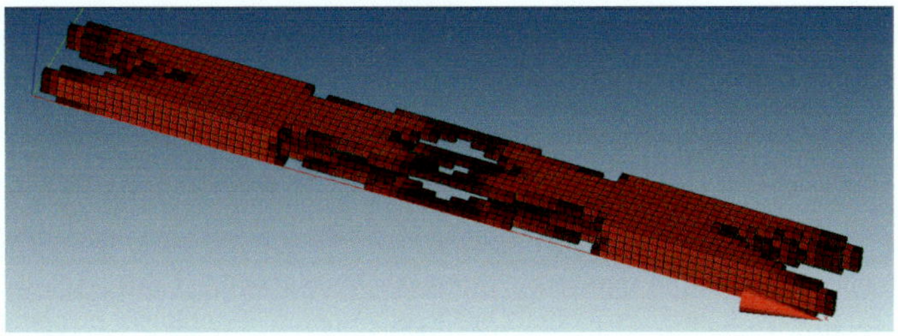

Fig. 2 Example of topology with the 3D MMC framework

Bifurcation detection and optimisation

A computational framework based on the harmonic balance method is employed to compute the dynamic response. Hill's stability analysis is performed to detect and locate the potential bifurcations. Similarly to [2], the objective function for the optimisation is based on a bifurcation measure, able to handle multiple bifurcations and to locate them at chosen frequency and/or amplitude levels. It is written as follows:

$$f(T, P(\theta)) = |T - P|\Psi(\theta) + \frac{1}{|T|}\sum_{\tau \in T}\prod_{\pi\theta \in P}\left|\frac{\pi(\theta) - \tau}{\tau}\right|^{1/|P|} \tag{3}$$

with T the set of target bifurcation, $P(\theta)$ the set of bifurcations predicted for the model parameters θ. The notation $|T|$ represents the number of elements in the set T.

Geometry Parameterisation

Traditional topology optimisation algorithms, such as SIMP or LSM, require the analytic definition of the gradient to ensure reasonable computational time. As this information is unavailable in our case, and as the problem is nonlinear, non-convex, with many local minima, the choice is made to go for global optimisation algorithms. As they are parametric, the Moving Morphable Component (MMC) framework is employed here to ensure the parametrisation of the beam. Coupled to the global optimisation algorithm, it was proven to be efficient in dealing with topology optimisation for nonlinear vibration mitigation [1]. A 3D individual component is characterised by 9 parameters, which define explicitly a Level-Set Function (LSF) [5]. The global geometry is defined as the union of the unitary LSF. By moving, rotating, and shrinking each component, complex topologies can be described. The LSF is then projected on the initial mesh to define the density field. An example of topology obtained with the MMC framework is given in Figure 2.

Global Optimisation

The optimisation problem is non-convex with many local minima. For this reason, a global optimisation algorithm is considered. The strategy proposed here couples two algorithms, namely the CMA-ES [9] and the PSO [10]. A preliminary study has been done on the case of linear vibrations, and it appears that the CMA-ES is very robust and efficient to reach approximately the global minima, but then requires many model calls to reach the precise solution. On the other hand, the PSO algorithm is very efficient when initialised near the optimal point. The solution proposed here couples the two algorithms.

Acknowledgments Yichang Shen would like to thank HORIZON MSCA Postdoctoral Fellowship action and the UKRI Postdoc Guarantee Fellowship provided by the EPSRC, with Grant Ref: EP/X026027/1.

References

1. Denimal, E., Renson, L., Wong, C., and Salles, L. "Topology optimisation of friction under-platform dampers using moving morphable components and the efficient global optimization algorithm". *Structural and Multidisciplinary Optimization*, 65(2):56 (2022).
2. Mélot, A., Denimal, E., and Renson, L. "Multi-parametric optimization for controlling bifurcation structures". *Proceedings of the Royal Society A*, 480(2283):20230505 (2024).
3. Mélot, A., Goy, E.D., and Renson, L. "Control of isolated response curves through optimization of codimension-1 singularities". *Computers & Structures*, 299:107394 (2024).

4. Dou, S., Strachan, B.S., Shaw, S.W., and Jensen, J.S. "Structural optimization for nonlinear dynamic response". *Philosophical Transactions of the Royal Society A: Mathematical, Physical and Engineering Sciences*, 373(2051):20140408 (2015).
5. Zhang, W., Li, D., Yuan, J., Song, J., and Guo, X. "A new three-dimensional topology optimization method based on moving morphable components (mmcs)". *Computational Mechanics*, 59:647–665 (2017).
6. Touzé, C., Vizzaccaro, A., and Thomas, O. "Model order reduction methods for geometrically nonlinear structures: a review of nonlinear techniques". *Nonlinear Dynamics*, 105:1141–1190 (2021).
7. Shen, Y., Béreux, N., Frangi, A., and Touzé, C. "Reduced order models for geometrically nonlinear structures: assessment of implicit condensation in comparison with invariant manifold approach". *European Journal of Mechanics A/Solids*, 86:104165 (2021).
8. Vizzaccaro, A., Shen, Y., Salles, L., Blahos, J., and Touzé, C. "Direct computation of nonlinear mapping via normal form for reduced-order models of finite element nonlinear structures". *Computer Methods in Applied Mechanics and Engineering*, 384:113957 (2021).
9. Nomura, M. and Shibata, M. "Cmaes: A simple yet practical python library for cma-es". *arXiv preprint arXiv:2402.01373* (2024).
10. Miranda, L.J. "Pyswarms: a research toolkit for particle swarm optimization in python". *Journal of Open Source Software*, 3(21):433 (2018).

Chapter 21
Coupled Harmonic Balance based approach for the non-linear dynamics of spur gear pairs

Giacomo Saletti, Giuseppe Battiato, and Stefano Zucca

Abstract Noise, Vibration and Harshness analyses are of great interest for the latest developments of the gearboxes of electric vehicles. Gearboxes are now the main source of vibrations since electric powertrains are much quieter than internal combustion engines. Traditionally, the simulation of the non-linear gears dynamics is studied by first performing a series of preliminary static analyses to compute the static transmission error (STE). The STE (i.e. in the form of varying mesh stiffness) is then accepted as the system's excitation source to compute the dynamic transmission error (DTE). This paper presents a novel approach to analyze the non-linear dynamics of gears which does not require any preliminary static analyses. The method consists of a frequency domain approach based on the Harmonic Balance Method (HBM) and the Alternating Frequency-Time (AFT) scheme, allowing for much faster simulations when compared to the widely used direct time integration (DTI). The contact between the teeth is modeled as intermittent, penalty based with varying gap. The time varying gap between the teeth is initially approximated to a step function that guarantees the design contact ratio.

The methodology introduced is tested on a lumped parameter model of a spur gear pair already proposed and simulated in the literature. The results obtained with the novel approach are compared with the DTI simulation of the model as reference. The excellent match between the different approaches validates the reliability of developed methodology.

Keywords Gear dynamics · Non-linear dynamics · Harmonic Balance Method · Alternating Frequency-Time · Transmission error

Introduction

Gears serve as critical components in various mechanical systems, yet they are frequently subjected to issues such as vibrations. These vibrations are detrimental as they can induce damage, generate noise, and lead to malfunctions. The primary cause of vibrations in gears is the variation in contact stiffness, which fluctuates due to the changing number of teeth in contact during gear rotation [1].

The issue is well known and has been studied systematically since the mid-1800s with the early work of Walker who proposed a coefficient relating the static and dynamic loads operating on gears as reported by [2]. In the last century several lumped parameters models have been developed for the characterization of gear dynamics [3], addressing torsional vibrations [4], specifically teeth dynamics [1] and more in general geared rotor dynamics [5] where gears are often modelled as rigid disks with inertia on flexible shafts.

More recent studies that focused on torsional gear dynamics are still utilizing simple models for their analyses, Parker et al. [6] compared a single DOF model and a FE model in approximating the experimental dynamic response of a spur gear pair, Ambarisha and Parker [7] developed an analytical model able to capture the complex dynamics of a planetary gear when compared to its FE counterpart. In 2019 Dai et al. [8] proposed a hybrid analytical-computational approach that integrates the accurate FE static analyses with a compact lumped parameters model that can both accurately simulate the torsional vibrations of the gears and correctly capture the stresses on the teeth.

However, even dynamic analyses on lumped parameter models can require significant computational time if they are performed by time integration of the equation of motion of the system. The response of the gears is measured once steady state is reached, this requires performing the simulation of the initial transient until the response can be considered periodic.

Giacomo Saletti · Giuseppe Battiato · Stefano Zucca
Department of Mechanical and Aerospace Engineering, Politecnico di Torino, Torino 10129, IT
e-mail: giacomo.saletti@polito.it; giuseppe.battiato@polito.it; stefano.zucca@polito.it

Ludovic Renson et al. (eds.), *Nonlinear Structures & Systems, Vol. 1*, Conference Proceedings of the Society for Experimental Mechanics Series, https://doi.org/10.13052/9788743801467-21

Trying to overcome the limitations of time integration, frequency domain approaches were introduced in the field of gear dynamics. Harmonic Balance Method (HBM) allows for the determination of the steady state response without having to solve for initial transient.

Some definitions are necessary for the following discussion in this paper. Transmission errors (TE) are defined as the difference between the positions that perfectly rigid gears would occupy during ideal motion and the effective positions occupied by flexible gears. The static transmission error (STE) is typically defined as the TE obtained through a sequence of low-speed or quasi-static analyses on loaded gears [9]. The dynamic transmission error (DTE) is the counterpart of the STE when calculated at high rotational speeds. Instead of the STE, some prefer to introduce the concept of mesh stiffness, which is related to the ratio of the torque applied to the gears and the STE [6]. In practice, the STE and mesh stiffness provide similar information.

Commonly the study of gear dynamics in the frequency domain is performed by first computing the STE and subsequently using it as a dynamic excitation to obtain the DTE [6, 10, 11]. Initially solving the system to obtain the STE can be cumbersome, for large systems especially, moreover it represents an approximation because it assumes that the static (or quasi-static) solution of the system is not affected by the dynamics of the system. In the field of turbo-machinery, where HBM has been extensively applied, it has been proven that the uncoupled solution static-dynamic can lead to inaccurate results. Coupled HBM was developed to allow for accurate predictions of dynamic responses of non-linear models with friction damping. Zucca et al. [12] used coupled HBM with an alternating frequency time (AFT) on lumped parameter models with friction contact. They argue that considering the static and dynamic problem as uncoupled leads to inaccurate solution when comparing to DTI simulations as reference.

In this paper the use of coupled HBM is extended to gear dynamics developing a method capable of overcoming the differences between STE and DTE. The transmission error is obtained in a single analysis considering the coupling between the static solution and the dynamic one. The simulation algorithm is enabled by the use of the Alternating Frequency Time (AFT) approach [13]. During the Newton-Raphson iteration to solve the non-linear set of equations expressed in the frequency domain, the AFT approach allows to momentarily move to the time domain to compute the non-linear contact forces arising from the interaction between teeth pairs.

The methodology was implemented in a MATLAB script and the test system was solved making use of a continuation method to help convergence around turning points, however, it was not possible to reach convergence over the whole frequency ranges chosen.

The following sections will focus on:

- in section , Methodology, a comprehensive description of the proposed approach is presented, details on the numerical solution strategy are also highlighted;
- in section , Results and Discussion, comparative analyses are carried out on a simple lumped parameter model. Non-linear dynamic analyses are performed with both DTI (reference) and the developed methodology.
- In section , the final conclusions are drawn.

Methodology

The described approach is referred to as Coupled HBM and has been introduced in [12] for systems with dry friction contacts. In this paper the method is applied to systems with clearance type nonlinearities, specifically to gear dynamics. Several advantages of the methodology can be listed:

- Frequency domain approaches are faster than time domain approaches
- No preliminary static analyses are required
- No approximations/assumptions are required on the mesh stiffness
- Suitable for the integration with different contact models

The approach also presents some limitations:

- because of the clearance type nonlinearities, several harmonics might be required to obtain a converged solution
- The stiffness matrix can be singular for gear dynamics problems, leading to bad conditioning of the dynamic stiffness and hard to obtain Newton-Raphson convergence.

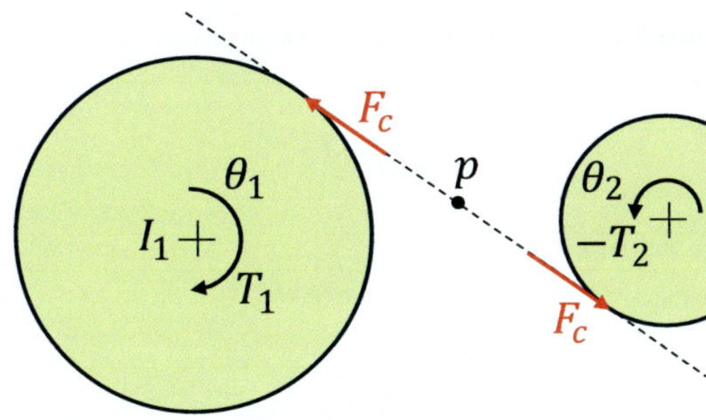

Fig. 1 Lumped parameters model

Single DOF model

The demonstration of the applicability of the method is performed through the introduction of a simple, lumped parameter model. The introduced one (see Figure 1) was already studied in [6], when trying to approximate some available experimental results. The model is composed of two gears with I_1 and I_2 moments of inertia, r_1 and r_2 base radii and T_1 and $T_2 = T_1 r_2/r_1$ the torques applied to the gear. Given that no torsional stiffness or damping constraint the rotational DOF of the gears, the equations of motion can be written as:

$$I_1\ddot{\theta}_1 + F_c r_1 = T_1 \tag{1}$$

$$I_2\ddot{\theta}_2 - F_c r_2 = -T_1 r_2/r_1 \tag{2}$$

From equations 1 and 2, the following set of equations can be easily obtained:

$$I_2 I_1\ddot{\theta}_1 r_1 + I_2 F_c r_1^2 = I_2 T_1 r_1 \tag{3}$$

$$I_1 I_2\ddot{\theta}_2 r_2 + I_1 F_c r_2^2 = -I_1 T_1 r_2^2/r_1 \tag{4}$$

Subtracting Eq. 3 from Eq. 4 is obtained

$$I_1 I_2\left(\ddot{\theta}_2 r_2 - \ddot{\theta}_1 r_1\right) - F_c\left(I_1 r_2^2 + I_2 r_1^2\right) = -T_1\left(\frac{I_1 r_2^2}{r_1} + I_2 r_1\right) \tag{5}$$

$$\frac{I_1 I_2}{I_1 r_2^2 + I_2 r_1^2}\left(\ddot{\theta}_2 r_2 - \ddot{\theta}_1 r_1\right) - F_c + T_1\frac{\frac{I_1 r_2^2}{r_1} + I_2 r_1}{I_1 r_2^2 + I_2 r_1^2} = 0 \tag{6}$$

Introducing the TE as the relative DOF between the gears $x = \theta_2 r_2 - \theta_1 r_1$ and introducing a damping term proportional to $\dot{x}$, the equation of motion can be expressed in the form:

$$\frac{I_1 I_2}{I_1 r_2^2 + I_2 r_1^2}\ddot{x} + C\dot{x} - F_c + T_1\frac{\frac{I_1 r_2^2}{r_1} + I_2 r_1}{I_1 r_2^2 + I_2 r_1^2} = 0 \tag{7}$$

$$M\ddot{x} + C\dot{x} - F_c + F_{ext} = 0 \tag{8}$$

The external excitation is collected in the F_{ext} term, it is directly dependent on the torque T_1 applied to the first gear. The external excitation is generally a periodic function known a priori.

F_c is the contact force being exchanged between the teeth of the gears. In the most general case, F_c is a non-linear function of time (t), the transmission error (x) and its derivative ($\dot{x}$).

Considering the gears rotating at a constant angular speed ω', once steady state is reached, every point of the gears will occupy the same position after a $T' = 2\pi/\omega'$ time period. However, considering the geometrical periodicity of the gears, already after a period of $T = 2\pi/z\omega'$ with z the number of teeth, the gears are in an equivalent position. Consequently, the contact forces are also periodic with the same period T and the effective pulsation of the system is $\omega = z\omega'$.

The displacement and forces in equation 8 can be expressed as Fourier series:

$$x(t) = \sum_{h=0}^{H} \tilde{x}_h e^{jh\omega t} \tag{9}$$

$$F_{ext}(t) = \sum_{h=0}^{H} \tilde{F}_{ext,h} e^{jh\omega t} \tag{10}$$

$$F_c(t, x, \dot{x}) = \sum_{h=0}^{H} \tilde{F}_{c,h} e^{jh\omega t} \tag{11}$$

The quantities $\tilde{x}_h$, $\tilde{F}_{ext,h}$ and $\tilde{F}_{c,h}$ are the complex Fourier coefficients associated with the hth order, j is the imaginary unit. The series are truncated at an arbitrary order H.

Equations 9, 10 and 11 are substituted in equation 8 and the HBM is enforced. The equation for a generic order h is:

$$\left(-h^2\omega^2 M + jh\omega C\right)\tilde{x}_h - \tilde{F}_{c,h} + \tilde{F}_{ext,h} = 0 \tag{12}$$

The generic model presented in Figure 1 can be simulated using direct time integration (DTI) solving equation 8 or it can be solved in its frequency domain expression (equation 12) using non-linear solution methods (e.g. Newton-Raphson) with Alternating Frequency Time scheme (AFT) [13].

The AFT scheme starts from a given guess value of the Fourier coefficients of the generalized displacements. The displacements are then expressed in time domain over a single period via the Inverse Fast Fourier Transform (IFFT). The contact model allows to obtain the contact force evolution over time given the evolution of the generalized displacements. The sequence of generalized contact forces is eventually expressed back in the frequency domain via the Fast Fourier Transform (FFT), thus obtaining the Fourier coefficients of the contact forces $\tilde{F}_{c,h}$ for the different harmonics retained in the HBM.

If the residual of the equation of motion is still too large, a new guess of the Fourier coefficient of the generalized displacements must be provided. The iterations proceed until the residual of the equation of motion can be considered sufficiently small.

Contact model

A contact model is required to express F_c as a function of time (t), the transmission error (x) and its derivative $\dot{x}$. This model can then be used as is in the different formulations of the problem (time domain or frequency domain).

Considering the rigid rotation and mating of the gears, the contact ratio $c_r > 1$ defines the average number of teeth in contact during a period of the motion. The varying number of teeth in contact leads to changes in the equivalent stiffness of the contact which represent the source of dynamic excitation in the studied model. Moreover, the vibrations induced by this phenomenon can lead the teeth on opposing gear to close or open the contact in a different instant than the rigid case. The proposed contact model wants to mimic this behavior.

The basic idea is that, during one period of the rigid rotation of the gears a couple of teeth will be either in contact or not. The contact status (open or close) depends on the relative distance between the teeth. This is distance is referred to as *gap*.

The gap between two teeth can be either zero or a positive value. The relative position of the contact points $x = \theta_2(t)r_2 - \theta_1(t)r_1$ is positive if they tend to separate and negative if they tend to interpenetrate. A clearance type contact is defined in the form of

$$\begin{cases} F_c(t) = -k_c\left(x(t) + gap(t)\right) & \text{if } x + gap < 0 \\ F_c(t) = 0 & \text{if } x + gap \geq 0 \end{cases} \tag{13}$$

where k_c is the contact stiffness. The contact status is fully dependent on the total relative position of the contact at every time step. This contact model allows the system to find an equilibrium considering both the static and dynamic components of the response.

Considering a model with contact ratio $c_r = 1.75$, during one (rigid) rotation period of the gears, three couples of teeth will be in contact with each other at a certain point. Considering the representation in Figure 2, the central couple of teeth will be in contact for the entire rotation period, the *high* couple of teeth will start as separated and will enter contact after some time. On the contrary, the *low* couple of teeth will start in contact and will separate as the rotation continues.

To consider the interactions between the teeth described, the system presented in Figure 1 is completed in Figure 3 including three contact elements representing the couples of teeth.

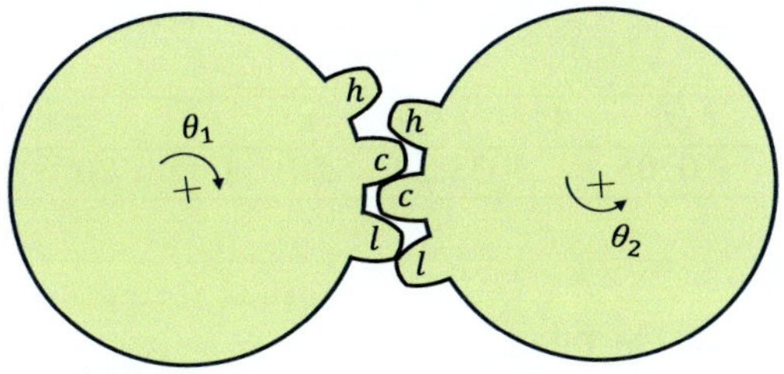

Fig. 2 Teeth in contact

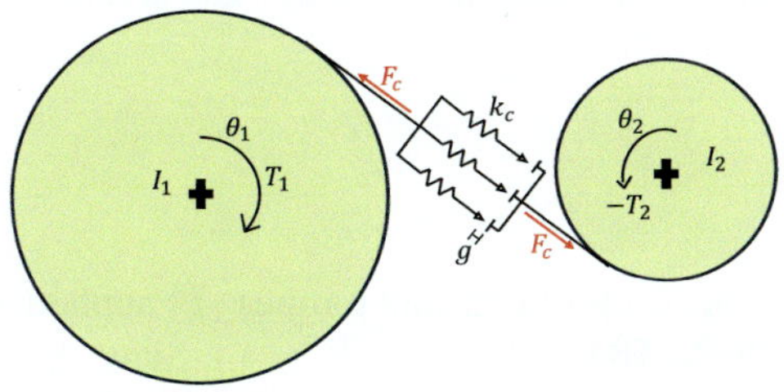

Fig. 3 Gears with contact elements

The gap operating in the different contact elements will vary over time. As a first approximation a step function is introduced (Figure 4), the gap for all the teeth is set to 0 when the teeth are in contact and to 1 when they are open. The gap jumps are set to enforce the design contact ratio. The approximation of the gap variation as a step function is not representative of the real behavior of the teeth contact, the results obtained with this model will not be quantitatively close to any experimental results possibly available. However, since the scope of this paper is to show the applicability of the coupled HBM to the simulation of gear dynamics having an accurate contact model is not a requirement. In future work, effort will be put into this aspect to improve the quality of the results and to investigate the applicability of the methodology to more representative models.

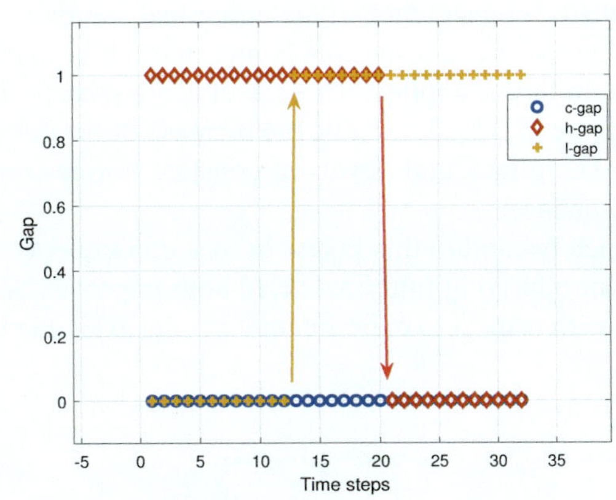

Fig. 4 Gap variation for the different contact elements

Results and discussion

The model considered for the following analyses considers a couple of identical gears. The equation 7 can then be simplified as follows:

$$\frac{I}{2r^2}\ddot{x} + C\dot{x} - F_c + T\frac{1}{r} = 0 \tag{14}$$

Table 1 System parameters.

$\mathbf{I}\,(Kg/m^2)$	$\mathbf{r}\,(m)$	$\mathbf{C}\,(Ns/m)$	$\mathbf{z}$	$\mathbf{c_r}$	φ	$\mathbf{k_c}\,(N/m)$
0.0158	0.0705	4038.2	50	1.75	20°	2.6e8

Where the parameters considered are collected in Table 1.

To validate the methodology proposed, the reference results are obtained by direct time integration of the system expressed in equation 14. The solutions are obtained implementing as a MATLAB algorithm the Newmark method with $\beta = 0.5$ and $\gamma = 0.25$ [14]. The simulation is carried out until steady state is reached.

Non-linear forced responses

Different analyses are performed for multiple levels of constant torques (T) applied, namely 100 Nm, 200 Nm and 300 Nm. The number of harmonics considered in the MHBM solution is $H = 14$. Since the system has only one degree of freedom (DOF), it is possible to simulate it using a large number of harmonics, although good results could be obtained by retaining fewer harmonics. It was also intention of the authors to properly capture the contact force behavior, considering that the unrealistic step gap function induces instant changes in the contact status leading to steep gradients of the contact stiffness during the rotation of the gears.

In Figure 5 the non-linear forced responses of the system are presented; the jumps phenomena are highlighted by black arrows. The HBM simulations are performed implementing a continuation method, however, it was hard to obtain the unstable branches of the response. The simulations for both DTI and HBM are carried out starting from the lowest frequency of the range (forward, "*forw*") and from the highest as well (backward, "*back*").

The different simulation methods are in excellent agreement with each other. All the peaks are predicted accurately both in terms of amplitude and frequency. The jump phenomena typical for this kind of systems are predicted accurately by the HBM simulations.

A slight difference seems to be present around the jump of the first super-harmonic (around 1250 Hz), in that region the continuation method struggled to obtain a converged solution due to the presence of an unstable branch of the response. The apparent difference is in fact due to the missed convergence of the numerical solution.

In Figure 6 a comparison of the non-linear forced responses is presented. It is worth noting that the different responses seem to be scaling linearly with the constant torque applied. This effect is not present in more accurate simulations or in the experimental results available in the literature [6, 15]. Typically, the increase of the constant torque level tends to "linearize" the system, resulting in less softened peaks. In fact, high levels of constant torque counters the complete separation of the teeth at excitation frequencies close to resonance.

The presented system does not completely capture this behavior as a consequence of the simplified contact modelling. More realistic contact modelling will be introduced in future work, to both improve convergence of the solution method and increase accuracy of the results obtained with respect to experimental results available in the literature.

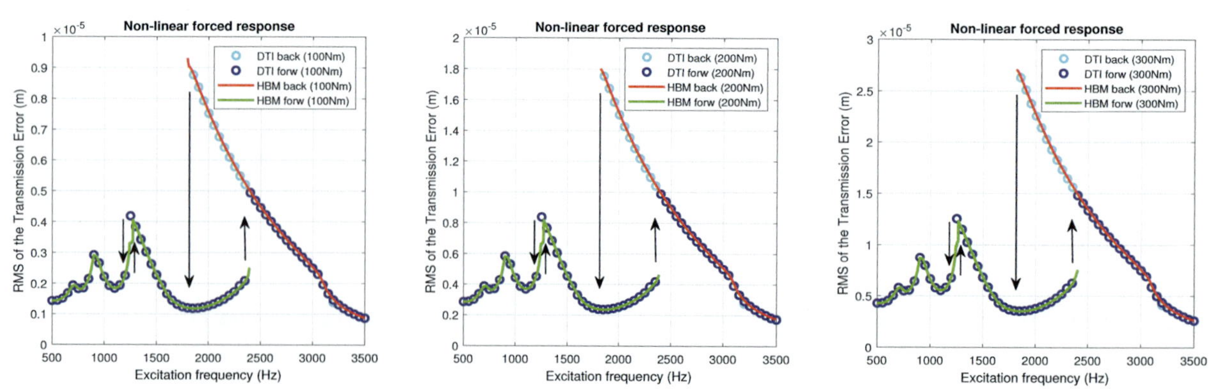

Fig. 5 Non-linear forced responses of the system simulated with DTI and HBM with different static torques applied

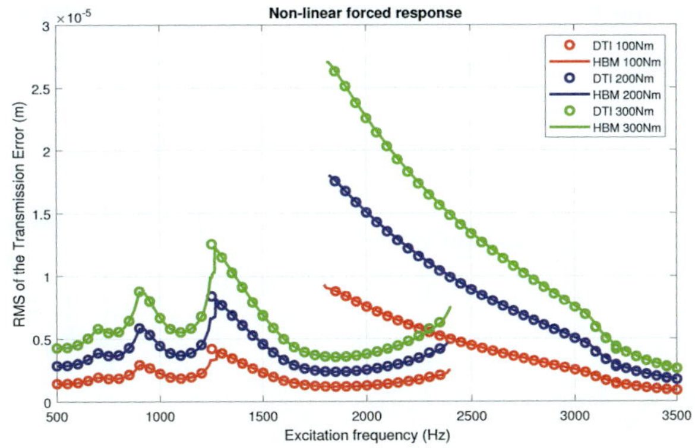

Fig. 6 Non-linear forced responses comparison for different torque levels

Contact force

In this sub-section, a specific focus on the contact quantities is presented. The accuracy of the proposed methodology is assessed by comparison of the contact force and equivalent contact stiffness in the HBM and DTI simulations. The following comparisons are performed for the case with $T = 100$ Nm constant torque applied, but equivalent results were obtained for the other cases.

In Figure 7 the plots associated with an excitation frequency of 3250 Hz are presented. Being relatively far from the resonances, the response at regime for a single period is close to a sinusoidal signal (on the left). The simulation carried out with the proposed methodology (HBM) and with the time integration (DTI) are in very good agreement. On the right side the computed contact force is presented, the steep changes are associated with the gap jumps shown in Figure 4. The instantaneous ratio between the contact force and the response is defined as the mesh stiffness, this quantity is plotted associated with the contact force. The mesh stiffness for this case is a step function that varies between k_c and $2k_c$, giving an indication on the number of contact elements (see Figure 3) operating at each time step. Both the mesh stiffness and the contact force are simulated remarkably well by the HBM solution method when compared to the DTI standard approach.

The same comparisons are presented for other significant excitation frequencies, in Figure 8 the focus is on 1250 Hz. Specifically, at the resonance frequency of the first super-harmonic, the displacement response presents two peaks. The mesh stiffness obtained for this excitation frequency shows three distinguished levels, for some time steps the stiffness is zero, which means that the two gears are completely separated and they are not exchanging any force. This effect is given by the dynamics of the system that is producing vibration levels high enough to open the contact.

In Figure 9 and Figure 10 the results at the main resonance's frequency are presented. At this frequency two stable results are possible, the response with the highest amplitude is presented in Figure 9, while the one with the lowest amplitude is presented in Figure 10. The response in Figure 9 is mainly mono-harmonic, the contact force reaches the highest values

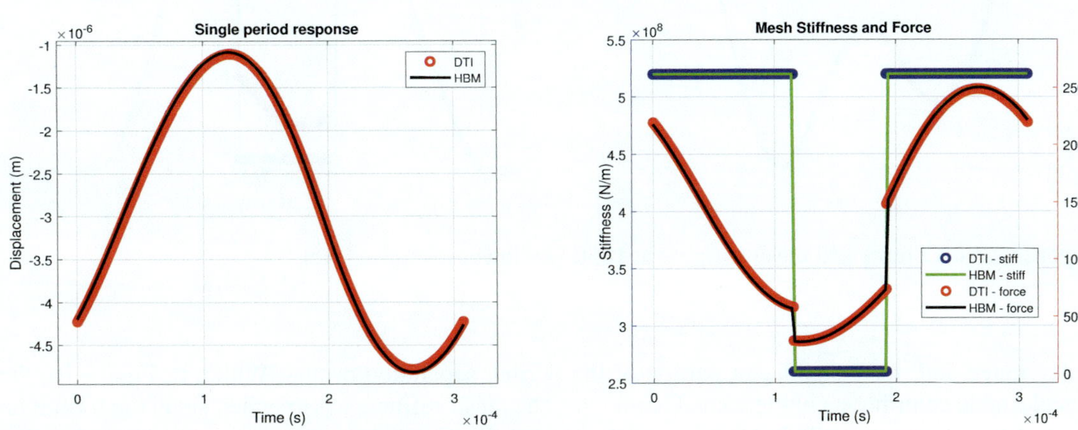

Fig. 7 Response, contact force and mesh stiffness at 3250 Hz

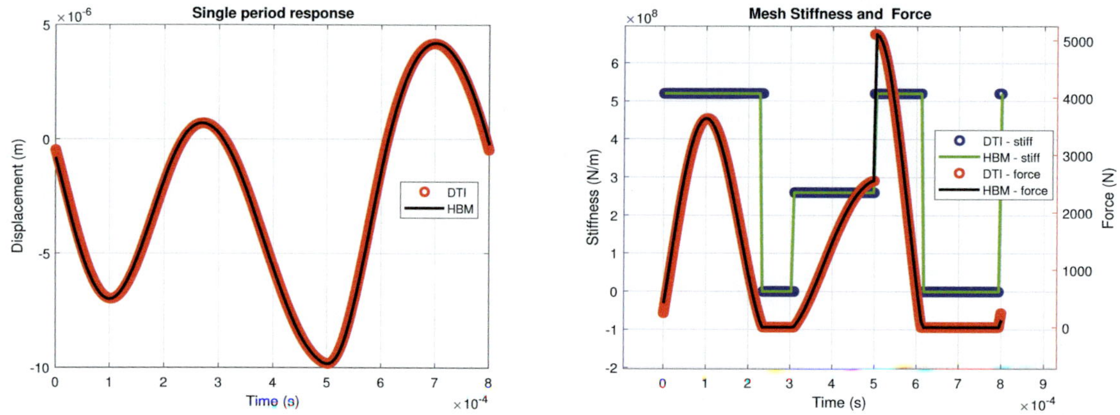

Fig. 8 Response, contact force and mesh stiffness at 1250 Hz

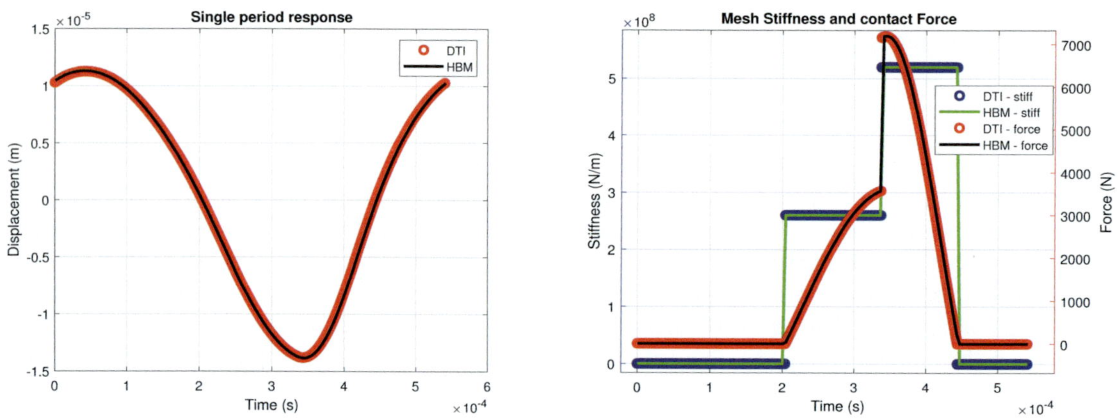

Fig. 9 Response, contact force and mesh stiffness at 1850 Hz, higher response level

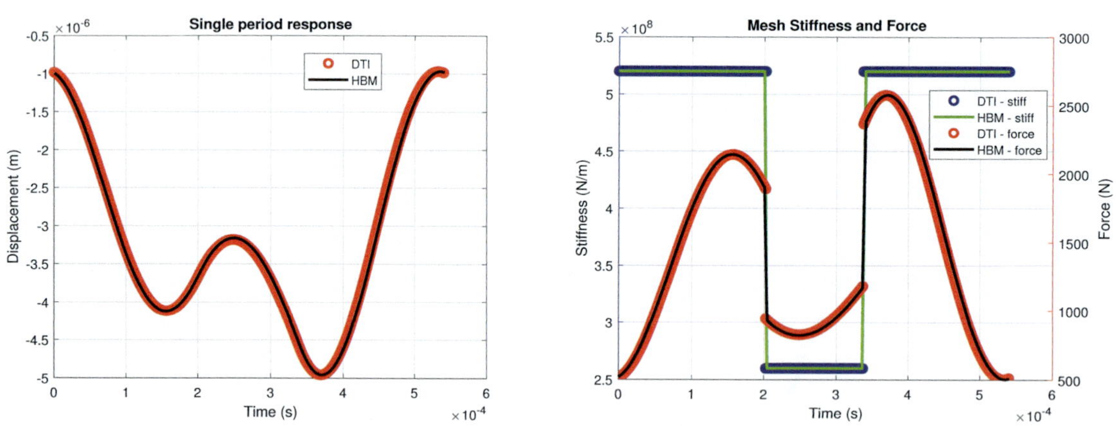

Fig. 10 Response, contact force and mesh stiffness at 1850 Hz, lower response level

of the frequency range and, for a significant portion of the period, separation occurs. While, in Figure 10, the response shows a non-neglectable contribution of the second harmonic. The mesh stiffness has assumed again the form it has far from resonance (as already seen in Figure 7), but in this case the contact force signal presents two peaks given that the excitation frequency of interest is closer to the first super-harmonic.

Conclusions

A new approach to the analyses of spur gear pairs is presented in this paper. The already well-established Harmonic Balance Method in its coupled static-dynamic formulation is extended to the equation of motion of a lumped parameters model of a spur gear pair. A solution scheme is proposed making use of the Alternating Frequency-Time method. The AFT allows to compute the non-linear contact forces moving temporarily to the time domain via a IFFT of the generalized displacements. Subsequently the Fourier coefficient of the contact forces are obtained with the FFT of the single period time signal of the contact forces.

A contact model is required to express the contact force as a function of the generalized displacements and its derivative. A basic penalty-based contact model is introduced defining a time varying gap between the teeth.

The formulation obtained allows to simulate the system in the frequency domain without the need for any preliminary static analysis.

In the results section, multiple non-linear responses are presented comparing the responses amplitudes obtained with the Newmark method and with the proposed methodology. The comparison holds for all the frequency range considered.

Contact force, mesh stiffness and one period responses are also compared at specific frequencies. HBM and DTI show overlapping results for all the cases.

Coupled HBM proved to be an effective alternative to DTI for the simulation of simple lumped parameters models. The results obtained are not close quantitatively to the experimental behavior of the modeled gears, however other studies (see [8]) have shown that lumped parameters model can simulate real components as long as an accurate contact model is introduced.

The proposed methodology allows for the incorporation of various contact models without affecting the overall framework. The authors plan to leverage this flexibility in future work by introducing more realistic contact models to enhance the accuracy of the simple lumped parameter model in relation to actual spur gear pairs.

References

1. Harris, S.L. "Dynamic Loads on the Teeth of Spur Gears". *Proceedings of the Institution of Mechanical Engineers*, 172(1):87–112 (1958).
2. Fisher, A. "Factors in calculating the load-carrying capacity of helical gears". *Machinery*, 98:545–552 (1961).
3. Özgüven, H.N. and Houser, D.R. "Mathematical models used in gear dynamics—a review". *Journal of sound and vibration*, 121(3):383–411 (1988).
4. Nakata, T. "The dynamic load on gear caused by the varying elasticity of the mating teeth". In *Proc 6ˆ¡ th¿ Japan National Congress for Applied Mechanics*, volume 493 (1956).
5. Lund, J.W. "Critical Speeds, Stability and Response of a Geared Train of Rotors". *Journal of Mechanical Design*, 100(3):535–538 (1978).
6. Parker, R., Vijayakar, S., and Imajo, T. "Non-linear dynamic response of a spur gear pair: modelling and experimental comparisons". *Journal of Sound and Vibration*, 237(3):435–455 (2000).
7. Ambarisha, V.K. and Parker, R.G. "Nonlinear dynamics of planetary gears using analytical and finite element models". *Journal of sound and vibration*, 302(3):577–595 (2007).
8. Dai, X., Cooley, C.G., and Parker, R.G. "An Efficient Hybrid Analytical-Computational Method for Nonlinear Vibration of Spur Gear Pairs". *Journal of Vibration and Acoustics*, 141(1):011006 (2019).
9. Özgüven, H. and Houser, D. "Dynamic analysis of high speed gears by using loaded static transmission error". *Journal of Sound and Vibration*, 125(1):71–83 (1988).
10. Blankenship, G. and Kahraman, A. "Steady state forced response of a mechanical oscillator with combined parametric excitation and clearance type non-linearity". *Journal of Sound and Vibration*, 185(5):743–765 (1995).
11. Velex, P. and Ajmi, M. "Dynamic tooth loads and quasi-static transmission errors in helical gears–approximate dynamic factor formulae". *Mechanism and Machine Theory*, 42(11):1512–1526 (2007).
12. Zucca, S. and Firrone, C.M. "Nonlinear dynamics of mechanical systems with friction contacts: Coupled static and dynamic Multi-Harmonic Balance Method and multiple solutions". *Journal of Sound and Vibration*, 333(3):916–926 (2014).
13. Cameron, T.M. and Griffin, J.H. "An Alternating Frequency/Time Domain Method for Calculating the Steady-State Response of Nonlinear Dynamic Systems". *Journal of Applied Mechanics*, 56(1):149–154 (1989).
14. Newmark, N.M. "A method of computation for structural dynamics.". *Journal of the engineering mechanics division*, 85(3):67–94 (1959).
15. Hotait, M. and Kahraman, A. "Experiments on the relationship between the dynamic transmission error and the dynamic stress factor of spur gear pairs". *Mechanism and Machine Theory*, 70:116–128 (2013).

Chapter 22
Multi-level Stochastic Model Identification of Complex Aeroelastic Systems Considering Aerodynamic Nonlinearities

Michael McGurk and Jie Yuan

Abstract This study investigates the nonlinearities in aeroelastic systems that lead to Limit Cycle Oscillations (LCO) below the linear flutter threshold. We employ an adaptive multi-level data-driven model updating framework within a Bayesian inference scheme, achieving up to a 20% improvement in accuracy over conventional methods. However, previous models struggled to fully capture experimental data, particularly at high angles of attack, due to reliance on a quasi-steady linear aerodynamic model. To enhance model fidelity, we incorporate the nonlinear Beddoes-Leishman (B-L) model, which effectively accounts for dynamic stall and vortex shedding. The resulting probabilistic bifurcation diagram successfully encompasses all experimental data within 95% confidence bands, demonstrating significant improvement in accuracy compared to the quasi-steady model. This study confirms the multi-level framework's effectiveness in accurately representing experimental behaviour through improved mathematical modelling.

Keywords Bayesian model updating · Beddoes-Leishman · LCO · Aeroelasticity

Introduction

Nonlinearities in aeroelastic systems can result in subcritical post-flutter responses, leading to self-sustaining oscillations, known as Limit Cycle Oscillations (LCO), at velocities below the linear flutter threshold. These behaviours have been experimentally observed in high-aspect-ratio flexible wings, and recent studies have identified subcritical LCO formation in highly flexible wings during low-speed flutter experiments. Accurately validating mathematical models to capture these nonlinear behaviours is, therefore, crucial. However, generating the bifurcation diagrams required to describe LCO behaviour is often computationally expensive.

In previous work, an adaptive multi-level data-driven model updating framework was demonstrated, in which data-driven models were developed and employed within a Bayesian inference scheme to obtain probabilistic estimates of nonlinear parameters and LCO behaviour [1]. This process was applied to a simple aerofoil test case with a structural nonlinearity in the pitch degree of freedom and a linear quasi-steady (Q-S) aerodynamic model. The proposed methodology provided up to a 20% improvement in accuracy compared to conventional deterministic methods, while significantly increasing computational efficiency in both the updating and uncertainty quantification processes when applied to a nonlinear aeroelastic test case with experimental data.

Despite these improvements, none of the probabilistic responses fully captured the experimental data, indicating potential noise or a mismatch between the mathematical model and experimental results, especially at high angles of attack. It is hypothesised that this discrepancy arises from the use of a quasi-steady linear aerodynamic model, which fails to account for nonlinear aerodynamic effects at high angles of attack. To address this, the nonlinear Beddoes-Leishman (B-L) model, which accounts for dynamic stall and vortex shedding, has been employed [2]. The aim is to improve the agreement between the probabilistic behaviour estimates and the experimental data. The B-L model has proven successful in low-airspeed analyses and has shown the capability to capture high-amplitude nonlinear oscillations, while remaining relatively computationally

Michael McGurk
Aerospace Centre of Excellence, University of Strathclyde, Montrose St, Glasgow, G1 1XJ, UK
e-mail: michael.mcgurk@strath.ac.uk

Jie Yuan
Computational Engineering Design Group, University of Southampton, Burgess Rd, Southampton, SO16 7QF, UK
e-mail: j.yuan@soton.ac.uk

Ludovic Renson et al. (eds.), *Nonlinear Structures & Systems, Vol. 1*, Conference Proceedings of the Society for Experimental Mechanics Series, https://doi.org/10.13052/9788743801467-22

efficient compared to alternative models [3]. In this study, the vortex shedding phase is of particular interest, as it can induce additional lift [4]. In theory, this could lead to higher amplitude LCO at greater pitch angles and improve the fit of the data to the bifurcation curves from the previous study. The new model also introduces more degrees of freedom, allowing this study to further demonstrate the improved computational efficiency of the multi-level updating framework. Additionally, the inclusion of aerodynamic nonlinearities enables the consideration of more nonlinear parameters in the model updating process.

Methodology

The Beddoes-Leishman model, as outlined in [2], was applied to the two-degree-of-freedom structural model used in the previous study [1]. This new model introduces additional complexity, expanding the system from a 6-degree-of-freedom problem to a 16-degree-of-freedom system, by adding twelve aerodynamic states (denoted x_1– x_{12}) alongside the four structural states. As in the previous study, a structural nonlinearity is introduced in the pitch degree of freedom (α), defined by the two nonlinear parameters $K_{\alpha 2}$ and $K_{\alpha 3}$

$$f_{nl} = K_{\alpha 2}\alpha^2 + K_{\alpha 3}\alpha^3$$

These were the only parameters estimated by the multilevel data-driven Bayesian framework in the previous case. However, with the introduction of a nonlinear aerodynamic model, we can now identify key nonlinear aerodynamic terms that influence vortex-induced lift and estimate these terms as well. The primary aerodynamic state influencing the vortex shedding phase in this study is x_{12}, which represents the normal force coefficient due to vortex shedding. This acts as an additional lift force during the vortex shedding phase and is evaluated with:

$$\dot{x}_{12} = \left(\frac{2V}{c}\right)\frac{-x_{12} + \dot{C}_v}{T_v}$$

Where V is the airspeed, c is the wing chord, and T_v is a time constant. $\dot{C}_v$ is a derivative dependent on the constant C_α, meaning the magnitude of the normal force due to lift is influenced by C_α. Another term that significantly affects the impact of the vortex-induced lift force is C_{n1}. This term essentially defines the points in heave and pitch where vortex shedding occurs. In summary, four key terms have been identified as having a significant influence on the shape of the bifurcation diagrams: the structural terms $K_{\alpha 2}$ and $K_{\alpha 3}$, and the aerodynamic terms C_α and C_{n1}.

Results

The multi-level data-driven Bayesian framework was applied to the new model with the aim of estimating $K_{\alpha 2}$, $K_{\alpha 3}$, C_α, and C_{n1} over three levels of model updating based on the experimental points shown in Figure 1a [5]. The resulting scatter

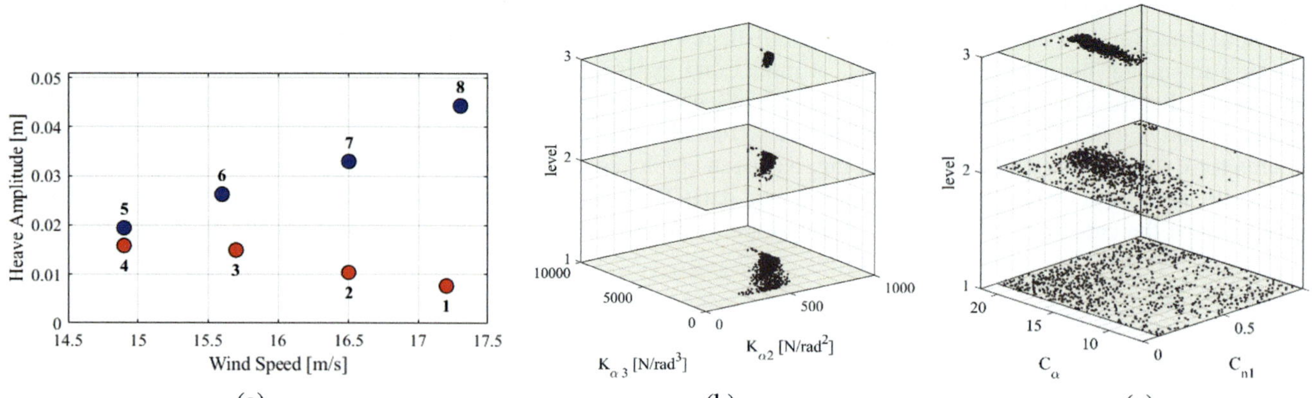

Fig. 1 a) Control based continuation experimental data adapted from [5], b) Structural parameter scatter plots, c) Aerodynamic parameters scatter plots

plots for each level are shown in Figure 1. The mean estimates for $K_{\alpha 2}$ and $K_{\alpha 3}$ were 722.35 N/rad^2 and 4762.8 N/rad^3, respectively, representing an increase of up to 23% compared to the results obtained using the quasi-steady aerodynamic model. The mean estimates for C_α and C_{n1} were 17.7 and 0.474, respectively. This reflects a 30% increase in C_α from the value reported in ref [1] and a 12% decrease in C_{n1}.

The resulting probabilistic bifurcation diagram is presented in Figure 2a, showing that all experimental points are captured within the 95% confidence bands. When compared to the probabilistic results from the quasi-steady study in Figure 2b, a significant improvement is evident, as 37.5% of the data in the previous model falls outside the confidence region. A comparison of the mean estimations from each model is provided in Figure 3. From Figure 3a, it is clear that the results from the B-L model offer a better fit to the experimental data, particularly in the upper portion of the diagram, as anticipated. A direct comparison of accuracy, shown in Figure 3b, demonstrates that the B-L model improves accuracy at every data point. Both models capture the system's behaviour well before the turning point; however, beyond this point, there is a noticeable decline in accuracy for both models, although the BL model performs significantly better.

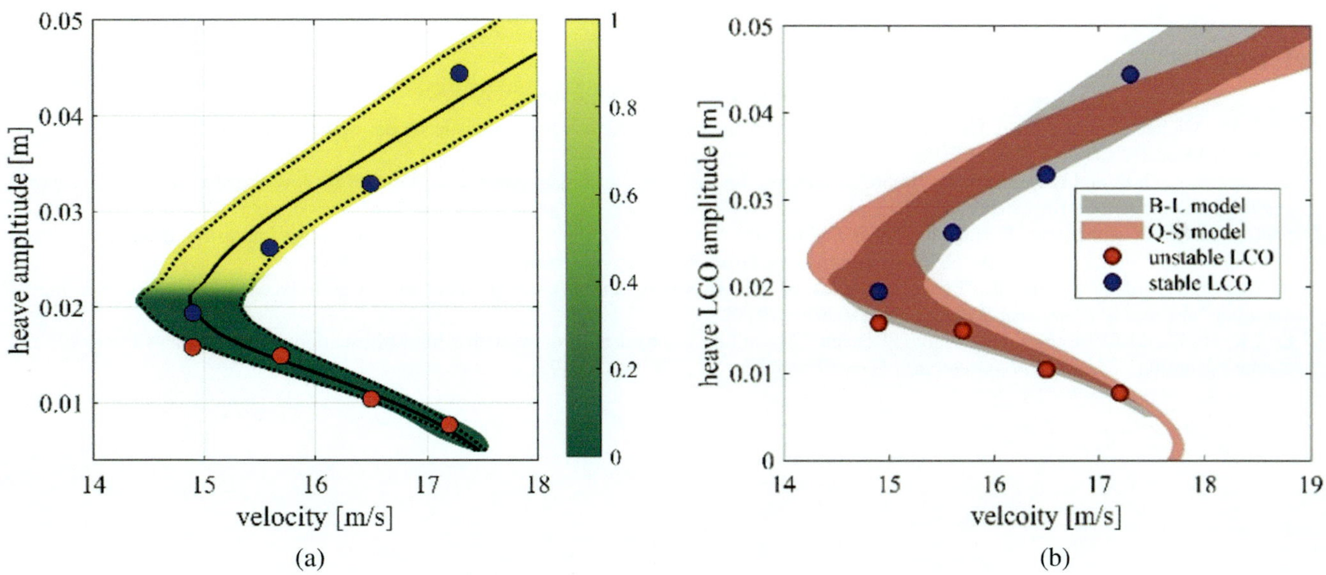

Fig. 2 a) Probabilistic bifurcation diagram, b) Probabilistic bifurcation diagram comparison

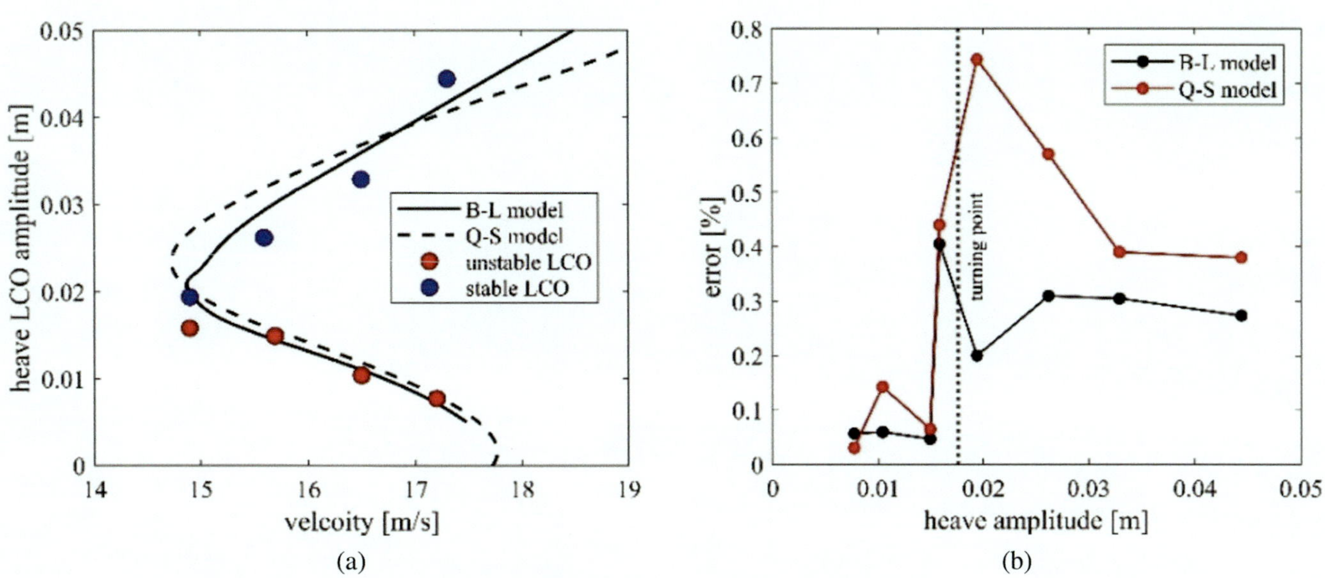

Fig. 3 a) Mean bifurcation diagrams, b) Mean bifurcation diagrams comparison

Conclusion

The findings of this study demonstrate that when applying the multi-level framework to the B-L model, the behaviour observed in the experimental data is captured with greater accuracy. The probabilistic bifurcation diagram successfully encompassed all the data using the B-L model, and the mean plot showed improved accuracy at every point when compared to the Q-S model. Further investigation is required to determine whether the aerodynamic parameters can be altered to such an extent and whether this reflects the true aerodynamics of the system. However, what this study does confirm is that if the mathematical model can represent the experimental behaviour, the multi-level framework will function effectively.

Acknoledgements

The authors acknowledge the experimental data provied by Dr Kyoung Hyun Lee available at https://github.com/Kyounghyunlee/.

References

1. McGurk, Michael, and Jie Yuan. "A multilevel data-driven Bayesian approach for stochastic model updating of nonlinear aeroelastic systems." *AIAA SCITECH 2024 Forum*. 2024.
2. dos Santos, L. G. P., and F. D. Marques. "Improvements on the Beddoes–Leishman dynamic stall model for low-speed applications." *Journal of Fluids and Structures* 106 (2021): 103375.
3. Santos, Carlos R., Muhammad R. Hajj, and Flavio D. Marques. "Structural nonlinearities influence on the energy harvesting from stall-induced oscillations." *2018 AIAA/ASCE/AHS/ASC Structures, Structural Dynamics, and Materials Conference*. 2018.
4. Konstantinidis, Efstathios, et al. "Phase dynamics of effective drag and lift components in vortex-induced vibration at low mass–damping." *Journal of Fluids and Structures* 96 (2020): 103028.
5. Lee, K. H., David AW Barton, and Ludovic Renson. "Modelling of physical systems with a hopf bifurcation using mechanistic models and machine learning." *Mechanical Systems and Signal Processing* 191 (2023): 110173.

Chapter 23
Resonance Tracking of Nonlinear Structures by Amplitude Controlled Sine-Dwell Testing

Gleb Kleyman, Hans-Jürgen Borutta, and Sebastian Schwarzendahl

Abstract Sine-dwell testing is a common methodology in high cycle fatigue (HCF) testing. During a sine-dwell test the structure is excited with a single sine tone exactly at a resonance frequency until structural failure or sufficient time/cycles passed. When the resonance frequency shifts due to accumulated damage or temperature changes the controller tracks the resonance frequency by adjusting the drive frequency to maintain a defined phase relation between two channels.

Nonlinear systems are known to have modal parameters varying with the excitation amplitude. The evolution of the resonance frequency over the excitation level is often referred to as the backbone-curve of a system. Backbone-curves can provide useful insights into the physical mechanisms involved in the nonlinearity.

For many nonlinear structures, the backbone-curve can be experimentally determined using the sine-dwell method combined with amplitude control. In this approach, the resonance frequency is tracked as the excitation level is increased step by step. This iterative phase control technique is an extension of traditional sine-dwell testing and is now available in the latest m+p VibControl software. It allows test engineers to better understand the nonlinear behavior of structures using familiar technology.

This paper discusses the application of this method to a cantilever beam excited within an amplitude range where it exhibits geometric nonlinear behavior. When excited using conventional amplitude-controlled sweeps, the system experiences a jump phenomenon, with both resonance frequency and damping varying. It is demonstrated that these amplitude-dependent modal parameters can be characterized through nonlinearity plots derived from data obtained during sine-dwell testing. These results can be directly compared to classical methods, offering a reduction in testing time.

Keywords Resonance Tracking · Sine Dwell · Backbone-Curve · Phase Resonance · Nonlinear Modal Analysis

Introduction

Nonlinearities are a common occurrence in vibration and modal testing. It is likely that all practical engineering structures exhibit some degree of nonlinearity. Typically, this nonlinearity results in modal parameters that change with load levels. These modes require careful consideration, as the resonance may shift to critical frequencies when the excitation level changes.

Various factors, such as structural joints, boundary conditions, and material properties, can be identified as sources of nonlinearity [1]. However, geometric nonlinearity is often overlooked. Mechanical systems are classified as geometrically nonlinear when the nonlinear effects arise solely from the system's kinematics. In practical applications, slender structures like beams tend to demonstrate geometrically nonlinear behavior when subjected to sufficiently large deformation amplitudes, even though the strains and stresses remain within the linear elastic range. It has been demonstrated that, disregarding modal interaction, plane beam systems can be represented as Duffing oscillators [2]:

$$m\ddot{q} + d\dot{q} + kq - \alpha q^3 = f, \tag{1}$$

Here, q represents the (modal) displacement, while m, d, and k denote the structural parameters. The term f corresponds to the excitation force, and α represents the cubic stiffness term. In this context, it is important to differentiate between the two primary types of geometrically nonlinear effects: (1) midplane stretching, which results from the coupling of transverse

Gleb Kleyman
m+p international Mess- und Rechnertechnik GmbH, Thurnithistraße 2, 30519 Hannover, Germany
e-mail: gleb.kleyman@mpihome.com; hans.borutta@mpihome.com; sebastian.schwarzendahl@mpihome.com

Ludovic Renson et al. (eds.), *Nonlinear Structures & Systems, Vol. 1*, Conference Proceedings of the Society for Experimental Mechanics Series, https://doi.org/10.13052/9788743801467-23

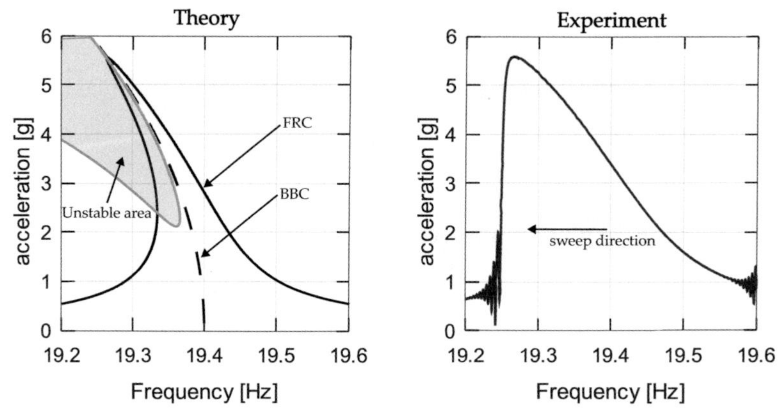

Fig. 1 Theoretical frequency responce curve vs. swept sine measurement of a cantilever beam

vibrations with longitudinal strain, and (2) nonlinear elasticity, arising from large cross-sectional rotations. The latter effect tends to exhibit a softening behavior, which here is reflected as a negative nonlinear stiffness term in the Duffing model. In the left subplot of Fig. 1, the frequency response curve (FRC) and backbone-curve (BBC) of a softening Duffing oscillator are shown, estimated using the Harmonic Balance method [3]. The FRC is depicted as a solid black line, while the BBC is represented by a dashed line. The unstable region is shaded in gray. For comparison, the right subplot presents the measured response of a base-excited cantilever beam, where the base acceleration was maintained at a constant excitation level during a swept sine test.

It can be observed that as the measured response approaches the upper bifurcation point, where the system typically loses stability, the curve abruptly jumps from the upper to the lower stable branch. A similar phenomenon occurs in the opposite direction at the lower turning point of the FRC (not shown in the figure). In both cases, the middle branch of unstable solutions remains undetected. Traditional experimental modal analysis methods are insufficient to fully characterize such systems due to the non-uniqueness of the frequency response. As a result, new approaches are required. Common alternatives include using the backbone-curve or plotting the resonance frequency and damping as functions of the excitation amplitude (nonlinearity plots).

With growing attention to the identification of nonlinear systems over the past two decades [4, 5], the automated extraction of backbone-curves has become increasingly desirable. In recent years, advanced control methods have been introduced [6, 7]. These approaches are primarily based on two key principles: (a) stabilizing the system when it nears an unstable oscillatory state, and (b) tracking the backbone-curve as the excitation level increases. It can be observed that the BBC approaches the unstable region at higher oscillation levels. Consequently, specialized nonlinear identification methods, such as the previously mentioned PLL and CBC, incorporate time-domain feedback control to maintain stable oscillations at higher levels. However, as noted in [8, 9], this classical control strategy is not always required, as the driving point position and shaker properties can influence the instability region. When this is the case, a simpler frequency-domain approach, like the one presented here, is sufficient for tracking the resonance frequency. In this method, a defined phase relationship between two channels is maintained by iteratively adjusting the excitation frequency.

Experimental Setup and control strategy

A photograph and schematic representation of the experimental setup are shown in Fig. 2. The specimen used was a cantilever beam with dimensions of 190x4x0.8 cm, base-excited in the horizontal direction. Both the acceleration of the base and the beam were measured. The control system consisted of m+p VibRunner hardware and m+p VibControl 2.17 software. The third bending mode of the beam was investigated, with a linear natural frequency of approximately 19.4 Hz. The beam is made from a single piece of steel, with the only bolted connection being at the fixture to the shaker's slip table, minimizing frictional contacts. Hence, the observed nonlinear behavior is likely due to the large deflection, as discussed in the previous section.

For phase control, a relative acceleration $\ddot{q}$ between the beam and the base was introduced. As explained in [10], a 90° phase shift between $\ddot{q}$ and $\ddot{q}_{\mathrm{base}}$ was selected as the control target. The relative phase is calculated from the fundamental harmonics of these signals, and if it deviates from the reference phase, the excitation frequency is adjusted until the target is achieved. The amplitude controller functions similarly. However, the reference value for the amplitude controller can be set in a dwell table, allowing different excitation levels to be sequentially targeted during testing. This control strategy does not

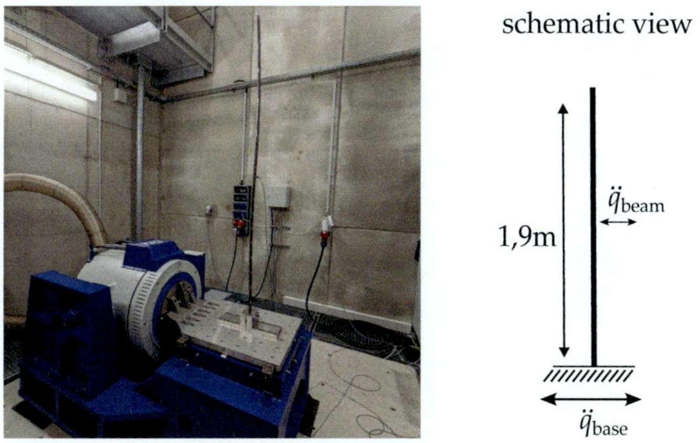

Fig. 2 Experimental setup of a base excited cantilever beam

inherently stabilize the system, as it operates in the frequency domain and cannot prevent the system from jumping between stable branches if a disturbance occurs. However, by keeping the amplitude increments small between different levels, it is possible to trace the backbone curve deep into the overhanging range, as the following results will demonstrate. Larger amplitudes than those shown in Fig. 4 were not targeted due to concerns about the structural integrity of the system.

Experimental Results

Two types of measurements were conducted: (1) resonance tracking using the sine-dwell testing mode and (2) classical amplitude-controlled swept sine excitation at various amplitude levels. The three subplots in Fig. 3 display the time records of the two acceleration signals, along with the corresponding phase and frequency values over the entire duration of the sine-dwell test. An initial frequency of $f_0 = 19.3\,\text{Hz}$ was set. Starting from this frequency, the control algorithm initiates a frequency sweep at the defined sweep rate to locate the targeted phase. Once the desired phase value is reached, the controller locks onto it and maintains tracking for the remainder of the test duration, which was set to 120 seconds in this instance. The relevant area is shaded in gray.

When the time period elapses, the next amplitude level is approached, with each level maintained for 90 seconds. This duration could be extended to minimize variations in the BBC measurement caused by transients; however, the current experiment was designed to minimize the overall testing time. After each increment, the resonance frequency is detuned, as indicated by the individual peaks in the phase curve. The controller then adjusts the excitation frequency to restore the phase resonance condition.

The left subplot of Fig. 4 displays the backbone curve (BBC), estimated from the results shown in Fig. 3, alongside the frequency response curves obtained from base acceleration-controlled sweeps. To calculate the BBC, measurement points were selected from the phase curve that fall within an interval of 88° to 92°. Each data point of the BBC represents an average over the corresponding interval at a specific amplitude level. The sweeps were conducted in a descending frequency

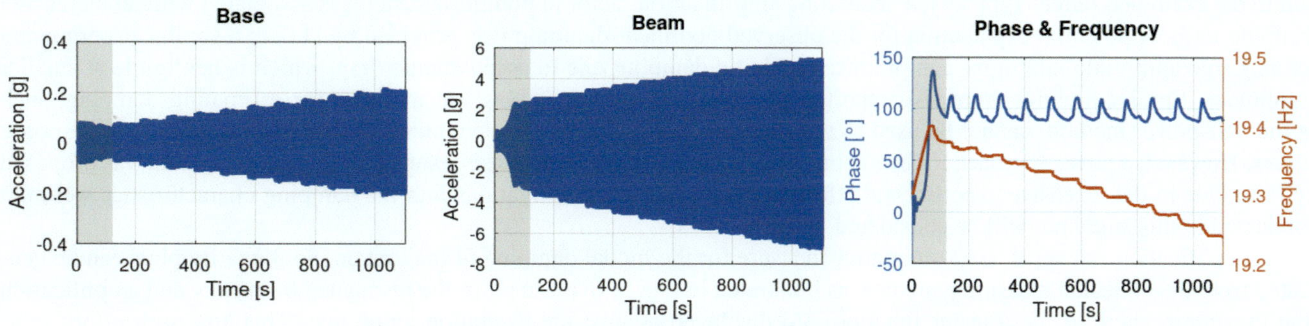

Fig. 3 Time domain data of a sine dwell test for resonance tracking

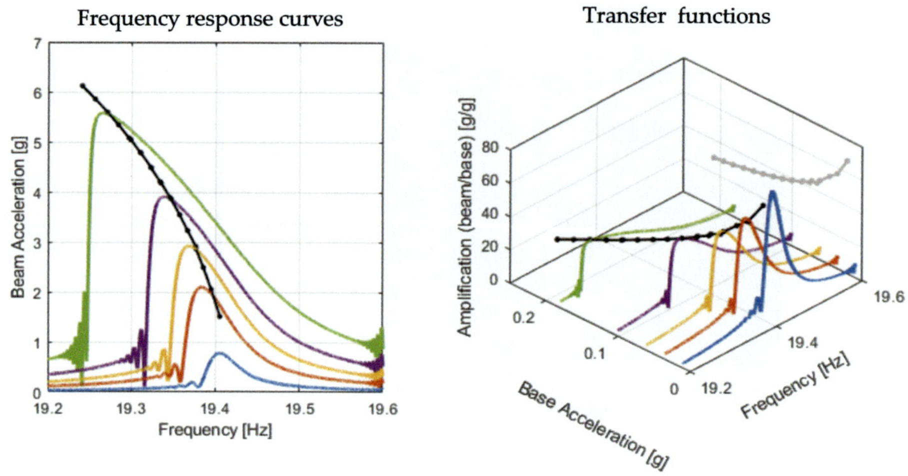

Fig. 4 Comparison of results from amplitude controlled swept-sine excitation and resonance tracking (backbone-curve)

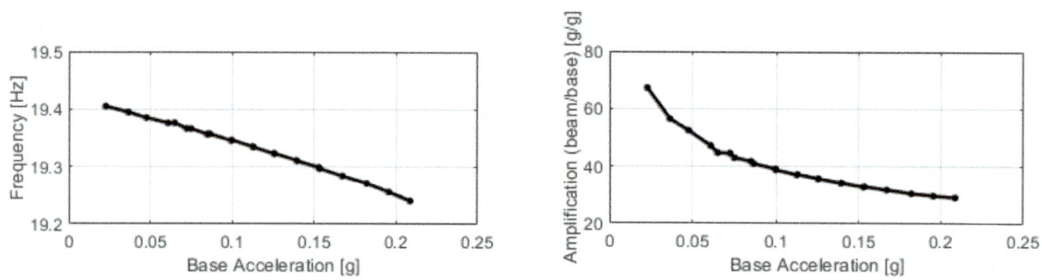

Fig. 5 Nonlinearity plots

direction at five different amplitude levels. It is evident that the BBC accurately captures the resonance peaks of the frequency response curves. Additionally, the resolution of the BBC is significantly finer due to the larger number of measurement points compared to the number of sweep levels. Notably, the measurement time for obtaining the BBC through sine-dwell testing is only half that of sequential sweep measurements.

The right subplot of Fig. 4 presents the transfer functions, obtained by dividing the output signals by their corresponding input values. In engineering applications, the transfer function is often considered the most important characteristic of a dynamic structure, as it defines the maximum response amplitude for a given excitation. Linear systems exhibit proportional behavior, which means their transfer functions remain constant for different excitation levels. Therefore, in a three-dimensional representation similar to Fig. 4, where different functions are plotted at the corresponding excitation levels, the locus of resonance peaks in the amplitude subspace (shown as a gray curve) would appear as a horizontal line. However, this is not the case for the nonlinear system in this example. Instead, amplification decreases with increasing excitation levels within the examined range. Typically, a decreasing amplification factor in nonlinear systems is associated with an increasing damping ratio. A potential explanation for the observed nonlinear damping was provided by [11] in a similar experimental setup. Large amplitude vibrations can produce quadratic damping due to aerodynamic drag, which is negligible at smaller amplitudes. Similar findings were also reported by [12]. Unfortunately, classical methods for estimating damping, such as the half-power method, cannot be used to validate this theory due to the asymmetrical shape of the frequency response curves. However, a novel approach for estimating nonlinear damping from phase-resonant base excitation tests was proposed by [10]. Due to the extensive experimental effort involved, further investigations into the damping characteristics were not conducted in this study but will be considered in future work.

If amplification can serve as an alternative measure for the modal damping of the system, nonlinearity plots can be generated from the collected measurement data, as illustrated in Fig. 5. In these plots, the resonance frequency and amplification (the maximum value of the transfer function) are displayed against the excitation amplitude. This approach offers valuable insights into the nonlinear structural properties of the mode under investigation. In this particular example, a nonlinear damping behavior emerged that was not considered in the mechanical model.

Conclusion

This study successfully demonstrates the application of the sine-dwell testing method combined with amplitude control to effectively investigate the nonlinear behavior of a cantilever beam. By tracking the resonance frequency as the excitation level is increased, the backbone-curve was measured, offering critical insights into the modal parameters that vary with excitation amplitude. The results indicate that traditional modal analysis techniques may be insufficient for fully characterizing nonlinear systems due to their complex response characteristics. The findings also reveal a notable nonlinear damping behavior that was not captured in the existing mechanical model, emphasizing the necessity for further refinement in future research. Overall, the research provides valuable methods and insights for engineers and researchers who want to better identify and model the complexity of nonlinear dynamic systems in practical applications.

References

1. Worden, K. and Tomlinson, G. *Nonlinearity in Structural Dynamics - Detection, Identification and Modelling*. IOP Publishing (2001).
2. Thomsen, J. *Vibration and Stability*. Springer-Verlag Berlin Heidelberg (2003).
3. Hagedorn, P. *Nichtlineare Schwingungen*. Akademische Verlagsgesellschaft (1978).
4. Kerschen, G., Worden, K., Vakakis, A.F., and Golinval, J.C. "Past, present and future of nonlinear system identification in structural dynamics". *Mechanical Systems and Signal Processing*, 20(3):505–592 (2006).
5. Noël, J. and Kerschen, G. "Nonlinear system identification in structural dynamics: 10 more years of progress". *Mechanical Systems and Signal Processing*, 83:2–35 (2017).
6. Hippold, P., Scheel, M., Renson, L., and Krack, M. "Robust and fast backbone tracking via phase-locked loops" (2024).
7. Tatzko, S., Kleyman, G., and Wallaschek, J. "Continuation methods for lab experiments of nonlinear vibrations". *GAMM-Mitteilungen*, 46(2):e202300009 (2023).
8. Stanbridge, A., Ewins, D., Sanliturk, K., and Ferreira, J. "Experimental investigation of dry friction damping and cubic stiffness nonlinearity". *Proceedings of the ASME Design Engineering Technical Conference*, 6:2141–2148 (2001).
9. Robbins, E., Kuether, R., Pacini, B., and Moreu, F. "Stabilizing a strongly nonlinear structure through shaker dynamics in fixed frequency voltage control tests". *Mechanical Systems and Signal Processing*, 190:110118 (2023).
10. Müller, F., Woiwode, L., Gross, J., Scheel, M., and Krack, M. "Nonlinear damping quantification from phase-resonant tests under base excitation". *Mechanical Systems and Signal Processing*, 177:109170 (2022).
11. Colin, M., Thomas, O., Grondel, S., and Cattan, E. "Very large amplitude vibrations of flexible structures: Experimental identification and validation of a quadratic drag damping model". *Journal of Fluids and Structures*, 97:103056 (2020).
12. Debeurre, M., Benacchio, S., Grolet, A., Grenat, C., Giraud-Audine, C., and Thomas, O. "Phase resonance testing of highly flexible structures: Measurement of conservative nonlinear modes and nonlinear damping identification". *Mechanical Systems and Signal Processing*, 215:111423 (2024).

For Product Safety Concerns and Information please contact our
EU representative GPSR@taylorandfrancis.com Taylor & Francis
Verlag GmbH, Kaufingerstraße 24, 80331 München, Germany